BIG IDEAS MATH®

Modeling Real Life

Grade 5

Volume 2

Ron Larson
Laurie Boswell

Erie, Pennsylvania
BigIdeasLearning.com

Big Ideas Learning, LLC
1762 Norcross Road
Erie, PA 16510-3838
USA

For product information and customer support, contact Big Ideas Learning at 1-877-552-7766 or visit us at BigIdeasLearning.com.

Cover Image
Valdis Torms, enmyo/Shutterstock.com

Printed in the U.S.A.

ISBN 13: 978-1-63598-894-9

3 4 5 6 7 8 9 10—22 21 20 19

About the Authors

Ron Larson

Ron Larson, Ph.D., is well known as the lead author of a comprehensive program for mathematics that spans school mathematics and college courses. He holds the distinction of Professor Emeritus from Penn State Erie, The Behrend College, where he taught for nearly 40 years. He received his Ph.D. in mathematics from the University of Colorado. Dr. Larson's numerous professional activities keep him actively involved in the mathematics education community and allow him to fully understand the needs of students, teachers, supervisors, and administrators.

Ron Larson

Laurie Boswell

Laurie Boswell, Ed.D., is the former Head of School at Riverside School in Lyndonville, Vermont. In addition to textbook authoring, she provides mathematics consulting and embedded coaching sessions. Dr. Boswell received her Ed.D. from the University of Vermont in 2010. She is a recipient of the Presidential Award for Excellence in Mathematics Teaching and is a Tandy Technology Scholar. Laurie has taught math to students at all levels, elementary through college. In addition, Laurie has served on the NCTM Board of Directors and as a Regional Director for NCSM. Along with Ron, Laurie has co-authored numerous math programs and has become a popular national speaker.

Laurie Boswell

Dr. Ron Larson and Dr. Laurie Boswell began writing together in 1992. Since that time, they have authored over four dozen textbooks. This successful collaboration allows for one voice from Kindergarten through Algebra 2.

Contributors, Reviewers, and Research

Big Ideas Learning would like to express our gratitude to the mathematics education and instruction experts who served as our advisory panel, contributing specialists, and reviewers during the writing of *Big Ideas Math: Modeling Real Life*. Their input was an invaluable asset during the development of this program.

Contributing Specialists and Reviewers

- **Sophie Murphy**, Ph.D. Candidate, Melbourne School of Education, Melbourne, Australia
  Learning Targets and Success Criteria Specialist and Visible Learning Reviewer
- **Linda Hall**, Mathematics Educational Consultant, Edmond, OK
  Advisory Panel
- **Michael McDowell**, Ed.D., Superintendent, Ross, CA
  Project-Based Learning Specialist
- **Kelly Byrne**, Math Supervisor and Coordinator of Data Analysis, Downingtown, PA
  Advisory Panel
- **Jean Carwin**, Math Specialist/TOSA, Snohomish, WA
  Advisory Panel
- **Nancy Siddens**, Independent Language Teaching Consultant, Las Cruces, NM
  English Language Learner Specialist
- **Kristen Karbon**, Curriculum and Assessment Coordinator, Troy, MI
  Advisory Panel
- **Kery Obradovich**, K–8 Math/Science Coordinator, Northbrook, IL
  Advisory Panel
- **Jennifer Rollins**, Math Curriculum Content Specialist, Golden, CO
  Advisory Panel
- **Becky Walker**, Ph.D., School Improvement Services Director, Green Bay, WI
  Advisory Panel and Content Reviewer
- **Deborah Donovan**, Mathematics Consultant, Lexington, SC
  Content Reviewer
- **Tom Muchlinski**, Ph.D., Mathematics Consultant, Plymouth, MN
  Content Reviewer and Teaching Edition Contributor
- **Mary Goetz**, Elementary School Teacher, Troy, MI
  Content Reviewer
- **Nanci N. Smith**, Ph.D., International Curriculum and Instruction Consultant, Peoria, AZ
  Teaching Edition Contributor
- **Robyn Seifert-Decker**, Mathematics Consultant, Grand Haven, MI
  Teaching Edition Contributor
- **Bonnie Spence**, Mathematics Education Specialist, Missoula, MT
  Teaching Edition Contributor
- **Suzy Gagnon**, Adjunct Instructor, University of New Hampshire, Portsmouth, NH
  Teaching Edition Contributor
- **Art Johnson**, Ed.D., Professor of Mathematics Education, Warwick, RI
  Teaching Edition Contributor
- **Anthony Smith**, Ph.D., Associate Professor, Associate Dean, University of Washington Bothell, Seattle, WA
  Reading and Writing Reviewer
- **Brianna Raygor**, Music Teacher, Fridley, MN
  Music Reviewer
- **Nicole Dimich Vagle**, Educator, Author, and Consultant, Hopkins, MN
  Assessment Reviewer
- **Janet Graham**, District Math Specialist, Manassas, VA
  Response to Intervention and Differentiated Instruction Reviewer
- **Sharon Huber**, Director of Elementary Mathematics, Chesapeake, VA
  Universal Design for Learning Reviewer

Student Reviewers

- T.J. Morin
- Alayna Morin
- Ethan Bauer
- Emery Bauer
- Emma Gaeta
- Ryan Gaeta
- Benjamin SanFrotello
- Bailey SanFrotello
- Samantha Grygier
- Robert Grygier IV
- Jacob Grygier
- Jessica Urso
- Ike Patton
- Jake Lobaugh
- Adam Fried
- Caroline Naser
- Charlotte Naser

Research

Ron Larson and Laurie Boswell used the latest in educational research, along with the body of knowledge collected from expert mathematics instructors, to develop the *Modeling Real Life* series. The pedagogical approach used in this program follows the best practices outlined in the most prominent and widely accepted educational research, including:

- *Visible Learning*
  John Hattie © 2009
- *Visible Learning for Teachers*
  John Hattie © 2012
- *Visible Learning for Mathematics*
  John Hattie © 2017
- *Principles to Actions: Ensuring Mathematical Success for All*
  NCTM © 2014
- *Adding It Up: Helping Children Learn Mathematics*
  National Research Council © 2001
- *Mathematical Mindsets: Unleashing Students' Potential through Creative Math, Inspiring Messages and Innovative Teaching*
  Jo Boaler © 2015
- *What Works in Schools: Translating Research into Action*
  Robert Marzano © 2003
- *Classroom Instruction That Works: Research-Based Strategies for Increasing Student Achievement*
  Marzano, Pickering, and Pollock © 2001
- *Principles and Standards for School Mathematics*
  NCTM © 2000
- *Rigorous PBL by Design: Three Shifts for Developing Confident and Competent Learners*
  Michael McDowell © 2017
- *Universal Design for Learning Guidelines*
  CAST © 2011
- Rigor/Relevance Framework®
  International Center for Leadership in Education
- *Understanding by Design*
  Grant Wiggins and Jay McTighe © 2005
- Achieve, ACT, and The College Board
- *Elementary and Middle School Mathematics: Teaching Developmentally*
  John A. Van de Walle and Karen S. Karp © 2015
- *Evaluating the Quality of Learning: The SOLO Taxonomy*
  John B. Biggs & Kevin F. Collis © 1982
- *Unlocking Formative Assessment: Practical Strategies for Enhancing Students' Learning in the Primary and Intermediate Classroom*
  Shirley Clarke, Helen Timperley, and John Hattie © 2004
- *Formative Assessment in the Secondary Classroom*
  Shirley Clarke © 2005
- *Improving Student Achievement: A Practical Guide to Assessment for Learning*
  Toni Glasson © 2009

Mathematical Processes and Proficiencies

Big Ideas Math: Modeling Real Life reinforces the Process Standards from NCTM and the Five Strands of Mathematical Proficiency endorsed by the National Research Council. With *Big Ideas Math*, students get the practice they need to become well-rounded, mathematically proficient learners.

Problem Solving/Strategic Competence

- *Think & Grow: Modeling Real Life* examples use problem-solving strategies, such as drawing a picture, circling knowns, and underlining unknowns. They also use a formal problem-solving plan: understand the problem, make a plan, and solve and check.
- Real-life problems are provided to help students learn to apply the mathematics that they are learning to everyday life.
- Real-life problems help students use the structure of mathematics to break down and solve more difficult problems.

Reasoning and Proof/Adaptive Reasoning

- *Explore & Grows* allow students to investigate math and make conjectures.
- Questions ask students to explain and justify their reasoning.

Communication

- Cooperative learning opportunities support precise communication.
- Exercises, such as *You Be The Teacher* and *Which One Doesn't Belong?,* provide students the opportunity to critique the reasoning of others.
- *Apply and Grow: Practice* exercises allow students to demonstrate their understanding of the lesson up to that point.
- *ELL Support* notes provide insights into how to support English learners.

Connections

- Prior knowledge is continually brought back and tied in with current learning.
- Performance Tasks tie the topics of a chapter together into one extended task.
- Real-life problems incorporate other disciplines to help students see that math is used across content areas.

Representations/Productive Disposition

- Real-life problems are translated into pictures, diagrams, tables, equations, and graphs to help students analyze relations and to draw conclusions.
- Visual problem-solving models help students create a coherent representation of the problem.
- Multiple representations are presented to help students move from concrete to representative and into abstract thinking.
- *Learning Targets* and *Success Criteria* at the start of each chapter and lesson help students understand what they are going to learn.
- Real-life problems incorporate other disciplines to help students see that math is used across content areas.

Conceptual Understanding

- *Explore & Grows* allow students to investigate math to understand the reasoning behind the rules.

Procedural Fluency

- Skill exercises are provided to continually practice fundamental skills.
- Prior knowledge is continually brought back and tied in with current learning.

Meeting Proficiency and Major Topics

Meeting Proficiency

As standards shift to prepare students for college and careers, the importance of focus, coherence, and rigor continues to grow.

FOCUS *Big Ideas Math: Modeling Real Life* emphasizes a narrower and deeper curriculum, ensuring students spend their time on the major topics of each grade.

COHERENCE The program was developed around coherent progressions from Kindergarten through eighth grade, guaranteeing students develop and progress their foundational skills through the grades while maintaining a strong focus on the major topics.

RIGOR *Big Ideas Math: Modeling Real Life* uses a balance of procedural fluency, conceptual understanding, and real-life applications. Students develop conceptual understanding in every *Explore and Grow*, continue that development through the lesson while gaining procedural fluency during the *Think and Grow*, and then tie it all together with *Think and Grow: Modeling Real Life*. Every set of practice problems reflects this balance, giving students the rigorous practice they need to be college- and career-ready.

Major Topics in Grade 5

Number and Operations in Base Ten

- Understand the place value system.
- Perform operations with multi-digit whole numbers and with decimals to hundredths.

Number and Operations—Fractions

- Use equivalent fractions as a strategy to add and subtract fractions.
- Apply and extend previous understandings of multiplication and division to multiply and divide fractions.

Measurement and Data

- Geometric measurement: understand concepts of volume and relate volume to multiplication and to addition.

Use the color-coded Table of Contents to determine where the major topics, supporting topics, and additional topics occur throughout the curriculum.

- Major Topic
- Supporting Topic
- Additional Topic

1 Place Value Concepts

2 Numerical Expressions

■ Major Topic
■ Supporting Topic
■ Additional Topic

Add and Subtract Decimals

Multiply Whole Numbers

Let's learn how to multiply whole numbers.

5 Multiply Decimals

6 Divide Whole Numbers

■ Major Topic
■ Supporting Topic
■ Additional Topic

7

Divide Decimals

Race Around the World: Division

Directions:

1. Players take turns.
2. On your turn, flip a Race Around the World: Division Card and find the quotient.
3. Move your piece to the next number on the board that is highlighted in the quotient.
4. The first player to make it back to North America wins!

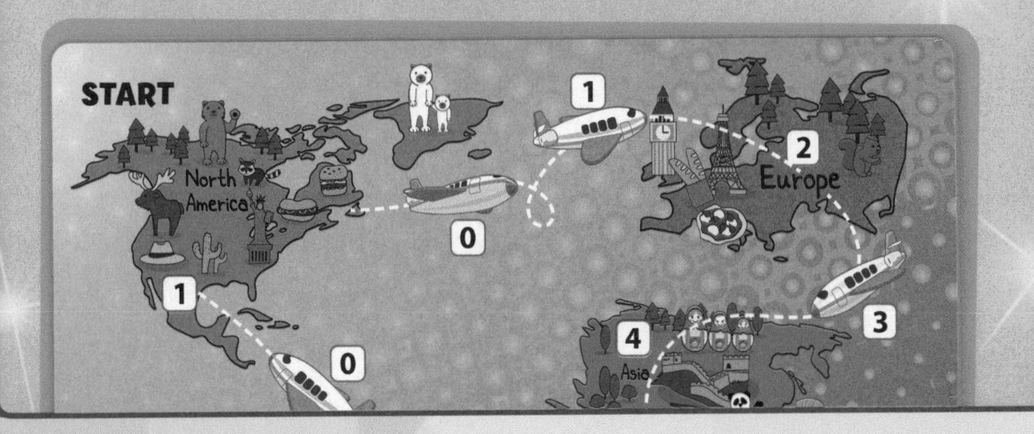

8 Add and Subtract Fractions

9 Multiply Fractions

■ Major Topic
■ Supporting Topic
■ Additional Topic

10 Divide Fractions

11 Convert and Display Units of Measure

Let's learn how to divide fractions!

12 Patterns in the Coordinate Plane

13 Understand Volume

- Major Topic
- Supporting Topic
- Additional Topic

14 Classify Two-Dimensional Shapes

Quadrilateral Lineup

Directions:

1. Players take turns spinning the spinner.
2. On your turn, cover a quadrilateral that matches your spin.
3. If you land on *Lose a Turn*, then do not cover a quadrilateral.
4. The first player to get four in a row twice, horizontally, vertically, or diagonally, wins!

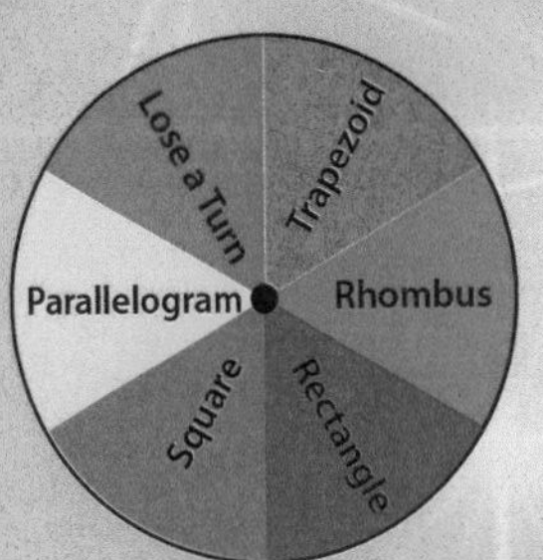

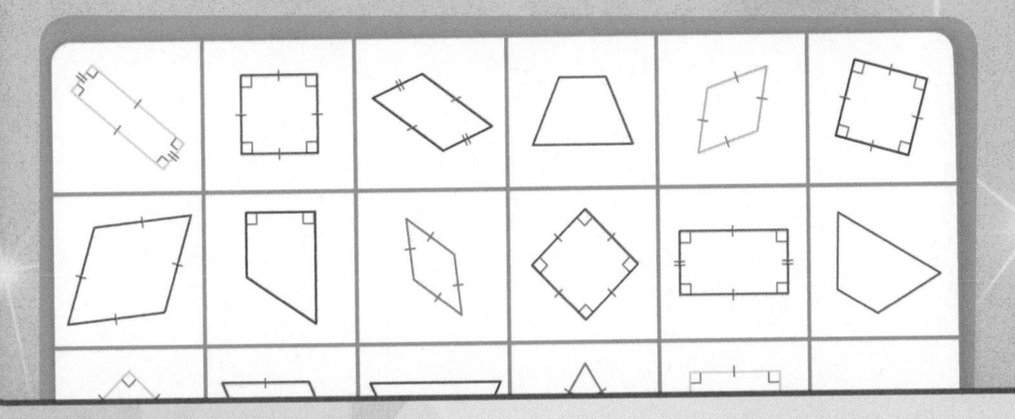

8 Add and Subtract Fractions

Chapter Learning Target:
Understand adding and subtracting fractions.

Chapter Success Criteria:

- I can find the factors of a number.
- I can write equivalent fractions.
- I can add and subtract fractions.
- I can solve a problem using fractions.

- The Mount Rushmore National Memorial is carved into the side of a mountain in South Dakota. Which four American presidents does the memorial depict?
- Three of the presidents' noses are each $6\frac{2}{3}$ yards long. George Washington's nose is $\frac{1}{3}$ yard longer. How can you find the length of George Washington's nose?

Name ______________________________

8 Vocabulary

Review Words
decimal fraction
mixed number
unit fraction

Organize It

Use the review words to complete the graphic organizer.

$\frac{1}{8}$	$\frac{3}{10}$	$4\frac{2}{3}$
$\frac{1}{15}$	$\frac{27}{100}$	$7\frac{5}{8}$
$\frac{1}{9}$	$\frac{98}{100}$	$2\frac{1}{5}$

Define It

Use your vocabulary cards to complete each definition.

1. simplest form: When the ______________ and denominator of a __________ have no __________ factors other than __________

2. common denominator: A ______________ that is the ______________ of __________ or more __________

3. proper fraction: A fraction __________ than __________

4. improper fraction: A fraction __________ than __________

Chapter 8 Vocabulary Cards

common denominator

improper fraction

proper fraction

simplest form

A fraction greater than 1

$\frac{3}{2}, \frac{6}{3}, \frac{9}{4}$

A number that is the denominator of two or more fractions

$\frac{3}{4} \quad \frac{2}{4}$

4 is a common denominator for $\frac{3}{4}$ and $\frac{2}{4}$.

When the numerator and denominator of a fraction have no common factors other than 1

$\frac{2}{6} = \frac{1}{3}$

↑

simplest form

A fraction less than 1

$\frac{1}{2}, \frac{2}{3}, \frac{3}{4}$

Name ___________________________

Simplest Form 8.1

Learning Target: Write fractions in simplest form.

Success Criteria:

- I can find the common factors of two numbers.
- I can write equivalent fractions.
- I can write a fraction in simplest form.

Explore and Grow

Use the model to write as many fractions as possible that are equivalent to $\frac{36}{72}$ but have numerators less than 36 and denominators less than 72.

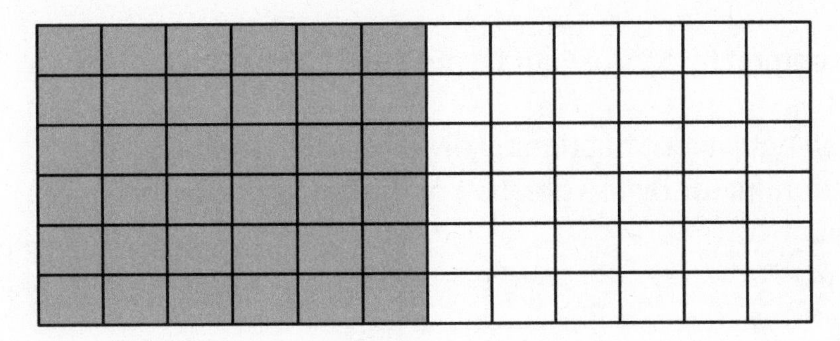

Which of your fractions has the fewest equal parts? Explain.

Construct Arguments When might it be helpful to write $\frac{48}{72}$ as $\frac{2}{3}$ in a math problem?

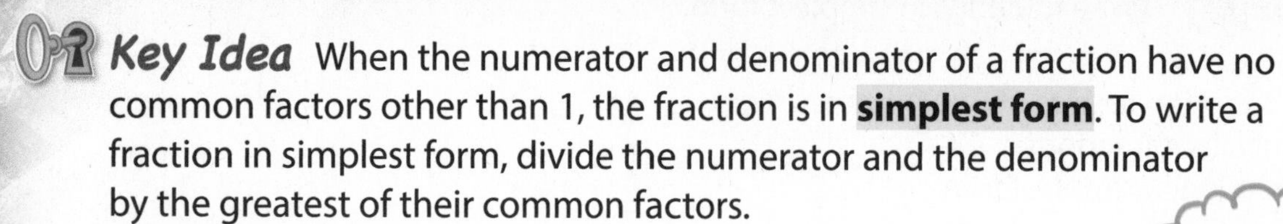

Think and Grow: Simplest Form

Key Idea When the numerator and denominator of a fraction have no common factors other than 1, the fraction is in **simplest form**. To write a fraction in simplest form, divide the numerator and the denominator by the greatest of their common factors.

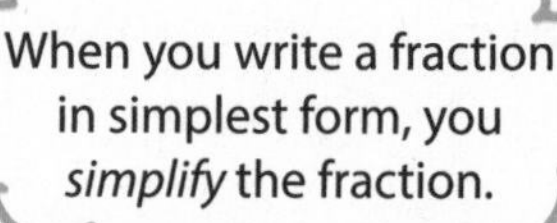

When you write a fraction in simplest form, you *simplify* the fraction.

Example Write $\frac{6}{8}$ in simplest form.

Step 1: Find the common factors of 6 and 8.

Factors of 6: (1), (2), 3, 6

Factors of 8: (1), (2), 4, 8

The common factors of 6 and 8 are _____ and _____.

Step 2: Write an equivalent fraction by dividing the numerator and the denominator by the greatest of the common factors.

Simplest form uses as few equal parts of a whole as possible to represent a fraction.

$$\frac{6}{8} = \frac{6 \div \square}{8 \div \square} = \frac{\square}{\square}$$

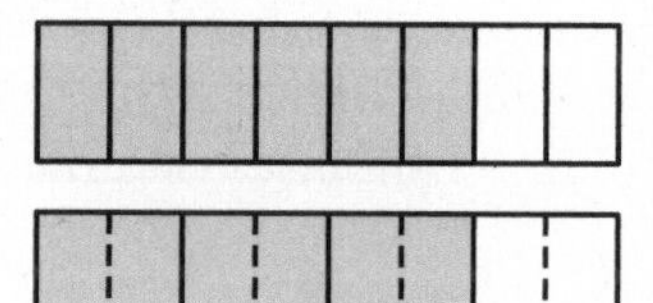

Because 3 and _____ have no common factors other than 1, _____ is in simplest form.

Show and Grow *I can do it!*

1. Use the model to write $\frac{2}{4}$ in simplest form.

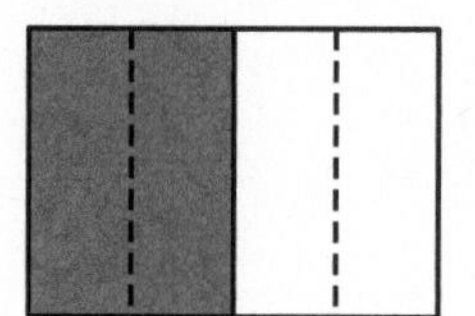

2. Write $\frac{8}{12}$ in simplest form.

Name ______________________________

Apply and Grow: Practice

Use the model to write the fraction in simplest form.

3. $\frac{8}{10} = \frac{\square}{\square}$

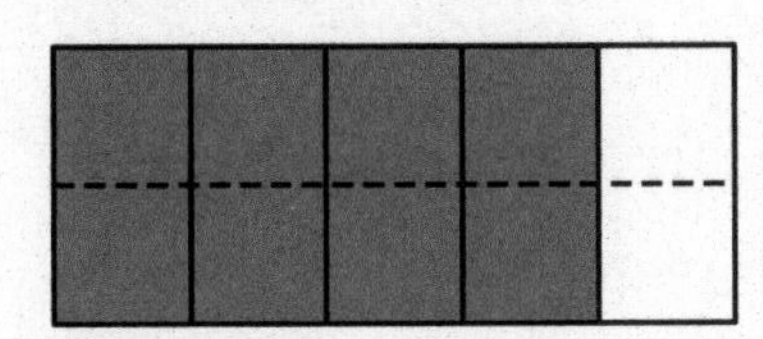

4. $\frac{5}{15} = \frac{\square}{\square}$

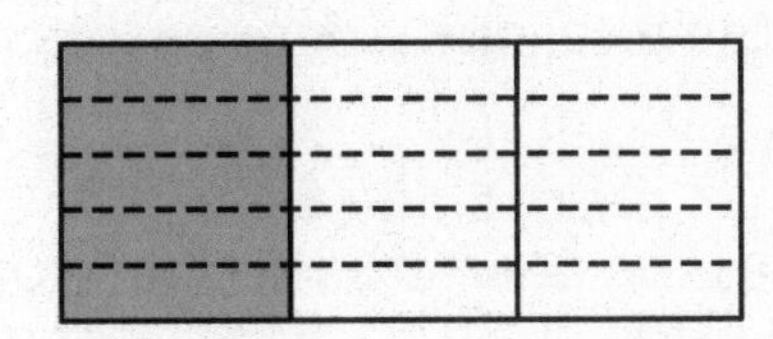

5. $\frac{10}{12} = \frac{\square}{\square}$

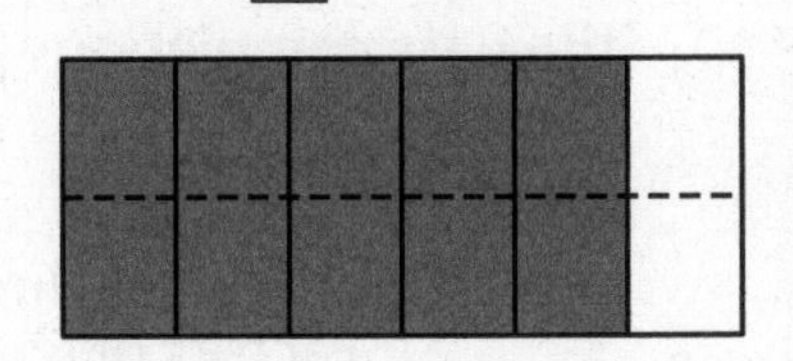

Write the fraction in simplest form.

6. $\frac{3}{6}$

7. $\frac{2}{10}$

8. $\frac{6}{8}$

9. $\frac{7}{14}$

10. $\frac{10}{100}$

11. $\frac{12}{4}$

12. Three out of nine baseball players are in the outfield. In simplest form, what fraction of the players are in the outfield?

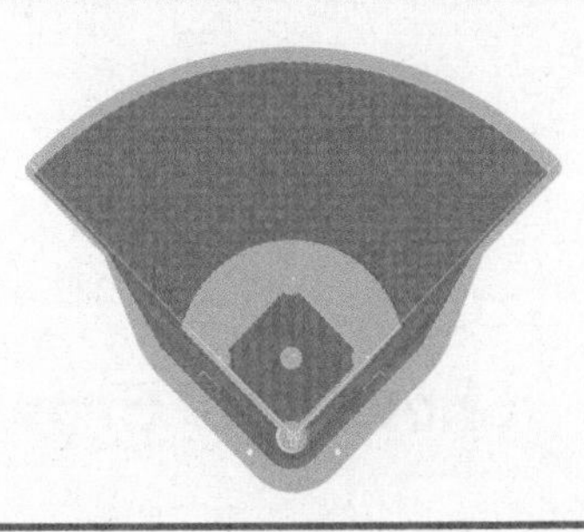

13. **YOU BE THE TEACHER** Your friend writes $\frac{2}{6}$ in simplest form. Is your friend correct? Explain.

$$\frac{2}{6} = \frac{2}{6 \div 2} = \frac{2}{3}$$

14. **MP Reasoning** The numerator and denominator of a fraction have 1, 2, and 4 as common factors. After you divide the numerator and denominator by 2, the fraction is still not in simplest form. Why?

Think and Grow: Modeling Real Life

Example A quarterback passes the ball 45 times during a game. The quarterback completes 35 passes. What fraction of the passes, in simplest form, does the quarterback *not* complete?

Find the number of passes that are not completed by subtracting the pass completions from the total number of passes.

$45 - 35 = 10$

Write a fraction for the passes the quarterback does not complete.

$\frac{10}{45}$ — 10 ← passes the quarterback does not complete; 45 ← total number of passes

Think: Why might you want to write the fraction in simplest form?

Find common factors of 10 and 45. Then write an equivalent fraction by dividing the numerator and the denominator by the greatest of the common factors.

Factors of 10: (1), 2, (5), 10

Factors of 45: (1), 3, (5), 9, 15, 45

$$\frac{10}{45} = \frac{10 \div \square}{45 \div \square} = \frac{\square}{\square}$$

The quarterback does not complete ______ of the passes.

Show and Grow *I can think deeper!*

15. There are 24 students in your class. Four of the students have blue eyes. What fraction of the class, in simplest form, do *not* have blue eyes?

16. **DIG DEEPER!** A student answers 4 out of 12 questions on a test incorrectly. What fraction of the questions, in simplest form, does the student answer incorrectly? Interpret the fraction.

Name Ethan #23 PM

Homework & Practice 8.1

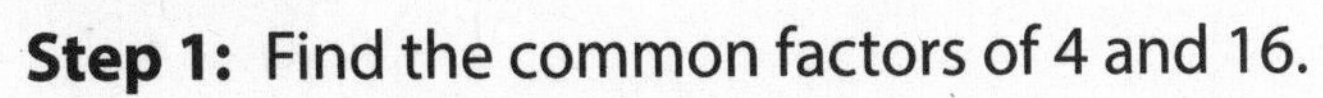

Learning Target: Write fractions in simplest form.

Example Write $\frac{4}{16}$ in simplest form.

Step 1: Find the common factors of 4 and 16.

Factors of 4: (1), (2), (4)

Factors of 16: (1), (2), (4), 8, 16

The common factors of 4 and 16 are 1, 2, and 4.

Step 2: Write an equivalent fraction by dividing the numerator and the denominator by the greatest of the common factors.

$$\frac{4}{16} = \frac{4 \div \boxed{4}}{16 \div \boxed{4}} = \frac{\boxed{1}}{\boxed{4}}$$

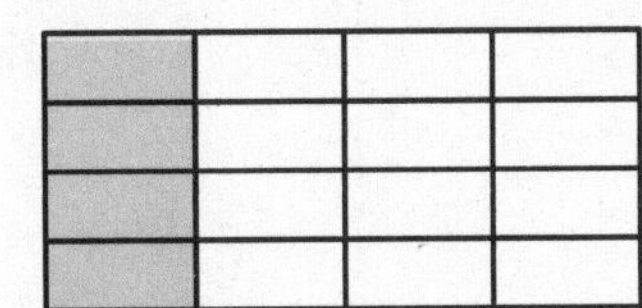

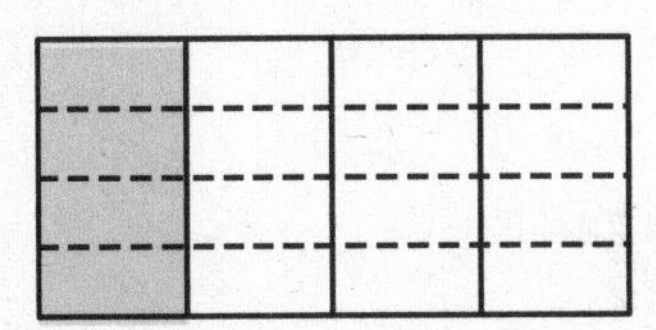

Because 1 and 4 have no common factors other than 1, $\frac{1}{4}$ is in simplest form.

Use the model to write the fraction in simplest form.

1. $\frac{6}{9} = \frac{\boxed{2}}{\boxed{3}}$

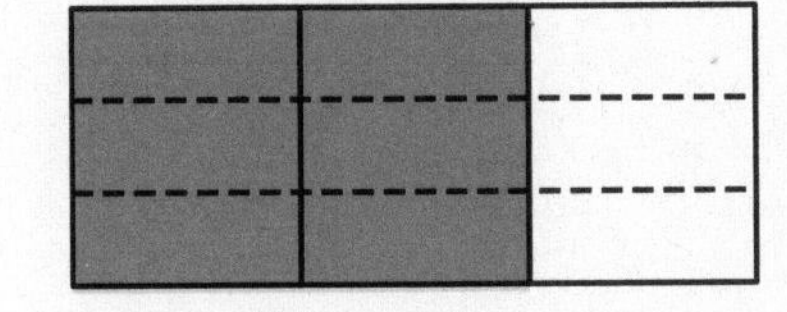

2. $\frac{3}{12} = \frac{\boxed{1}}{\boxed{4}}$

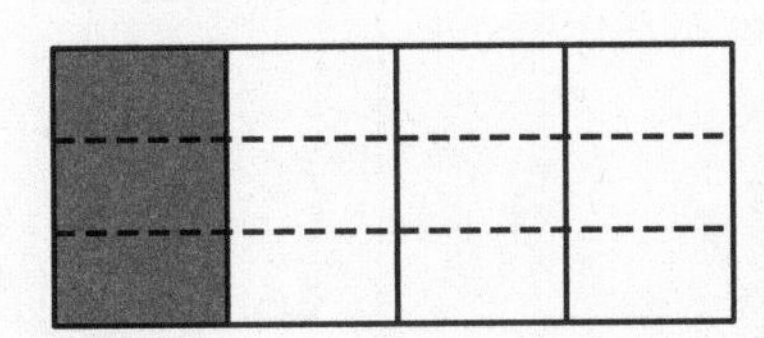

3. $\frac{5}{10} = \frac{\boxed{1}}{\boxed{2}}$

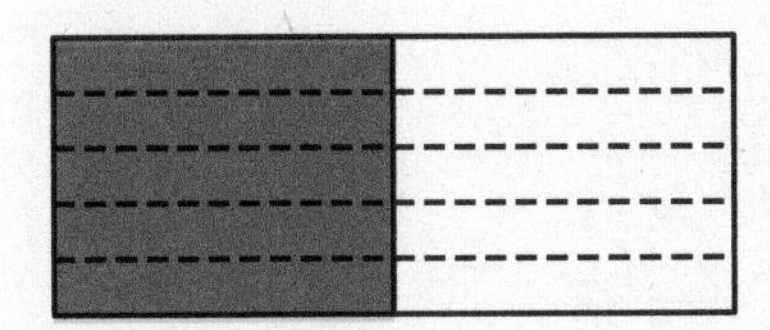

Write the fraction in simplest form.

4. $\frac{4}{8}$ $\frac{1}{2}$

5. $\frac{5}{100}$ $\frac{1}{20}$

6. $\frac{20}{15}$ $1\frac{1}{3}$

7. There are 18 students in your class. Six of the students pack their lunch. In simplest form, what fraction of the students in your class pack their lunch?

8. MP **Reasoning** Why do you have to divide a numerator and a denominator by the greatest of their common factors to write a fraction in simplest form?

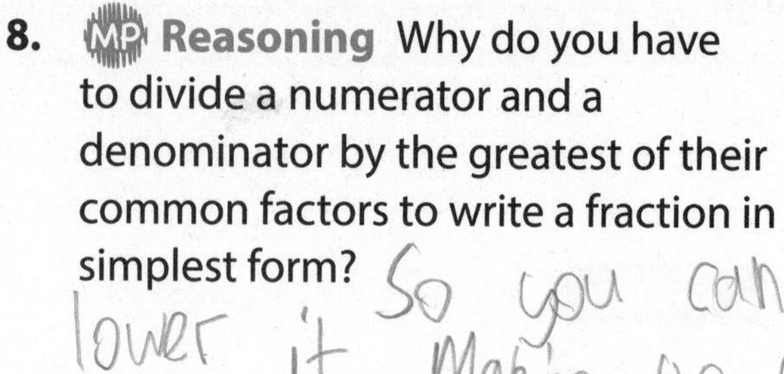

9. **Writing** Explain how you know when a fraction is in simplest form.

10. **Open-Ended** Write a fraction in which the numerator and the denominator have 1, 2, 4, and 8 as common factors. Then write the fraction in simplest form.

11. **Modeling Real Life** A flight attendant has visited 30 of the 50 states. What fraction of the states, in simplest form, has he *not* visited?

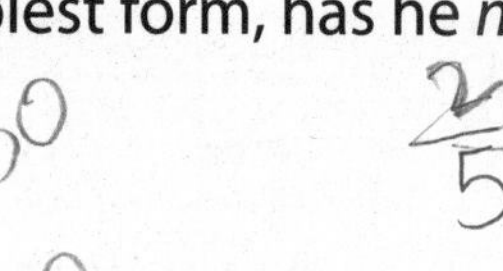

12. DIG DEEPER! A bin has red, orange, yellow, green, blue, and purple crayons. There are 4 of each color in the bin. In simplest form, what fraction of the crayons are red, orange, yellow, or green?

Review & Refresh

Estimate the sum or difference.

13. $598.44 - 45.61 =$ ______

14. $93.8 + 4.3 =$ ______

Name ________________________________

Estimate Sums and Differences of Fractions

8.2

Learning Target: Estimate sums and differences of fractions.

Success Criteria:

- I can use a number line and benchmarks to estimate a fraction.
- I can use mental math and benchmarks to estimate a fraction.
- I can use benchmarks to estimate sums and differences of fractions.

Explore and Grow

Plot $\frac{7}{12}$, $\frac{5}{6}$, and $\frac{1}{10}$ on the number line.

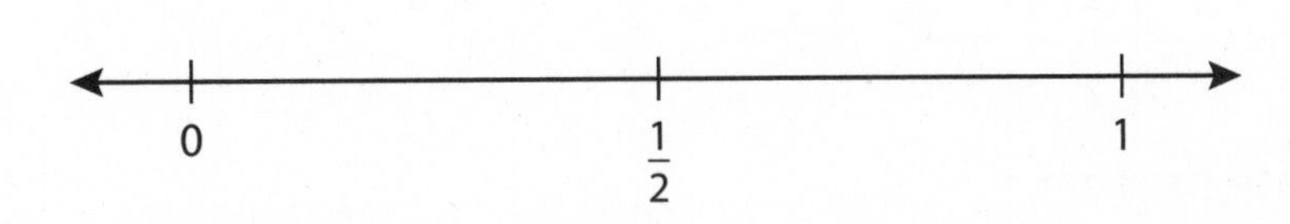

How can you estimate $\frac{7}{12} + \frac{5}{6}$?

How can you estimate $\frac{2}{3} - \frac{1}{10}$?

Reasoning Write two fractions that have a sum of about $\frac{1}{2}$. Then write two fractions that have a difference of about $\frac{1}{2}$. Explain your reasoning.

Think and Grow: Estimate Sums and Differences

You have used the benchmarks $\frac{1}{2}$ and 1 to compare fractions. You can use the benchmarks 0, $\frac{1}{2}$, and 1 to estimate sums and differences of fractions.

Example Estimate $\frac{1}{6} + \frac{5}{8}$.

Step 1: Use a number line to estimate each fraction.

$\frac{1}{6}$ is between 0 and $\frac{1}{2}$, but is closer to ______.

$\frac{5}{8}$ is between $\frac{1}{2}$ and 1, but is closer to ______.

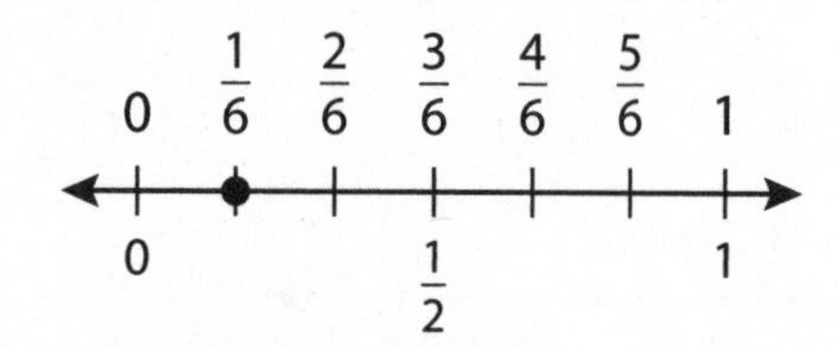

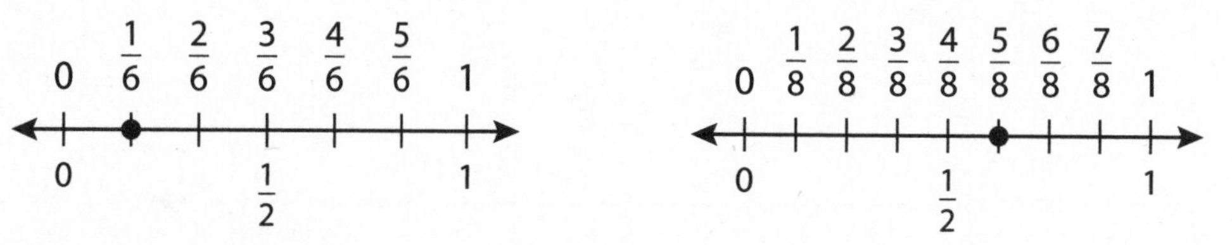

Step 2: Estimate the sum.

An estimate of $\frac{1}{6} + \frac{5}{8}$ is ______ + ______ = ______.

Example Estimate $\frac{9}{10} - \frac{2}{5}$.

Step 1: Use mental math to estimate each fraction.

$\frac{9}{10}$ is about ______. Think: The numerator is about the same as the denominator.

$\frac{2}{5}$ is about ______. Think: The numerator is about half of the denominator.

Compare the numerators to the denominators.

Step 2: Estimate the difference.

An estimate of $\frac{9}{10} - \frac{2}{5}$ is ______ − ______ = ______.

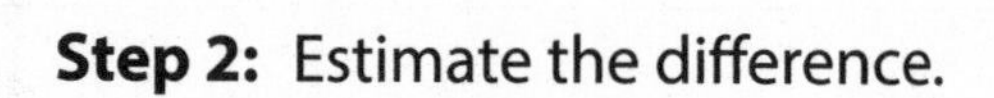

Show and Grow *I can do it!*

Estimate the sum or difference.

1. $\frac{1}{3} + \frac{11}{12}$

2. $\frac{3}{5} + \frac{5}{6}$

3. $\frac{15}{16} - \frac{7}{8}$

Name ___

Apply and Grow: Practice

Estimate the sum or difference.

4. $\frac{1}{6} + \frac{3}{5}$

5. $\frac{4}{5} - \frac{5}{12}$

6. $\frac{13}{16} + \frac{5}{6}$

7. $\frac{3}{6} - \frac{1}{8}$

8. $\frac{1}{14} + \frac{98}{100}$

9. $\frac{11}{12} - \frac{1}{8}$

10. You walk $\frac{1}{10}$ mile to your friend's house and then you both walk $\frac{2}{5}$ mile. Estimate how much farther you walk with your friend than you walk alone.

11. A carpenter has two wooden boards. One board is $\frac{3}{4}$ foot long and the other board is $\frac{1}{6}$ foot long. To determine whether the total length of the boards is 1 foot, should the carpenter use an estimate, or is an exact answer required? Explain.

12. MP **Number Sense** A fraction has a numerator of 1 and a denominator greater than 4. Is the fraction closer to 0, $\frac{1}{2}$, or 1? Explain.

Think and Grow: Modeling Real Life

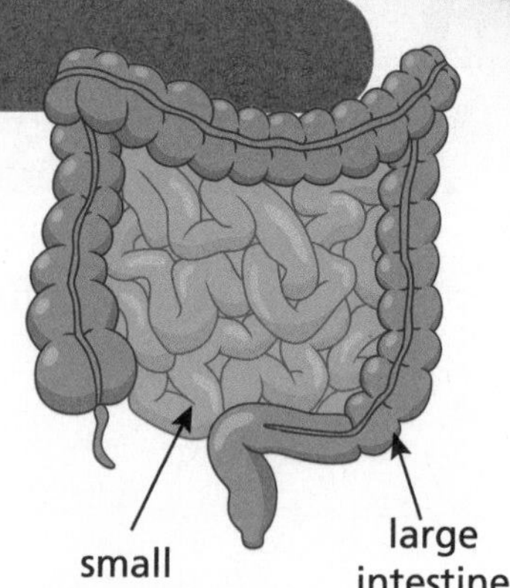

Example In the human body, the small intestine is about $20\frac{1}{12}$ feet long. The large intestine is about $4\frac{5}{6}$ feet long. About how long are the intestines in the human body?

To find the total length of the intestines, estimate $20\frac{1}{12} + 4\frac{5}{6}$.

Step 1: Use mental math to round each mixed number to the nearest whole number.

$20\frac{1}{12}$ is about _____. Think: $\frac{1}{12}$ is closer to 0 than it is to 1.

$4\frac{5}{6}$ is about _____. Think: $\frac{5}{6}$ is closer to 1 than it is to 0.

When the fractional part of a mixed number is close to 0 or 1, you can round the mixed number to the nearest whole number.

Step 2: Estimate the sum.

An estimate of $20\frac{1}{12} + 4\frac{5}{6}$ is _____ + _____ = _____.

So, the intestines in the human body are about _____ feet long.

Show and Grow *I can think deeper!*

13. A bullfrog jumps $5\frac{11}{12}$ feet. A leopard frog jumps $4\frac{1}{3}$ feet. About how much farther does the bullfrog jump than the leopard frog?

14. **DIG DEEPER!** A cell phone has 32 gigabytes of storage. The amounts of storage used by photos, songs, and apps are shown. About how many gigabytes of storage are left?

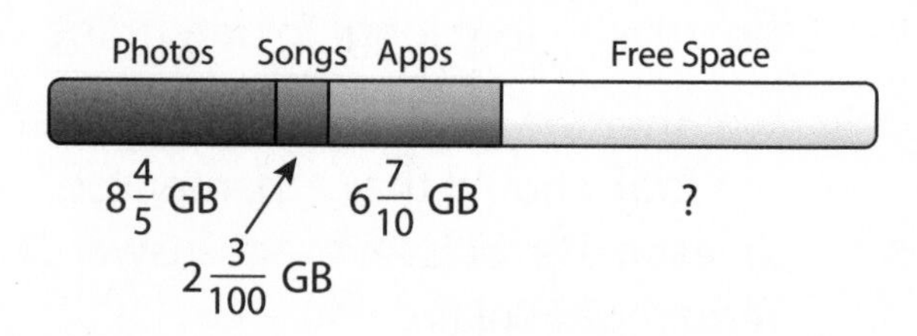

15. **DIG DEEPER!** Use two different methods to estimate how many cups of nut medley the recipe makes. Which estimate do you think is closer to the actual answer? Explain.

Nut Medley

- $1\frac{3}{8}$ cups almonds
- $\frac{5}{8}$ cup cashews
- $2\frac{1}{3}$ cups peanuts

Name ______________________________

Homework & Practice 8.2

Learning Target: Estimate sums and differences of fractions.

Example Estimate $\frac{5}{12} + \frac{8}{10}$.

Step 1: Use mental math to estimate each fraction.

$\frac{5}{12}$ is about $\underline{\frac{1}{2}}$. Think: The numerator is about half of the denominator.

$\frac{8}{10}$ is about $\underline{1}$. Think: The numerator is about the same as the denominator.

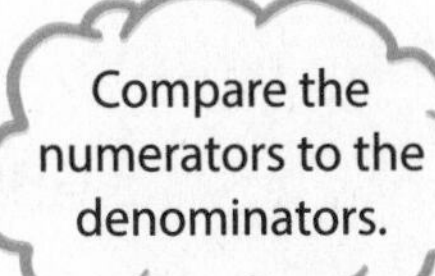
Compare the numerators to the denominators.

Step 2: Estimate the sum.

An estimate of $\frac{5}{12} + \frac{8}{10}$ is $\underline{\frac{1}{2}} + \underline{1} = \underline{1\frac{1}{2}}$.

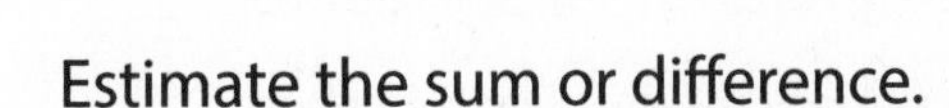

Estimate the sum or difference.

1. $\frac{11}{12} - \frac{5}{6}$

2. $\frac{17}{20} + \frac{13}{10}$

3. $\frac{3}{8} - \frac{1}{6}$

4. $\frac{7}{12} + \frac{2}{5}$

5. $\frac{4}{5} - \frac{7}{12}$

6. $\frac{1}{5} + 1\frac{10}{21}$

7. $3\frac{5}{8} - \frac{1}{10}$

8. $6\frac{1}{3} + 2\frac{4}{6}$

9. $5\frac{7}{8} - 4\frac{49}{100}$

10. You make a bag of trail mix with $\frac{2}{3}$ cup of raisins and $\frac{9}{8}$ cups of peanuts. About how much trail mix do you make?

11. You have $\frac{2}{3}$ cup of flour in a bin and $\frac{7}{8}$ cup of flour in a bag. To determine whether you have enough flour for a recipe that needs $1\frac{3}{4}$ cups of flour, should you use an estimate, or is an exact answer required? Explain.

12. **Writing** Explain how you know $\frac{9}{10} - \frac{3}{5}$ is about $\frac{1}{2}$.

13. MP **Precision** Your friend says $\frac{5}{8} + \frac{7}{12}$ is about 2. Find a closer estimate. Explain why your estimate is closer.

14. **Modeling Real Life** About how much taller is Robot A than Robot B?

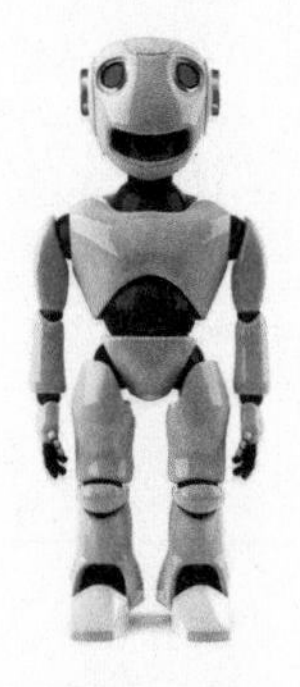

Robot A:

$1\frac{6}{10}$ meters tall

Robot B:

$1\frac{1}{5}$ meters tall

15. **Modeling Real Life** A class makes a paper chain that is $5\frac{7}{12}$ feet long. The class adds another $3\frac{5}{6}$ feet to the chain. About how long is the chain now?

Review & Refresh

Find the product. Check whether your answer is reasonable.

16. $509 \times 5 =$ _____

17. $7,692 \times 6 =$ _____

18. $31,435 \times 7 =$ _____

Name ______________________________

Find Common Denominators 8.3

Learning Target: Write fractions using a common denominator.

Success Criteria:

- I can list multiples of numbers.
- I can find a common denominator for two fractions.
- I can write fractions using a common denominator.

Explore and Grow

You cut a rectangular pan of vegetable lasagna into equal-sized pieces. You serve $\frac{1}{2}$ of the lasagna to a large table and $\frac{1}{3}$ of the lasagna to a small table. Draw a diagram that shows how you cut the lasagna.

What fraction of the lasagna does each piece represent? How does the denominator of the fraction compare to the denominators of $\frac{1}{2}$ and $\frac{1}{3}$?

Reasoning Is there another way you can cut the lasagna? Explain your reasoning.

Think and Grow: Find Common Denominators

Key Idea Fractions that have the same denominator are said to have a **common denominator**. You can find a common denominator either by finding a common multiple of the denominators or by finding the product of the denominators.

Example Use a common denominator to write equivalent fractions for $\frac{1}{2}$ and $\frac{5}{8}$.

List multiples of the denominators.

Multiples of 2: 2, 4, 6, (8), 10, 12, 14, (16), . . .

Multiples of 8: (8), (16), 24, 32, 40, . . .

• Because 8 is a multiple of 2, 8 is a common multiple of 2 and 8.
• The product of the denominators is always a common multiple!

Use a common denominator of 8. Write a fraction equivalent to $\frac{1}{2}$ with a denominator of 8.

$\frac{1}{2} = \frac{1 \times \square}{2 \times \square} = \frac{\square}{8}$

One way to write the fractions with a common denominator is ______ and $\frac{5}{8}$.

Example Use a common denominator to write equivalent fractions for $\frac{2}{3}$ and $\frac{1}{4}$.

Use the product of the denominators: $3 \times 4 =$ ______.

Write equivalent fractions with denominators of 12.

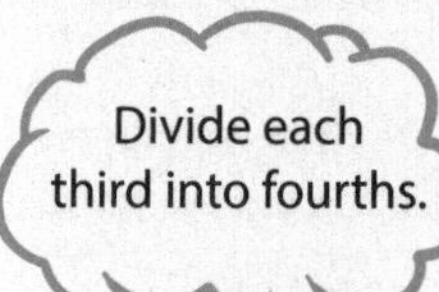

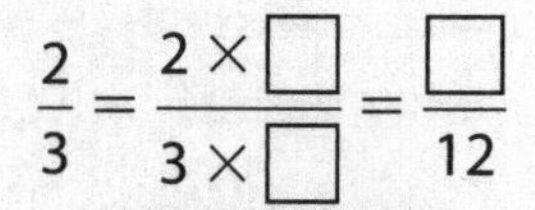

Divide each third into fourths.

$\frac{2}{3} = \frac{2 \times \square}{3 \times \square} = \frac{\square}{12}$

$\frac{1}{4} = \frac{1 \times \square}{4 \times \square} = \frac{\square}{12}$

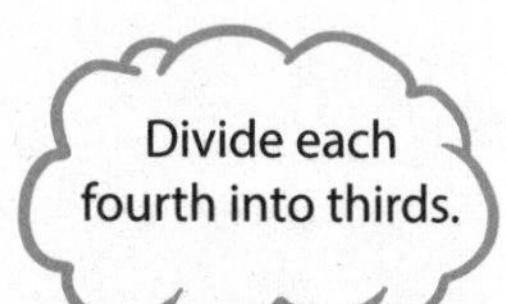

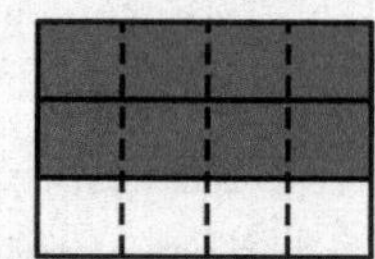

Divide each fourth into thirds.

One way to write the fractions with a common denominator is ______ and ______.

Show and Grow *I can do it!*

Use a common denominator to write an equivalent fraction for each fraction.

1. $\frac{2}{3}$ and $\frac{1}{6}$

2. $\frac{5}{6}$ and $\frac{3}{4}$

Name ________________________________

Apply and Grow: Practice

Use a common denominator to write an equivalent fraction for each fraction.

3. $\frac{2}{3}$ and $\frac{5}{6}$

4. $\frac{3}{4}$ and $\frac{1}{2}$

5. $\frac{5}{9}$ and $\frac{2}{3}$

6. $\frac{8}{21}$ and $\frac{3}{7}$

7. $\frac{1}{5}$ and $\frac{1}{2}$

8. $\frac{3}{4}$ and $\frac{1}{6}$

9. $\frac{3}{7}$ and $\frac{2}{9}$

10. $\frac{3}{8}$ and $\frac{5}{11}$

11. You walk your dog $\frac{3}{4}$ mile on Saturday and $\frac{5}{8}$ mile on Sunday. Use a common denominator to write an equivalent fraction for each fraction.

12. **Writing** Explain how to use the models to find a common denominator for for $\frac{1}{2}$ and $\frac{3}{5}$. Then write an equivalent fraction for each fraction.

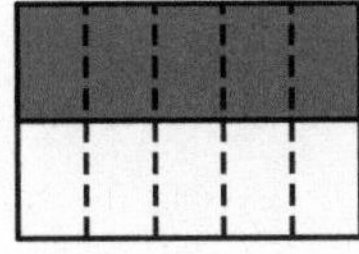

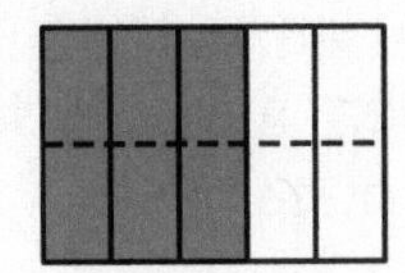

13. MP **Number Sense** Which pairs of fractions are equivalent to $\frac{1}{2}$ and $\frac{2}{3}$?

$\frac{2}{4}$ and $\frac{3}{4}$

$\frac{6}{12}$ and $\frac{8}{12}$

$\frac{3}{6}$ and $\frac{4}{6}$

$\frac{1}{18}$ and $\frac{12}{18}$

Think and Grow: Modeling Real Life

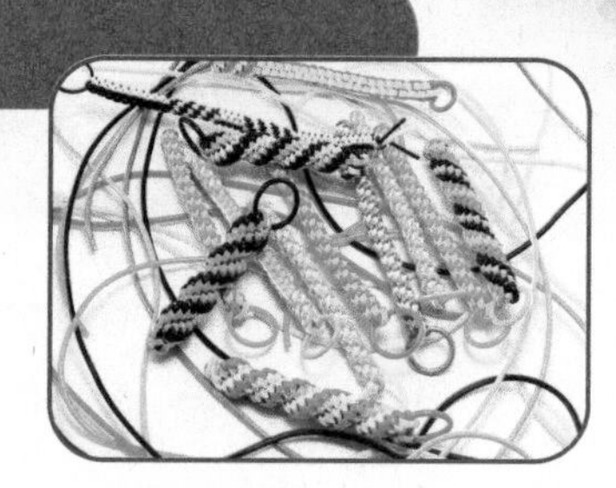

Example You and your friend make woven key chains. Your key chain is $\frac{2}{4}$ foot long. Your friend's is $\frac{3}{6}$ foot long. Are the key chains the same length?

Use a common denominator to write equivalent fractions for the lengths of the key chains. Use the product of the denominators.

$4 \times 6 =$ _____

Write equivalent fractions with denominators of 24.

$\frac{2}{4} = \frac{2 \times \square}{4 \times \square} = \frac{\square}{24}$ $\frac{3}{6} = \frac{3 \times \square}{6 \times \square} = \frac{\square}{24}$

Compare the lengths of the key chains.

So, the key chains ______________ the same length.

Show and Grow *I can think deeper!*

14. Your hamster weighs $\frac{13}{16}$ ounce. Your friend's hamster weighs $\frac{6}{8}$ ounce. Do the hamsters weigh the same amount?

15. **DIG DEEPER!** You have three vegetable pizzas of the same size. One has 4 equal slices. The second has 8 equal slices. The third has 6 equal slices. You cut the pizzas until all of them have the same number of slices. How many slices does each pizza have?

Name ______________________________

Homework & Practice 8.3

Learning Target: Write fractions using a common denominator.

Example Use a common denominator to write equivalent fractions for $\frac{1}{3}$ and $\frac{4}{9}$.

List multiples of the denominators.

Multiples of 3: 3, 6, (9), 12, 15, (18), 21, 24, (27), . . .

Multiples of 9: (9), (18), (27), . . .

- Because 9 is a multiple of 3, 9 is a common multiple of 3 and 9.
- The product of the denominators is always a common multiple!

Use a common denominator of 9. Write a fraction equivalent to $\frac{1}{3}$ with a denominator of 9.

$$\frac{1}{3} = \frac{1 \times \boxed{3}}{3 \times \boxed{3}} = \frac{\boxed{3}}{9}$$

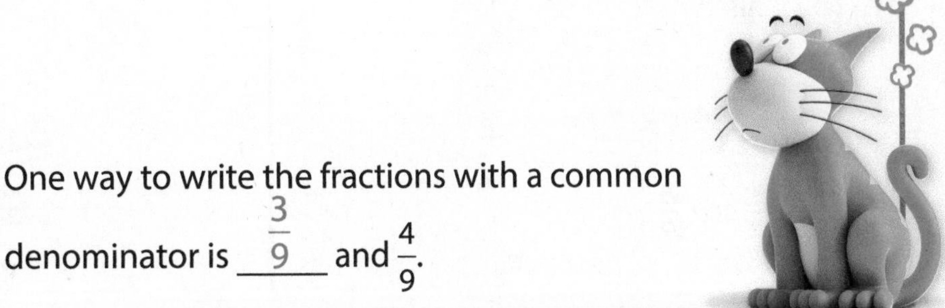

One way to write the fractions with a common denominator is $\underline{\frac{3}{9}}$ and $\frac{4}{9}$.

Use a common denominator to write an equivalent fraction for each fraction.

1. $\frac{1}{2}$ and $\frac{3}{8}$

2. $\frac{7}{9}$ and $\frac{2}{3}$

3. $\frac{5}{6}$ and $\frac{1}{2}$

4. $\frac{3}{4}$ and $\frac{5}{16}$

5. $\frac{18}{24}$ and $\frac{5}{6}$

6. $\frac{1}{3}$ and $\frac{1}{5}$

7. $\frac{3}{5}$ and $\frac{4}{7}$

8. $\frac{5}{8}$ and $\frac{2}{9}$

9. A mint plant grows $\frac{7}{8}$ inch in 1 week and $\frac{13}{16}$ inch the next week. Use a common denominator to write an equivalent fraction for each fraction.

10. Which One Doesn't Belong? Which pair of fractions is *not* equivalent to $\frac{2}{5}$ and $\frac{1}{10}$?

$\frac{12}{30}$ and $\frac{3}{30}$

$\frac{6}{15}$ and $\frac{5}{15}$

$\frac{20}{50}$ and $\frac{10}{100}$

$\frac{8}{20}$ and $\frac{2}{20}$

11. YOU BE THE TEACHER Your friend says she used a common denominator to find fractions equivalent to $\frac{2}{3}$ and $\frac{8}{9}$. Is your friend correct? Explain.

$\frac{2}{3} = \frac{2 \times 4}{3 \times 4} = \frac{8}{12}$

$\frac{8}{12}$ and $\frac{8}{9}$

12. Modeling Real Life Some friends spend $\frac{1}{3}$ hour collecting sticks and $\frac{5}{6}$ hour building a fort. Do they spend the same amount of time on each? Explain.

13. DIG DEEPER! Use a common denominator to write an equivalent fraction for each fraction. Which two students are the same distance from the school? Are they closer to or farther from the school than the other student?

Student	Distance from School (mile)	Equivalent Distance (mile)
A	$\frac{10}{12}$	
B	$\frac{7}{8}$	
C	$\frac{5}{6}$	

Review & Refresh

Find the value of the expression.

14. 10^2

15. 8×10^4

16. 6×10^3

17. 9×10^5

Name ______________________________

Add Fractions with Unlike Denominators 8.4

Learning Target: Add fractions with unlike denominators.

Success Criteria:

- I can write fractions using a common denominator.
- I can add fractions with like denominators.
- I can add fractions with unlike denominators.

Explore and Grow

Use a model to find the sum.

$\frac{2}{5} + \frac{7}{10} = ?$

$\square + \square = ?$

Explain how you can use a model to add fifths and tenths.

Construct Arguments How can you add two fractions with unlike denominators without using a model? Explain why your method makes sense.

Think and Grow: Add Fractions with Unlike Denominators

You can use equivalent fractions to add fractions that have unlike denominators.

Example Find $\frac{1}{4} + \frac{3}{8}$.

Use equivalent fractions to write the fractions with a common denominator. Then find the sum.

Think: 8 is a multiple of 4, so rewrite $\frac{1}{4}$ with a denominator of 8.

$\frac{1}{4} + \frac{3}{8} = \frac{\square}{8} + \frac{3}{8}$ Rewrite $\frac{1}{4}$ as $\frac{1 \times 2}{4 \times 2} = \frac{2}{8}$.

$= \frac{\square + \square}{8}$

$= \frac{\square}{\square}$

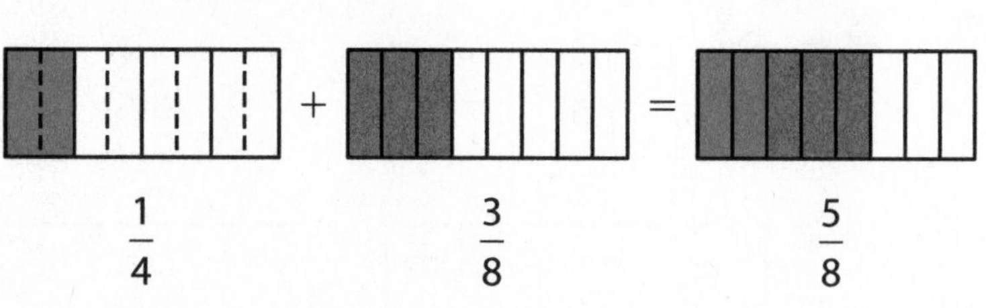

Example Find $\frac{7}{8} + \frac{1}{6}$. Estimate ______

Use equivalent fractions to write the fractions with a common denominator. Then find the sum.

Think: 8 is not a multiple of 6, so rewrite each fraction with a denominator of $8 \times 6 = 48$.

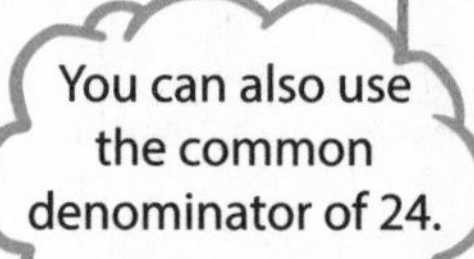

You can also use the common denominator of 24.

$\frac{7}{8} + \frac{1}{6} = \frac{\square}{48} + \frac{\square}{48}$ Rewrite $\frac{7}{8}$ as $\frac{7 \times 6}{8 \times 6} = \frac{42}{48}$ and $\frac{1}{6}$ as $\frac{1 \times 8}{6 \times 8} = \frac{8}{48}$.

$= \frac{\square + \square}{48}$

$= \frac{\square}{48}$, or $\frac{\square}{\square}$ Reasonable? ______ is close to ______. ✓

Show and Grow *I can do it!*

Add.

1. $\frac{5}{6} + \frac{2}{3} =$ ______

2. $\frac{1}{5} + \frac{3}{4} =$ ______

3. $\frac{1}{6} + \frac{1}{4} =$ ______

Name ____

Apply and Grow: Practice

Add.

4. $\frac{5}{8} + \frac{1}{4} =$ ______

5. $\frac{2}{3} + \frac{7}{12} =$ ______

6. $\frac{2}{5} + \frac{10}{15} =$ ______

7. $\frac{1}{6} + \frac{4}{8} =$ ______

8. $\frac{11}{12} + \frac{3}{5} =$ ______

9. $\frac{2}{9} + \frac{4}{3} + \frac{5}{9} =$ ______

10. Your friend buys $\frac{1}{8}$ pound of green lentils and $\frac{3}{4}$ pound of brown lentils. What fraction of a pound of lentils does she buy?

11. MP **Reasoning** Newton and Descartes find $\frac{1}{2} + \frac{1}{6}$. Newton says the sum is $\frac{4}{6}$. Descartes says the sum is $\frac{2}{3}$. Who is correct? Explain.

12. DIG DEEPER! Write two fractions that have a sum of 1 and have different denominators.

Think and Grow: Modeling Real Life

Example About $\frac{17}{25}$ of Earth's surface is covered by ocean water. About $\frac{3}{100}$ of Earth's surface is covered by other water sources. About how much of Earth's surface is covered by water?

Add $\frac{17}{25}$ and $\frac{3}{100}$ to find the fraction of Earth's surface that is covered by water.

Estimate _______

You can add the fractions because they each represent a part of the same whole: Earth.

Think: 100 is a multiple of 25, so rewrite $\frac{17}{25}$ with a denominator of 100.

$$\frac{17}{25} + \frac{3}{100} = \frac{\square}{100} + \frac{3}{100}$$

Rewrite $\frac{17}{25}$ as $\frac{17 \times 4}{25 \times 4} = \frac{68}{100}$.

$$= \frac{\square + \square}{100}$$

$$= \frac{\square}{100}$$

Reasonable? _____ is close to _____. ✓

So, about _____ of Earth's surface is covered by water.

Show and Grow *I can think deeper!*

13. The George Washington Bridge links Manhattan, NY, to Fort Lee, NJ. The part of the bridge in New Jersey is about $\frac{1}{2}$ mile long. The part in New York is about $\frac{2}{5}$ mile long. About how long is the George Washington Bridge?

The George Washington Bridge is the busiest motor vehicle bridge in the world.

---

14. **DIG DEEPER!** Your goal is to practice playing the saxophone for at least 2 hours in 1 week. Do you reach your goal? Explain.

Day	Monday	Wednesday	Friday
Practice Time	$\frac{3}{4}$ hour	$\frac{1}{2}$ hour	$\frac{2}{3}$ hour

Name ______________________________

Homework & Practice 8.4

Learning Target: Add fractions with unlike denominators.

Example Find $\frac{2}{5} + \frac{1}{3}$.   Estimate $\underline{\frac{1}{2} + \frac{1}{2} = 1}$

Use equivalent fractions to write the fractions with a common denominator. Then find the sum.

Think: 5 is not a multiple of 3, so rewrite each fraction with a denominator of $5 \times 3 = 15$.

$\frac{2}{5} + \frac{1}{3} = \frac{\boxed{6}}{15} + \frac{\boxed{5}}{15}$ Rewrite $\frac{2}{5}$ as $\frac{2 \times 3}{5 \times 3} = \frac{6}{15}$ and $\frac{1}{3}$ as $\frac{1 \times 5}{3 \times 5} = \frac{5}{15}$.

$= \frac{\boxed{6} + \boxed{5}}{15}$

$= \frac{\boxed{11}}{15}$ Reasonable? $\underline{\frac{11}{15}}$ is close to $\underline{1}$. ✓

Add.

1. $\frac{1}{9} + \frac{2}{3} =$ ______

2. $\frac{1}{2} + \frac{3}{4} =$ ______

3. $\frac{4}{6} + \frac{5}{12} =$ ______

4. $\frac{1}{3} + \frac{1}{4} =$ ______

5. $\frac{3}{2} + \frac{4}{5} =$ ______

6. $\frac{6}{8} + \frac{9}{10} + \frac{1}{8} =$ ______

7. You use beads to make a design. Of the beads, $\frac{1}{3}$ are red and $\frac{1}{6}$ are blue. The rest are white. What fraction of the beads are red or blue?

8. **YOU BE THE TEACHER** Your friend says the sum of $\frac{1}{5}$ and $\frac{9}{10}$ is $\frac{10}{15}$. Is your friend correct? Explain.

9. **MP Reasoning** Which expressions are equal to $\frac{14}{15}$?

$\frac{8}{8} + \frac{6}{7}$

$\frac{3}{5} + \frac{1}{3}$

$\frac{1}{5} + \frac{11}{15}$

$\frac{11}{10} + \frac{3}{5}$

10. **Modeling Real Life** There are 100 senators in the 115th Congress. Democrats make up $\frac{46}{100}$ of the senators, and Republicans make up $\frac{13}{25}$ of the senators. The rest are Independents. What fraction of the senators are Democrat or Republican?

11. **Modeling Real Life** Your friend needs 1 cup of homemade orange juice. He squeezes $\frac{1}{2}$ cup of orange juice from one orange and $\frac{3}{8}$ cup from another orange. Does your friend need to squeeze another orange? Explain.

12. **DIG DEEPER!** Of all the atoms in caffeine, $\frac{1}{12}$ are oxygen atoms, $\frac{1}{6}$ are nitrogen atoms, and $\frac{1}{3}$ are carbon atoms. The rest of the atoms are hydrogen. What fraction of the atoms in caffeine are oxygen, nitrogen, or hydrogen?

Review & Refresh

Use properties to find the sum or product.

13. 5×84

14. $521 + 0 + 67$

15. $25 \times 8 \times 4$

Name ____________________

Subtract Fractions with Unlike Denominators

8.5

Learning Target: Subtract fractions with unlike denominators.

Success Criteria:

- I can write fractions using a common denominator.
- I can subtract fractions with like denominators.
- I can subtract fractions with unlike denominators.

Explore and Grow

Use a model to find the difference.

$$\frac{7}{12} - \frac{1}{4} = ?$$

$$\square = ?$$

Explain how you can use a model to subtract fourths from twelfths.

Construct Arguments How can you subtract two fractions with unlike denominators without using a model? Explain why your method makes sense.

Think and Grow: Subtract Fractions with Unlike Denominators

You can use equivalent fractions to subtract fractions that have unlike denominators.

Example Find $\frac{9}{10} - \frac{1}{2}$.

Use equivalent fractions to write the fractions with a common denominator. Then find the difference.

Think: 10 is a multiple of 2, so rewrite $\frac{1}{2}$ with a denominator of 10.

$\frac{9}{10} - \frac{1}{2} = \frac{9}{10} - \frac{\square}{10}$

$= \frac{\square - \square}{10}$

$= \frac{\square}{\square}$, or $\frac{\square}{\square}$

Rewrite $\frac{1}{2}$ as $\frac{1 \times 5}{2 \times 5} = \frac{5}{10}$.

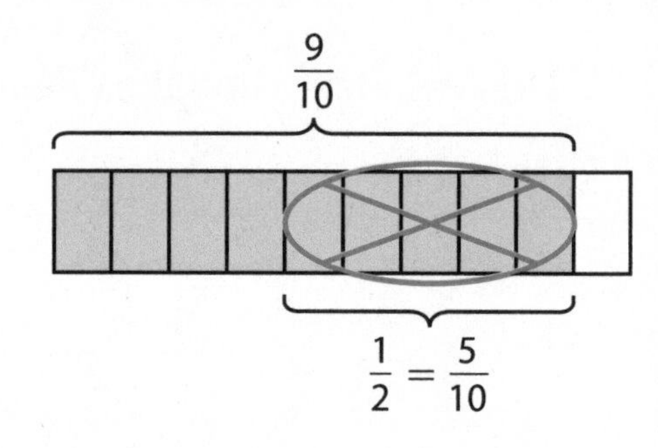

Example Find $\frac{4}{3} - \frac{1}{4}$. Estimate ______

Use equivalent fractions to write the fractions with a common denominator. Then find the difference.

Think: 4 is not a multiple of 3, so rewrite each fraction with a denominator of $3 \times 4 = 12$.

$\frac{4}{3} - \frac{1}{4} = \frac{\square}{12} - \frac{\square}{12}$

$= \frac{\square - \square}{12}$

$= \frac{\square}{12}$

Rewrite $\frac{4}{3}$ as $\frac{4 \times 4}{3 \times 4} = \frac{16}{12}$ and $\frac{1}{4}$ as $\frac{1 \times 3}{4 \times 3} = \frac{3}{12}$.

Reasonable? ______ is close to ______. ✓

Show and Grow *I can do it!*

Subtract.

1. $\frac{1}{2} - \frac{1}{4} =$ ______

2. $\frac{7}{9} - \frac{2}{3} =$ ______

3. $\frac{6}{5} - \frac{3}{8} =$ ______

Name ____________________

Apply and Grow: Practice

Subtract.

4. $\frac{10}{12} - \frac{3}{4} =$ _____

5. $\frac{1}{3} - \frac{1}{6} =$ _____

6. $\frac{9}{10} - \frac{2}{5} =$ _____

7. $\frac{5}{4} - \frac{2}{5} =$ _____

8. $\frac{13}{16} - \frac{3}{16} - \frac{5}{8} =$ _____

9. $\frac{8}{9} - \left(\frac{2}{3} + \frac{1}{6}\right) =$ _____

10. You have $\frac{3}{4}$ yard of wire. You use $\frac{1}{3}$ yard to make an electric circuit. How much wire do you have left?

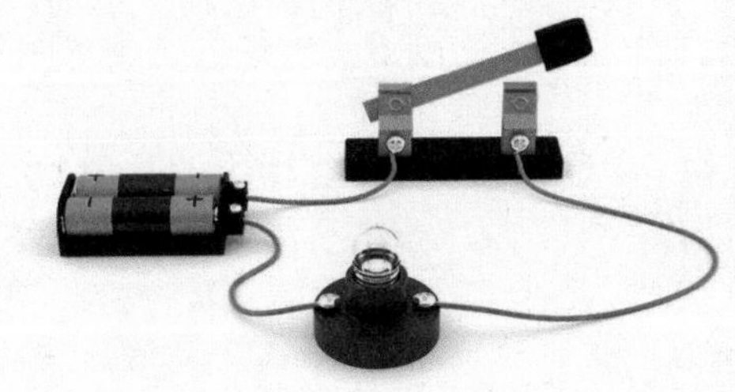

11. MP **Precision** Your friend finds $\frac{5}{8} - \frac{2}{5}$. Explain why his answer is unreasonable. What did he do wrong?

$$\frac{5}{8} - \frac{2}{5} = \frac{3}{3} = 1$$

12. **Number Sense** Which two fractions have a difference of $\frac{1}{8}$?

$\frac{1}{2}$ $\frac{2}{10}$ $\frac{1}{6}$ $\frac{3}{8}$

Think and Grow: Modeling Real Life

Quartz Sand

Volcanic Sand

Example A geologist needs $\frac{1}{2}$ cup of volcanic sand to perform an experiment. She has $\frac{3}{2}$ cups of quartz sand. She has $\frac{2}{3}$ cup more quartz sand than volcanic sand. Can she perform the experiment?

Find how many cups of volcanic sand the geologist has by subtracting $\frac{2}{3}$ from $\frac{3}{2}$.

Use equivalent fractions to write the fractions with a common denominator. Then find the difference.

Think: 3 is not a multiple of 2, so rewrite each fraction with a denominator of $2 \times 3 = 6$.

$$\frac{3}{2} - \frac{2}{3} = \frac{\square}{6} - \frac{\square}{6}$$

$$= \frac{\square - \square}{6}$$

$$= \frac{\square}{6}$$

Rewrite $\frac{3}{2}$ as $\frac{3 \times 3}{2 \times 3} = \frac{9}{6}$ and $\frac{2}{3}$ as $\frac{2 \times 2}{3 \times 2} = \frac{4}{6}$.

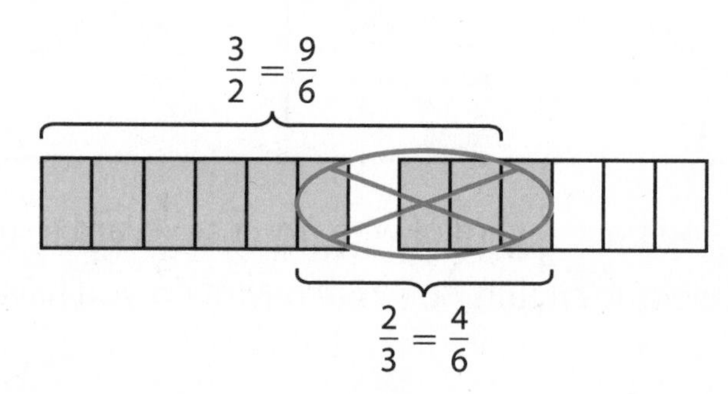

Compare the difference to $\frac{1}{2}$.

$$\frac{\square}{6} \overset{?}{>} \frac{1}{2}$$

So, the geologist _______ have enough volcanic sand to perform the experiment.

Show and Grow *I can think deeper!*

13. The world record for the longest dog tail is $\frac{77}{100}$ meter. The previous record was $\frac{1}{20}$ meter shorter than the current record. Was the previous record longer than $\frac{3}{4}$ meter?

14. **DIG DEEPER!** A woodworker has 1 gallon of paint for a tree house. He uses $\frac{3}{8}$ gallon to paint the walls and $\frac{1}{5}$ gallon to paint the ladder. He needs $\frac{1}{4}$ gallon to paint the roof. Does he have enough paint? Explain.

Name ____________________

Homework & Practice 8.5

Learning Target: Subtract fractions with unlike denominators.

Example Find $\frac{5}{8} - \frac{1}{6}$. Estimate $\frac{1}{2} - 0 = \frac{1}{2}$

Use equivalent fractions to write the fractions with a common denominator. Then find the difference.

Think: 8 is not a multiple of 6, so rewrite each fraction with a denominator of $6 \times 8 = 48$.

$\frac{5}{8} - \frac{1}{6} = \frac{30}{48} - \frac{8}{48}$ Rewrite $\frac{5}{8}$ as $\frac{5 \times 6}{8 \times 6} = \frac{30}{48}$ and $\frac{1}{6}$ as $\frac{1 \times 8}{6 \times 8} = \frac{8}{48}$.

$= \frac{30 - 8}{48}$

$= \frac{22}{48}$, or $\frac{11}{24}$

Reasonable? $\frac{11}{24}$ is close to $\frac{1}{2}$. ✓

Subtract.

1. $\frac{3}{4} - \frac{1}{8} =$ ______

2. $\frac{4}{5} - \frac{6}{15} =$ ______

3. $\frac{1}{2} - \frac{1}{8} =$ ______

4. $\frac{5}{3} - \frac{3}{4} =$ ______

5. $\frac{6}{8} - \frac{7}{10} =$ ______

6. $\frac{5}{6} - \frac{1}{4} - \frac{3}{12} =$ ______

7. You eat $\frac{1}{12}$ of a vegetable casserole. Your friend eats $\frac{1}{6}$ of the same casserole. How much more does your friend eat than you?

8. **Writing** Why do fractions need a common denominator before you can add or subtract them?

9. MP **Logic** Find a.

$$\frac{7}{10} - a = \frac{1}{2}$$

10. **DIG DEEPER!** Write and solve an equation to find the difference between Length A and Length B on the ruler.

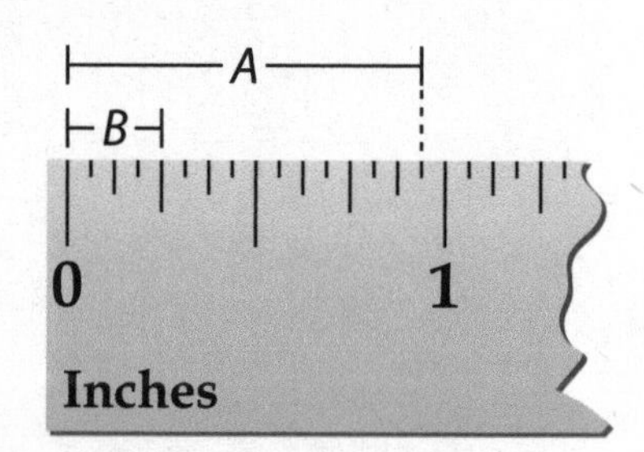

11. **Modeling Real Life** You want to stack cups in $\frac{1}{4}$ minute. Your first attempt takes $\frac{1}{2}$ minute. Your second attempt takes $\frac{3}{10}$ minute less than your first attempt. Do you meet your goal?

12. **Modeling Real Life** You and your friend each have a canvas of the same size. You divide your canvas into 5 sections and paint 3 of them. Your friend divides her canvas into 7 sections and paints 4 of them. Who paints more? How much more?

Review & Refresh

Evaluate. Check whether your answer is reasonable.

13. $1.7 + 5 + 4.3 =$ _____

14. $15.24 + 6.13 - 7 =$ _____

Name ______________________________

Add Mixed Numbers 8.6

Learning Target: Add mixed numbers with unlike denominators.

Success Criteria:

- I can add fractional parts and whole number parts of mixed numbers with unlike denominators.
- I can use equivalent fractions to add mixed numbers with unlike denominators.

Explore and Grow

Use a model to find the sum.

$$1\frac{3}{4} + 2\frac{1}{8}$$

Construct Arguments How can you add mixed numbers with unlike denominators without using a model? Explain why your method makes sense.

Think and Grow: Add Mixed Numbers

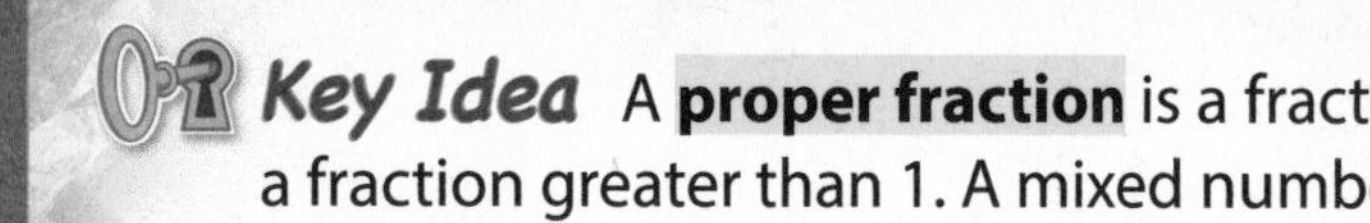

Key Idea A **proper fraction** is a fraction less than 1. An **improper fraction** is a fraction greater than 1. A mixed number represents the sum of a whole number and a proper fraction. You can use equivalent fractions to add mixed numbers.

Example Find $1\frac{1}{2} + 2\frac{5}{6}$.

One Way: Add the fractional parts and add the whole number parts.

To add the fractional parts, use a common denominator.

$$\begin{array}{r} 1\frac{1}{2} \\ +\ 2\frac{5}{6} \\ \hline \end{array} \longrightarrow \begin{array}{r} 1\frac{\square}{6} \\ +\ 2\frac{5}{6} \\ \hline 3\frac{8}{6} \end{array}, \text{ or } \square\frac{\square}{\square}$$

Another Way: Write the mixed numbers as improper fractions with a common denominator, then add.

$1\frac{1}{2} = 1 + \frac{1}{2} = \frac{2}{2} + \frac{1}{2} = \frac{3}{2} = \frac{\square}{6}$

$2\frac{5}{6} = 2 + \frac{5}{6} = \frac{12}{6} + \frac{5}{6} = \frac{17}{6}$

$\frac{9}{6} + \frac{17}{6} = \frac{\square}{\square}$, or $\square\frac{\square}{\square}$

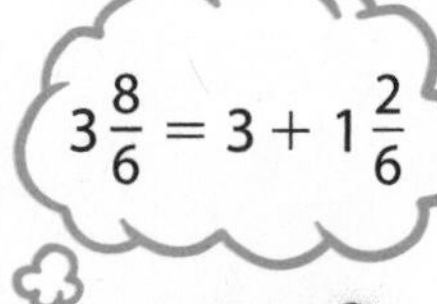

$\frac{26}{6} = \frac{24}{6} + \frac{2}{6}$

So, $1\frac{1}{2} + 2\frac{5}{6} = \square\frac{\square}{\square}$.

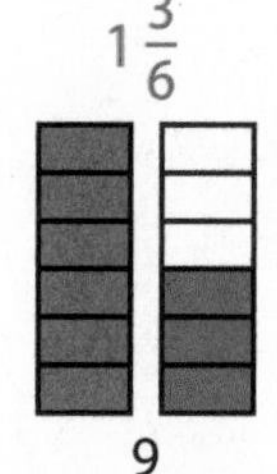

$1\frac{3}{6}$ + $2\frac{5}{6}$ = $3\frac{8}{6} = 4\frac{2}{6} = 4\frac{1}{3}$

$\frac{9}{6}$ + $\frac{17}{6}$ = $\frac{26}{6} = 4\frac{2}{6} = 4\frac{1}{3}$

Show and Grow *I can do it!*

Add.

1. $2\frac{2}{3} + 2\frac{1}{6} =$ ________

2. $1\frac{5}{12} + 3\frac{3}{4} =$ ________

Name ______________________________

Apply and Grow: Practice

Add.

3. $5\frac{4}{9} + 1\frac{2}{3} =$ ______

4. $3\frac{1}{2} + \frac{5}{12} =$ ______

5. $4\frac{5}{6} + 3\frac{5}{12} =$ ______

6. $\frac{4}{5} + 8\frac{7}{20} =$ ______

7. $2\frac{1}{3} + \frac{1}{6} + 3\frac{2}{3} =$ ______

8. $5\frac{1}{2} + 4\frac{3}{4} + 6\frac{5}{8} =$ ______

9. Your science class makes magic milk using $1\frac{1}{8}$ cups of watercolor paint and $1\frac{3}{4}$ cups of milk. How many cups of magic milk does your class make?

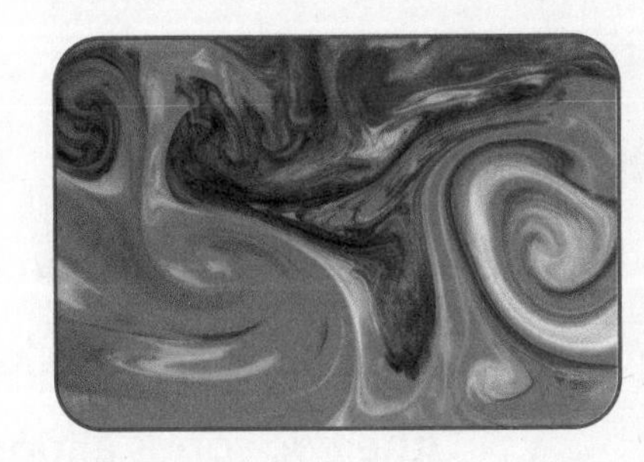

10. **MP Structure** Find $2\frac{3}{10} + 4\frac{2}{5}$ two different ways.

11. **DIG DEEPER!** Find the missing numbers.

$$2\frac{3}{4} + \square\frac{\square}{8} = 4\frac{3}{8}$$

Think and Grow: Modeling Real Life

Example You kayak $1\frac{8}{10}$ miles and then take a break. You kayak $1\frac{1}{4}$ more miles. How many miles do you kayak altogether?

Add the distances you kayak before and after you take a break.

Estimate ______

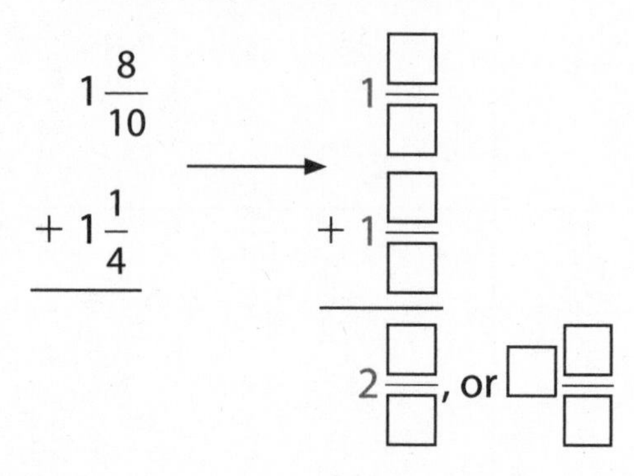

$$1\frac{8}{10} + 1\frac{1}{4} \rightarrow 1\frac{\square}{\square} + 1\frac{\square}{\square} = 2\frac{\square}{\square}, \text{ or } \square\frac{\square}{\square}$$

Reasonable? ______ is close to ______. ✓

You kayak ______ miles altogether.

Show and Grow *I can think deeper!*

12. You listen to a song that is $2\frac{3}{4}$ minutes long. Then you listen to a song that is $3\frac{1}{3}$ minutes long. How many minutes do you spend listening to the two songs altogether?

13. **DIG DEEPER!** A beekeeper collects $3\frac{3}{4}$ more pounds of honey from Hive 3 than Hive 1. Which hive produces the most honey? Explain.

Hive	Honey Collected (pounds)
1	$23\frac{5}{8}$
2	$27\frac{1}{2}$
3	?

Name ______________________

Homework & Practice 8.6

Learning Target: Add mixed numbers with unlike denominators.

Example Find $1\frac{3}{4} + 4\frac{5}{8}$.

One Way: Add the fractional parts and add the whole number parts.

To add the fractional parts, use a common denominator.

$$\begin{array}{r} 1\frac{3}{4} \\ +\ 4\frac{5}{8} \\ \hline \end{array} \longrightarrow \begin{array}{r} 1\frac{\boxed{6}}{8} \\ +\ 4\frac{5}{8} \\ \hline 5\frac{11}{8} \end{array}, \text{ or } \boxed{6}\frac{\boxed{3}}{\boxed{8}}$$

Another Way: Write the mixed numbers as improper fractions with a common denominator, then add.

$$1\frac{3}{4} = 1 + \frac{3}{4} = \frac{4}{4} + \frac{3}{4} = \frac{7}{4} = \frac{\boxed{14}}{8}$$

$$4\frac{5}{8} = 4 + \frac{5}{8} = \frac{32}{8} + \frac{5}{8} = \frac{37}{8}$$

$$\frac{14}{8} + \frac{37}{8} = \frac{\boxed{51}}{\boxed{8}}, \text{ or } \boxed{6}\frac{\boxed{3}}{\boxed{8}}$$

So, $1\frac{3}{4} + 4\frac{5}{8} = \boxed{6}\frac{\boxed{3}}{\boxed{8}}$.

Add.

1. $6\frac{2}{5} + 1\frac{3}{10} =$ ______

2. $2\frac{2}{3} + 5\frac{3}{6} =$ ______

3. $\frac{1}{4} + 3\frac{2}{5} =$ ______

4. $9\frac{5}{7} + \frac{2}{3} =$ ______

5. $2\frac{1}{2} + 1\frac{3}{4} + \frac{1}{2} =$ ______

6. $2\frac{2}{3} + 4\frac{1}{2} + 3\frac{5}{6} =$ ______

7. A veterinarian spends $3\frac{3}{4}$ hours helping cats and $5\frac{1}{2}$ hours helping dogs. How many hours does she spend helping cats and dogs altogether?

8. **Writing** How is adding mixed numbers with unlike denominators similar to adding fractions with unlike denominators? How is it different?

9. **MP Logic** Can you add two mixed numbers and get a sum of 2? Explain.

10. **MP Structure** Shade the model to represent the sum. Then write an equation to represent your model.

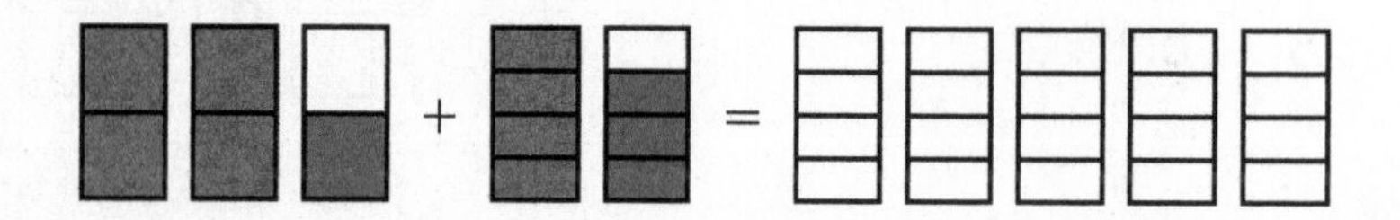

11. **Modeling Real Life** An emperor tamarin has a body length of $9\frac{5}{10}$ inches and a tail length of $14\frac{1}{4}$ inches. How long is the emperor tamarin?

12. **DIG DEEPER!** A long jumper jumps $1\frac{2}{3}$ feet farther on her third attempt than her second attempt. On which attempt does she jump the farthest? Explain.

Attempt	First	Second	Third
Distance (feet)	$15\frac{5}{6}$	$13\frac{3}{4}$	?

Review & Refresh

Find the product. Check whether your answer is reasonable.

13. 354×781

14. $4{,}029 \times 276$

15. 950×326

Name ______________________________

Subtract Mixed Numbers 8.7

Learning Target: Subtract mixed numbers with unlike denominators.

Success Criteria:

- I can subtract fractional parts and whole number parts of mixed numbers with unlike denominators.
- I can use equivalent fractions to subtract mixed numbers with unlike denominators.

Explore and Grow

Use a model to find the difference.

$$3\frac{5}{6} - 2\frac{1}{3}$$

Construct Arguments How can you subtract mixed numbers with unlike denominators without using a model? Explain why your method makes sense.

Think and Grow: Subtract Mixed Numbers

You can use equivalent fractions to subtract mixed numbers that have fractional parts with unlike denominators.

Example Find $3\frac{1}{4} - 1\frac{1}{2}$.

One Way: Subtract the fractional parts and subtract the whole number parts.

To subtract the fractional parts, use a common denominator.

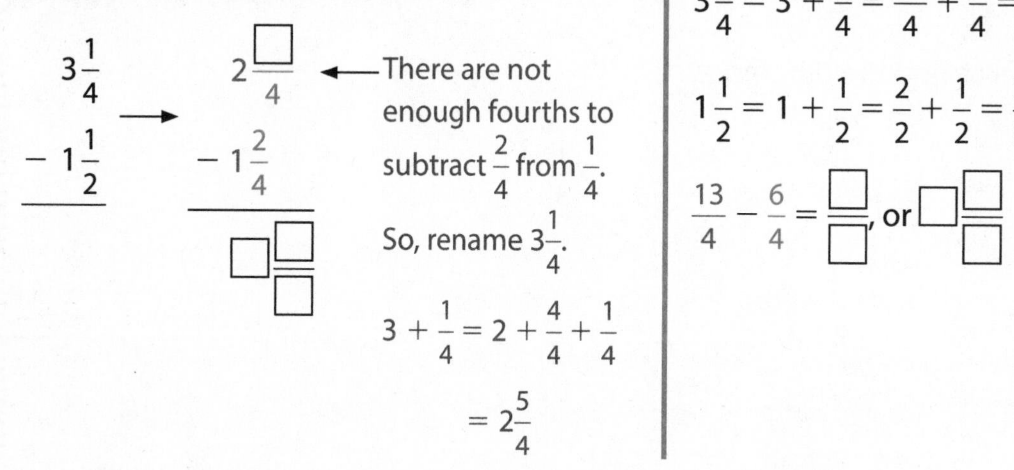

$3\frac{1}{4}$ → $2\frac{\square}{4}$
$-\,1\frac{1}{2}$ → $-\,1\frac{2}{4}$
$\square\frac{\square}{\square}$
There are not enough fourths to subtract $\frac{2}{4}$ from $\frac{1}{4}$.
So, rename $3\frac{1}{4}$.
$3 + \frac{1}{4} = 2 + \frac{4}{4} + \frac{1}{4}$
$= 2\frac{5}{4}$

Another Way: Write the mixed numbers as improper fractions with a common denominator, then subtract.

$3\frac{1}{4} = 3 + \frac{1}{4} = \frac{12}{4} + \frac{1}{4} = \frac{13}{4}$

$1\frac{1}{2} = 1 + \frac{1}{2} = \frac{2}{2} + \frac{1}{2} = \frac{3}{2} = \frac{\square}{4}$

$\frac{13}{4} - \frac{6}{4} = \frac{\square}{\square}$, or $\square\frac{\square}{\square}$

So, $3\frac{1}{4} - 1\frac{1}{2} = \square\frac{\square}{\square}$.

$3\frac{1}{4} - 1\frac{1}{2} = 1\frac{3}{4}$

$\frac{13}{4} - \frac{6}{4} = \frac{7}{4}$

Show and Grow *I can do it!*

Subtract.

1. $1\frac{4}{5} - 1\frac{3}{10} =$ _____

2. $5\frac{7}{12} - 3\frac{5}{6} =$ _____

Name ______________________________

Apply and Grow: Practice

Subtract.

3. $8\frac{11}{12} - 5\frac{2}{3} =$ ______

4. $6 - 4\frac{3}{4} =$ ______

5. $21\frac{2}{9} - 10\frac{1}{3} =$ ______

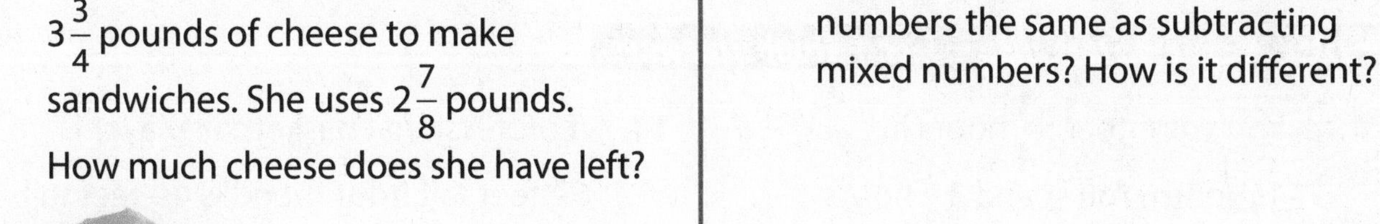

6. $7\frac{1}{2} - \frac{5}{8} =$ ______

7. $9\frac{7}{20} - 1\frac{3}{5} =$ ______

8. $7\frac{5}{6} - 1\frac{1}{6} - 2\frac{2}{3} =$ ______

9. A volunteer at a food bank buys $3\frac{3}{4}$ pounds of cheese to make sandwiches. She uses $2\frac{7}{8}$ pounds. How much cheese does she have left?

10. **Writing** How is adding mixed numbers the same as subtracting mixed numbers? How is it different?

11. **MP Number Sense** Write the words as an expression. Then evaluate.

Subtract the sum of four and three-fourths and two and five-eighths from eleven and seven-eighths.

12. **DIG DEEPER!** Find the missing number.

$$3\frac{1}{4} - 1\frac{\square}{12} = 2\frac{1}{6}$$

Think and Grow: Modeling Real Life

Example A dragonfly is $1\frac{1}{2}$ inches long. How much longer is the walking leaf than the dragonfly?

Walking Leaf

$2\frac{2}{3}$ in.

To find how much longer the walking leaf is than the dragonfly, subtract the length of the dragonfly from the length of the walking leaf.

$$\begin{array}{r} 2\frac{2}{3} \\ -\ 1\frac{1}{2} \\ \hline \end{array} \longrightarrow \begin{array}{r} 2\frac{\square}{\square} \\ -\ 1\frac{\square}{\square} \\ \hline \square\frac{\square}{\square} \end{array}$$

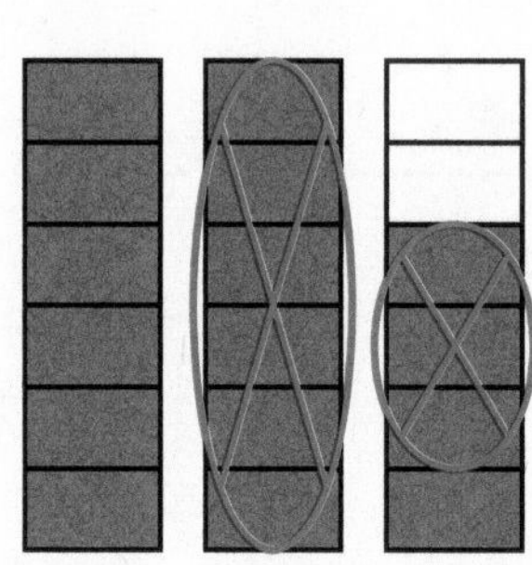

The walking leaf is _____ inches longer than the dragonfly.

Show and Grow *I can think deeper!*

13. You volunteer $5\frac{3}{4}$ hours in 1 month. You spend $3\frac{1}{3}$ hours volunteering at an animal shelter. You spend the remaining hours picking up litter on the side of the road. How many hours do you spend picking up litter?

14. A professional basketball player is $6\frac{3}{4}$ feet tall. Your friend is $4\frac{5}{6}$ feet tall. How much taller is the basketball player than your friend?

15. **DIG DEEPER!** Your rain gauge has $2\frac{1}{2}$ inches of water. After a rainstorm, your rain gauge has $1\frac{3}{4}$ more inches of water. It is sunny for a week. Now your rain gauge has $2\frac{2}{3}$ inches of water. How many inches of water evaporated?

Name ____________________

Homework & Practice 8.7

Learning Target: Subtract mixed numbers with unlike denominators.

Example Find $5\frac{3}{10} - 2\frac{4}{5}$.

One Way: Subtract the fractional parts and subtract the whole number parts.

To subtract the fractional parts, use a common denominator.

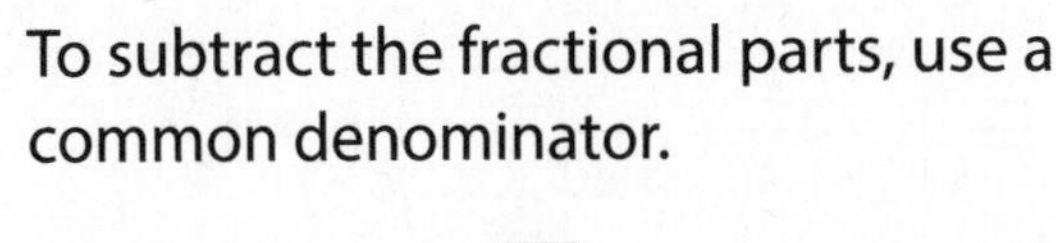

$$5\frac{3}{10} - 2\frac{4}{5} \rightarrow 4\frac{\boxed{13}}{10} - 2\frac{8}{10} = \boxed{2}\frac{\boxed{5}}{\boxed{10}}, \text{ or } \boxed{2}\frac{\boxed{1}}{\boxed{2}}$$

There are not enough tenths to subtract $\frac{8}{10}$ from $\frac{3}{10}$. So, rename $5\frac{3}{10}$.

$5 + \frac{3}{10}$

$= 4 + \frac{10}{10} + \frac{3}{10}$

$= 4\frac{13}{10}$

Another Way: Write the mixed numbers as improper fractions with a common denominator, then subtract.

$5\frac{3}{10} = 5 + \frac{3}{10} = \frac{50}{10} + \frac{3}{10} = \frac{53}{10}$

$2\frac{4}{5} = 2 + \frac{4}{5} = \frac{10}{5} + \frac{4}{5} = \frac{14}{5} = \frac{\boxed{28}}{10}$

$\frac{53}{10} - \frac{28}{10} = \frac{\boxed{25}}{\boxed{10}}$, or $\boxed{2}\frac{\boxed{1}}{\boxed{2}}$

So, $5\frac{3}{10} - 2\frac{4}{5} = \boxed{2}\frac{\boxed{1}}{\boxed{2}}$.

Subtract.

1. $9\frac{5}{6} - 4\frac{1}{2} =$ ______

2. $3\frac{2}{3} - \frac{1}{9} =$ ______

3. $6\frac{1}{3} - 1\frac{11}{12} =$ ______

4. $12\frac{5}{6} - 7\frac{3}{10} =$ ______

5. $5 - 2\frac{3}{4} =$ ______

6. $4\frac{1}{5} - 2\frac{1}{4} =$ ______

Subtract.

7. $7\frac{5}{8} - 1\frac{5}{6} =$ ______

8. $8\frac{1}{9} - 6\frac{7}{8} =$ ______

9. $1\frac{6}{7} + 5\frac{13}{14} - 2\frac{1}{2} =$ ______

10. **YOU BE THE TEACHER** Your friend says the difference of 8 and $3\frac{7}{10}$ is $5\frac{7}{10}$. Is your friend correct? Explain.

11. **DIG DEEPER!** Use a symbol card to complete the equation. Then find *b*.

$$4\frac{1}{4} - 1\frac{17}{20} \; \square \; b = 1\frac{1}{2}$$

$+$ $-$

12. **Modeling Real Life** The world record for the heaviest train pulled with a human beard is $2\frac{3}{4}$ metric tons. The world record for the heaviest train pulled by human teeth is $4\frac{1}{5}$ metric tons. How much heavier is the train pulled by teeth than the train pulled with a beard?

13. **Modeling Real Life** Your friend's hair is $50\frac{4}{5}$ centimeters long. Your hair is $8\frac{9}{10}$ centimeters long. How much longer is your friend's hair than yours?

Review & Refresh

14. Round 6.294.

Nearest whole number: ______

Nearest tenth: ______

Nearest hundredth: ______

15. Round 10.571.

Nearest whole number: ______

Nearest tenth: ______

Nearest hundredth: ______

Name ______________________________

Problem Solving: Fractions 8.8

Learning Target: Solve multi-step word problems involving fractions and mixed numbers.

Success Criteria:
- I can understand a problem.
- I can make a plan to solve.
- I can solve a problem using an equation.

Explore and Grow

Make a plan to solve the problem.

At a state park, every $\frac{1}{10}$ mile of a walking trail is marked. Every $\frac{1}{4}$ mile of a separate biking trail is marked. The table shows the number of mileage markers you and your friend pass while walking and biking on the trails. Who travels farther? How much farther?

	Markers Passed	
	Walking Trail	**Biking Trail**
You	8	9
Your Friend	12	6

Make Sense of Problems You decide to walk farther and you pass 4 more mileage markers on the walking trail. Does this change your plan to solve the problem? Explain.

Think and Grow: Problem Solving: Fractions

Example To repair a skate ramp, you cut a piece of wood from a $9\frac{1}{2}$-foot-long board. Then you cut the remaining piece in half. Each half is $3\frac{5}{12}$ feet long. How long is the first piece you cut?

Understand the Problem

What do you know?	What do you need to find?
• The board is $9\frac{1}{2}$ feet long. • You cut a piece from the board. • You cut the rest into two pieces that are each $3\frac{5}{12}$ feet long.	• You need to find the length of the first piece you cut.

Make a Plan

How will you solve?

Write and solve an equation: Subtract the sum of the lengths of the last two pieces you cut from the total length of the board.

Solve

Length of the first piece you cut = Total length of the board − (Length of one-half of remaining piece + Length of one-half of remaining piece)

Let g represent the length of the first piece you cut.

$$g = 9\frac{1}{2} - \left(3\frac{5}{12} + 3\frac{5}{12}\right)$$

$$= 9\frac{1}{2} - \square\frac{\square}{\square}$$

$$= \square\frac{\square}{\square}$$

So, the length of the first piece you cut is ______ feet.

Show and Grow *I can do it!*

1. Explain how you can check your answer in the example above.

Name ____________________

Apply and Grow: Practice

Understand the problem. What do you know? What do you need to find? Explain.

2. A racehorse eats $38\frac{1}{2}$ pounds of food each day. He eats $22\frac{3}{4}$ pounds of hay and $7\frac{1}{2}$ pounds of grains. How many pounds of his daily diet is *not* hay or grains?

3. In 2015, American Pharoah won all of the horse races shown in the table. How many kilometers did American Pharoah run in the races altogether?

Race	Length (kilometers)
Kentucky Derby	2
Preakness Stakes	$1\frac{9}{10}$
Belmont Stakes	$2\frac{2}{5}$

Understand the problem. Then make a plan. How will you solve? Explain.

4. You have $2\frac{1}{2}$ cups of blueberries. You use $1\frac{1}{4}$ cups for pancakes and $\frac{1}{3}$ cup for muffins. What fraction of a cup of blueberries do you have left?

5. A customer orders 2 pounds of cheese at a deli. The deli worker places some cheese in a bowl and weighs it. The scale shows $1\frac{1}{4}$ pounds. The bowl weighs $\frac{1}{8}$ pound. What fraction of a pound of cheese does the worker need to add to the bowl?

6. MP **Reasoning** Student A is $8\frac{1}{2}$ inches shorter than Student B. Student B is $3\frac{1}{4}$ inches taller than Student C. Student C is $56\frac{3}{8}$ inches tall. How tall is Student A? Student B?

7. **DIG DEEPER!** A police dog spends $\frac{1}{8}$ of his workday in a police car, $\frac{3}{4}$ of his workday in public, and the rest of his workday at the police station. What fraction of the dog's day is spent at the police station?

Think and Grow: Modeling Real Life

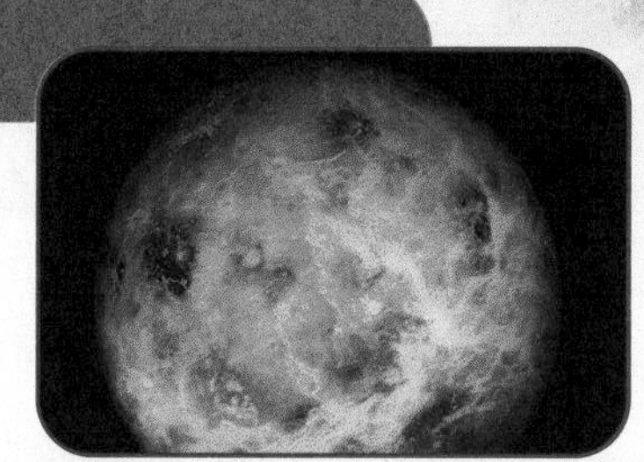

Magellan was an unmanned spacecraft whose primary mission was to gather images of the surface of Venus.

Example The *Magellan* spacecraft, launched by the United States, spent $5\frac{5}{12}$ years in space before it burned in Venus's atmosphere. Its first 4 cycles around Venus each lasted $\frac{2}{3}$ year. The remaining cycles around Venus lasted a total of $1\frac{1}{2}$ years. How long did it take *Magellan* to travel from Earth to Venus?

Think: What do you know? What do you need to find? How will you solve?

Length of time from Earth to Venus	+	Number of cycles	×	Length of one cycle around Venus	+	Length of remaining cycles	=	Total time spent in space

Let *m* represent the length of time from Earth to Venus.

$$m + 4 \times \frac{2}{3} + 1\frac{1}{2} = 5\frac{5}{12}$$

$$m + \square\frac{\square}{\square} + 1\frac{1}{2} = 5\frac{5}{12}$$

$$m + \square\frac{\square}{\square} = 5\frac{5}{12}$$

$$m = 5\frac{5}{12} - \square\frac{\square}{\square}$$ Write the related subtraction equation.

$$m = \square\frac{\square}{\square}, \text{ or } \square\frac{\square}{\square}$$

So, it took *Magellan* $\square\frac{\square}{\square}$ years to travel from Earth to Venus.

Show and Grow *I can think deeper!*

8. You have one of each euro coin shown. Your friend has four euro coins that have a total weight of $21\frac{3}{10}$ grams. Whose coins weigh more? How much more?

Euro Coin	1 cent	20 cent	50 cent	1 euro
Weight (g)	$2\frac{3}{10}$	$5\frac{3}{4}$	$7\frac{4}{5}$	$7\frac{1}{2}$

Name ______________________________

Homework & Practice 8.8

Learning Target: Solve multi-step word problems involving fractions and mixed numbers.

Example Newton makes a model canoe. He cuts a board into three pieces. Two of the pieces are each $1\frac{1}{3}$ feet long for the oars, and one of the pieces is $2\frac{3}{4}$ feet long for the canoe. How long was the board Newton started with?

Think: What do you know? What do you need to find? How will you solve?

Length of board he started with	=	Length of one oar piece	+	Length of one oar piece	+	Length of canoe piece

Let g represent the length of the board he started with.

$$g = 1\frac{1}{3} + 1\frac{1}{3} + 2\frac{3}{4}$$

$$= 1\frac{4}{12} + 1\frac{4}{12} + 2\frac{9}{12}$$

$$= 4\frac{17}{12}, \text{ or } 5\frac{5}{12}$$

So, the length of the board he started with was $5\frac{5}{12}$ feet.

Understand the problem. What do you know? What do you need to find? Explain.

1. Your goal is to exercise for 15 hours this month. You exercise for $3\frac{1}{2}$ hours the first week and $3\frac{3}{4}$ hours the next week. How many more hours do you need to exercise to reach your goal?

2. A taxi driver travels $4\frac{5}{8}$ miles to his first stop. He travels $1\frac{3}{4}$ miles less to his second stop. How many miles does the taxi driver travel for the two stops?

Understand the problem. Then make a plan. How will you solve? Explain.

3. During the U.S. Civil War, $\frac{5}{9}$ of the states fought for the Union, and $\frac{11}{36}$ of the states fought for the Confederacy. The rest of the states were border states. What fraction of the states were border states?

4. You have $6\frac{3}{4}$ pounds of clay. You use $4\frac{1}{6}$ pounds to make a medium-sized bowl and $1\frac{1}{2}$ pounds to make a small bowl. How many pounds of clay do you have left?

5. DIG DEEPER! Newton and Descartes have a 70-day summer vacation. They go to camp for $\frac{23}{70}$ of their vacation, and they travel for $\frac{6}{35}$ of their vacation. They stay home the rest of their vacation. How many weeks do Newton and Descartes spend at home?

6. **Modeling Real Life** A farmer plants beets in a square garden with side lengths of $12\frac{2}{3}$ feet. He plants squash in a garden with a perimeter of $50\frac{1}{2}$ feet. Which garden has a greater perimeter? How much greater is it?

7. DIG DEEPER! Which grade uses more leafy greens daily for its classroom rabbits? How much more does it use?

4th Grade	5th Grade
4 rabbits	2 rabbits
Leafy greens each rabbit eats:	Leafy greens each rabbit eats:
A.M. $\frac{3}{4}$ cup	A.M. $1\frac{1}{4}$ cups
P.M. 1 cup	P.M. $1\frac{1}{3}$ cups

Review & Refresh

Find the product.

8. $0.43 \times 1{,}000 =$ ______

9. $25.8 \times 0.1 =$ ______

Name ______________________________

Performance Task 8

Many historic landmarks are located in Washington, D.C.

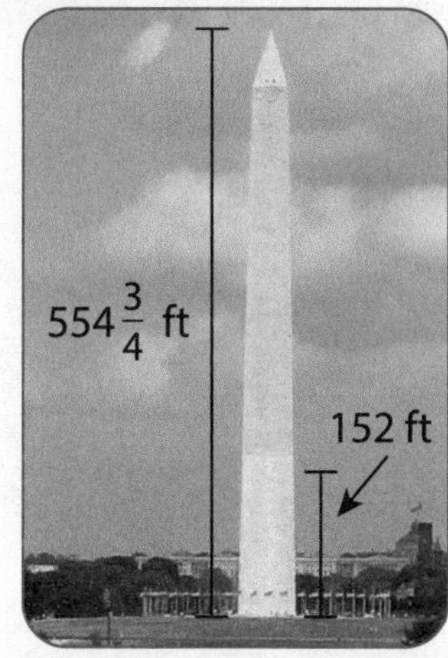

The Washington Monument is made out of marble and has an aluminum pyramid at the top.

1. Initial construction of the Washington Monument began in 1848. When the height of the monument reached 152 feet, construction halted due to lack of funds. How many feet were added to the height of the monument when construction resumed 23 years later?

2. You visit several historic landmarks. You start at the Capitol Building and walk to the Washington Monument, then the Lincoln Memorial, then the White House, and then back to the Capitol Building.

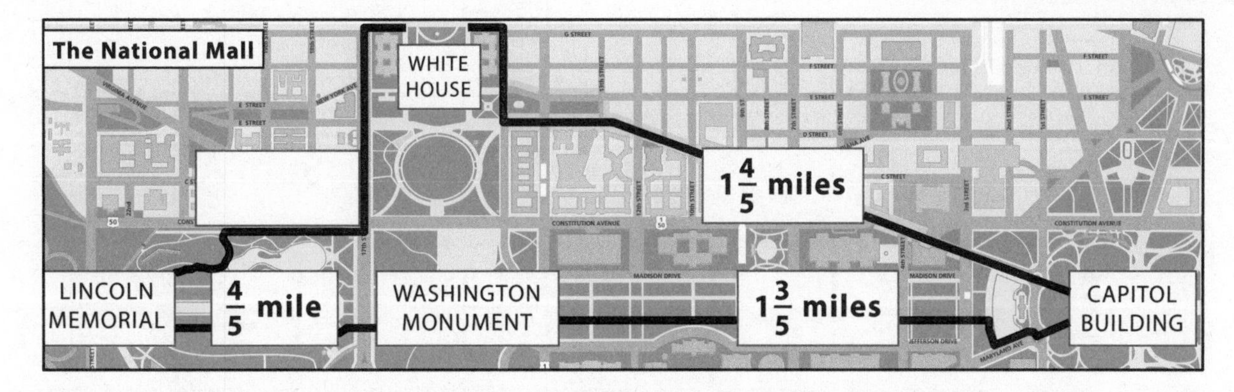

a. You walk 3 miles each hour. It takes you $\frac{1}{2}$ hour to walk from the Lincoln Memorial to the White House. What is the distance from the Lincoln Memorial to the White House? Label the map.

b. What is the total distance you walk visiting the landmarks?

3. A law in Washington, D.C., restricts a new building's height to no more than 20 feet taller than the width of the street it faces. You design a building with stories that are each 15 feet tall for a street that is $88\frac{2}{3}$ feet wide. What is the greatest number of stories your building can have? How much shorter is your building than the height restriction?

Mixed Number Subtract and Add

Directions:

1. Each player flips four Mixed Number Cards.
2. Each player arranges the cards to create two differences that will have the greatest possible sum.
3. Each player records the two differences, and then adds the differences.
4. Players repeat Steps 1–3.
5. Each player adds Sum A and Sum B to find the total. The player with the greatest total wins!

Greater Number	–	Lesser Number	=	Score
$\square\frac{\square}{\square}$	–	$\square\frac{\square}{\square}$	=	$\square\frac{\square}{\square}$
$\square\frac{\square}{\square}$	–	$\square\frac{\square}{\square}$	=	$+\ \square\frac{\square}{\square}$
			Sum A	$\square\frac{\square}{\square}$
$\square\frac{\square}{\square}$	–	$\square\frac{\square}{\square}$	=	$\square\frac{\square}{\square}$
$\square\frac{\square}{\square}$	–	$\square\frac{\square}{\square}$	=	$+\ \square\frac{\square}{\square}$
			Sum B	$\square\frac{\square}{\square}$

Sum A	+	Sum B	=	Total
$\square\frac{\square}{\square}$	+	$\square\frac{\square}{\square}$	=	$\square\frac{\square}{\square}$

Name ________________________________

Chapter Practice 8

8.1 Simplest Form

Write the fraction in simplest form.

1. $\frac{2}{12}$

2. $\frac{15}{30}$

3. $\frac{16}{24}$

4. $\frac{18}{36}$

5. $\frac{8}{32}$

6. $\frac{25}{10}$

8.2 Estimate Sums and Differences of Fractions

Estimate the sum or difference.

7. $\frac{7}{8} - \frac{1}{5}$

8. $\frac{5}{6} + \frac{9}{10}$

9. $\frac{11}{12} - \frac{89}{100}$

10. MP **Precision** Your friend says $\frac{7}{8} - \frac{5}{12}$ is about 0. Find a closer estimate. Explain why your estimate is closer.

8.3 Find Common Denominators

Use a common denominator to write an equivalent fraction for each fraction.

11. $\frac{1}{4}$ and $\frac{1}{2}$

12. $\frac{2}{3}$ and $\frac{2}{9}$

13. $\frac{2}{3}$ and $\frac{5}{6}$

14. $\frac{4}{5}$ and $\frac{1}{3}$

15. $\frac{5}{6}$ and $\frac{8}{9}$

16. $\frac{4}{5}$ and $\frac{3}{4}$

8.4 Add Fractions with Unlike Denominators

Add.

17. $\frac{2}{15} + \frac{2}{3} =$ _____

18. $\frac{3}{4} + \frac{1}{8} =$ _____

19. $\frac{7}{2} + \frac{1}{6} =$ _____

20. $\frac{5}{9} + \frac{1}{2} =$ _____

21. $\frac{7}{10} + \frac{5}{6} =$ _____

22. $\frac{1}{6} + \frac{11}{12} + \frac{4}{6} =$ _____

8.5 Subtract Fractions with Unlike Denominators

Subtract.

23. $\frac{1}{4} - \frac{1}{8} =$ _____

24. $\frac{3}{2} - \frac{7}{10} =$ _____

25. $\frac{15}{16} - \frac{7}{8} =$ _____

26. $\frac{4}{3} - \frac{2}{5} =$ _____

27. $\frac{5}{6} - \frac{3}{4} =$ _____

28. $\frac{7}{10} - \frac{2}{5} + \frac{11}{20} =$ _____

8.6 Add Mixed Numbers

Add.

29. $1\frac{3}{4} + 7\frac{5}{8} =$ _____

30. $3\frac{3}{10} + 2\frac{7}{20} =$ _____

31. $\frac{1}{3} + 6\frac{4}{5} =$ _____

32. $5\frac{8}{9} + \frac{5}{6} =$ _____

33. $2\frac{2}{3} + \frac{4}{9} + 4\frac{1}{3} =$ _____

34. $5\frac{1}{2} + 2\frac{5}{8} + 3\frac{3}{4} =$ _____

8.7 Subtract Mixed Numbers

Subtract.

35. $8\frac{7}{10} - 1\frac{2}{5} =$ _____

36. $15\frac{97}{100} - 10\frac{7}{20} =$ _____

37. $4 - 3\frac{5}{6} =$ _____

38. $5\frac{1}{3} - 2\frac{1}{2} =$ _____

39. $9\frac{2}{5} - 6\frac{3}{4} =$ _____

40. $2\frac{3}{8} + 7\frac{1}{2} - 1\frac{11}{16} =$ _____

41. **Modeling Real Life** A family adopts a puppy that weighs $7\frac{7}{8}$ pounds. They take him to the vet 2 weeks later, and he weighs $12\frac{3}{16}$ pounds. How much weight did the puppy gain?

8.8 Problem Solving: Fractions

42. A radio station plays three commercials between two songs. The commercials play for 2 minutes altogether. The first commercial is $\frac{1}{2}$ minute, and the second commercial is $1\frac{1}{4}$ minutes. How long is the third commercial?

43. Your friend plants a tree seedling on Earth Day that is $1\frac{1}{3}$ feet tall. In 1 year, the tree grows $1\frac{5}{6}$ feet. After 2 years, the tree is $4\frac{11}{12}$ feet tall. How much did the tree grow in the second year?

9 Multiply Fractions

- The elevation of a location on Earth is its height above sea level. What do you think this means?
- The highest point at the Grand Canyon is about $1\frac{2}{3}$ miles above sea level. How can you find its elevation in feet?

Chapter Learning Target:
Understand multiplying fractions.

Chapter Success Criteria:
- I can identify a fraction as a sum of unit fractions.
- I can write a fraction as a sum of unit fractions.
- I can multiply fractions.
- I can solve a problem using fractions.

Name ____________________

9 Vocabulary

Review Words

common factors
decimal fraction
improper fraction
numerator
simplest form
unit fraction

Organize It

Use review words to complete the graphic organizer.

[ ]

When the [ ] and denominator of a fraction have no [ ] other than 1

$\frac{15}{16}$ factors of 15: (1), 3, 5, 15
factors of 16: (1), 2, 4, 8, 16

Define It

Match each review word to a fraction.

1. decimal fraction
2. improper fraction
3. unit fraction

$\frac{1}{8}$

$\frac{17}{100}$

$\frac{13}{5}$

Name ___________________________

Multiply Whole Numbers by Fractions

9.1

Learning Target: Multiply whole numbers by fractions.

Success Criteria:

- I can use a model to multiply a whole number by a fraction.
- I can write a multiplication expression as a repeated addition expression.
- I can write a multiple of a fraction as a multiple of a unit fraction.

Explore and Grow

Write any proper fraction that is *not* a unit fraction. Draw a model to represent your fraction.

Draw a model to find a multiple of your fraction.

Reasoning How can you use a model to multiply a whole number by a fraction? Explain.

Think and Grow: Multiply Whole Numbers by Fractions

Example Find $3 \times \frac{2}{5}$.

One Way: Use repeated addition.

$$3 \times \frac{2}{5} = \frac{2}{5} + \frac{2}{5} + \frac{2}{5}$$

$$= \frac{\square + \square + \square}{5}$$

$$= \frac{\square}{\square}$$

Another Way: Rewrite the expression as a multiple of a unit fraction.

$$3 \times \frac{2}{5} = 3 \times \left(\_\_\_\_ \times \frac{1}{5}\right)$$

$$= (3 \times \_\_\_\_) \times \frac{1}{5}$$ Associative Property of Multiplication

$$= \_\_\_\_ \times \frac{1}{5}$$

$$= \frac{\square}{\square}$$

So, $3 \times \frac{2}{5} =$ ______.

$\frac{1}{5}$	$\frac{1}{5}$	$\frac{1}{5}$	$\frac{1}{5}$	$\frac{1}{5}$
$\frac{1}{5}$	$\frac{1}{5}$	$\frac{1}{5}$	$\frac{1}{5}$	$\frac{1}{5}$
$\frac{1}{5}$	$\frac{1}{5}$	$\frac{1}{5}$	$\frac{1}{5}$	$\frac{1}{5}$

$3 \times \frac{2}{5}$

$\frac{6}{5}$

Show and Grow *I can do it!*

Multiply.

1. $2 \times \frac{3}{4} =$ ______

2. $4 \times \frac{5}{8} =$ ______

Name ________________

Apply and Grow: Practice

Multiply.

3. $5 \times \frac{7}{10} =$ ______

4. $8 \times \frac{2}{3} =$ ______

5. $7 \times \frac{5}{6} =$ ______

6. $9 \times \frac{1}{2} =$ ______

7. $6 \times \frac{3}{100} =$ ______

8. $15 \times \frac{4}{7} =$ ______

9. $10 \times \frac{5}{3} =$ ______

10. $4 \times \frac{5}{2} =$ ______

11. $3 \times \frac{11}{8} =$ ______

Find the unknown number.

12. $\square \times \frac{3}{20} = \frac{9}{20}$

13. $\square \times \frac{4}{9} = \frac{24}{9}$

14. $\square \times \frac{5}{12} = \frac{50}{12}$

15. A recipe calls for $\frac{3}{4}$ cup of dried rice noodles. You make 4 batches of the recipe. How many cups of dried rice noodles do you use?

Thai Rice Noodle Salad

16. 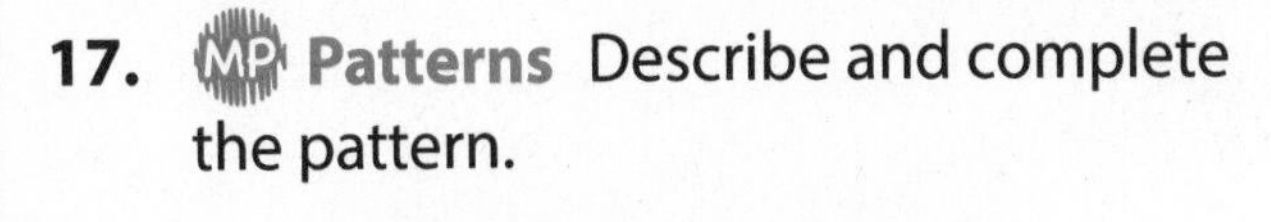

Your friend says that $5 \times \frac{3}{5}$ is equal to $\frac{3}{5} + \frac{3}{5} + \frac{3}{5} + \frac{3}{5} + \frac{3}{5}$. Is your friend correct? Explain.

17. MP **Patterns** Describe and complete the pattern.

Expression	Product
$9 \times \frac{1}{4}$	$\frac{9}{4}$
$9 \times \frac{2}{4}$	
$9 \times \frac{3}{4}$	
$9 \times \frac{4}{4}$	

Think and Grow: Modeling Real Life

Example Your goal is to make a waterslide that is at least 10 meters long. You make the waterslide using 10 plastic mats that are each $\frac{3}{2}$ meters long. Do you reach your goal?

Find the length of the waterslide by multiplying the number of mats by the length of each mat.

$10 \times \frac{3}{2} = 10 \times \left(\_\_\_\_ \times \frac{1}{2}\right)$

$= (10 \times \_\_\_\_) \times \frac{1}{2}$ Associative Property of Multiplication

$= \_\_\_\_ \times \frac{1}{2}$

$= \_\_\_\_$

You ________ reach your goal.

Show and Grow *I can think deeper!*

18. An excavator is moving 4 piles of dirt that are the same size. Each pile requires $\frac{3}{4}$ hour to move. Can the excavator move all of the piles in 2 hours?

19. You walk dogs $\frac{5}{4}$ miles two times each day. How far do you walk the dogs in 1 week?

20. **DIG DEEPER!** You have 10 feet of string. You need $\frac{5}{3}$ feet of string to make 1 necklace. You make 5 necklaces. Do you have enough string to make another necklace? Explain.

Name ______________________

Homework & Practice 9.1

Learning Target: Multiply whole numbers by fractions.

Example Find $2 \times \frac{5}{6}$.

One Way: Use repeated addition.

$2 \times \frac{5}{6} = \frac{5}{6} + \frac{5}{6}$

$= \frac{\boxed{5} + \boxed{5}}{6}$

$= \frac{\boxed{10}}{\boxed{6}}$, or $\frac{\boxed{5}}{\boxed{3}}$

Another Way: Rewrite the expression as a multiple of a unit fraction.

$2 \times \frac{5}{6} = 2 \times \left(\underline{\ 5\ } \times \frac{1}{6}\right)$

$= (2 \times \underline{\ 5\ }) \times \frac{1}{6}$ Associative Property of Multiplication

$= \underline{\ 10\ } \times \frac{1}{6} = \frac{\boxed{10}}{\boxed{6}}$, or $\frac{\boxed{5}}{\boxed{3}}$

So, $2 \times \frac{5}{6} = \underline{\ \frac{5}{3}\ }$.

$\frac{10}{6}$

$\frac{1}{6}$	$\frac{1}{6}$	$\frac{1}{6}$	$\frac{1}{6}$	$\frac{1}{6}$	$\frac{1}{6}$
$\frac{1}{6}$	$\frac{1}{6}$	$\frac{1}{6}$	$\frac{1}{6}$	$\frac{1}{6}$	$\frac{1}{6}$

$2 \times \frac{5}{6}$

Multiply.

1. $5 \times \frac{2}{3} =$ ______

2. $9 \times \frac{7}{8} =$ ______

3. $4 \times \frac{11}{12} =$ ______

4. $8 \times \frac{35}{100} =$ ______

5. $3 \times \frac{1}{2} =$ ______

6. $7 \times \frac{2}{5} =$ ______

7. $6 \times \frac{7}{4} =$ ______

8. $12 \times \frac{8}{7} =$ ______

9. $25 \times \frac{10}{9} =$ ______

Find the unknown number.

10. $\square \times \frac{5}{7} = \frac{25}{7}$

11. $\square \times \frac{9}{10} = \frac{63}{10}$

12. $\square \times \frac{3}{5} = \frac{27}{5}$

13. You make 5 servings of pancakes. You top each serving with $\frac{1}{4}$ cup of strawberries. How many cups of strawberries do you use?

14. **Which One Doesn't Belong?** Which one does *not* belong with the other three?

$4 \times \frac{3}{8}$

$3 \times \frac{1}{8}$

$\frac{3}{8} + \frac{3}{8} + \frac{3}{8} + \frac{3}{8}$

$1 + \frac{1}{2}$

15. **Modeling Real Life** You complete $\frac{5}{2}$ inches of a weaving each day for 5 days. The weaving needs to be at least 11 inches long. Is your weaving complete?

16. **DIG DEEPER!** You spend $\frac{7}{2}$ hours playing drums each day for 2 days. Your friend spends $\frac{5}{4}$ hours playing drums each day for 6 days. Who spends more time playing drums? How much more?

Review & Refresh

Find the product.

17. 0.6×0.4

18. 2.37×1.9

19. 52.8×0.75

Name ______________________________

9.2 Use Models to Multiply Fractions by Whole Numbers

Learning Target: Multiply fractions by whole numbers.

Success Criteria:

- I can divide a whole into equal parts.
- I can use a model to find part of a group.
- I can use a model to multiply a fraction by a whole number.

Explore and Grow

You need to give water to $\frac{2}{3}$ of the dogs at a shelter. There are 12 dogs at the shelter. How many dogs need water? Draw a model to support your answer.

Reasoning How can you use a model to multiply a fraction by a whole number? Explain.

Think and Grow: Multiply Fractions by Whole Numbers

You can use models to multiply a fraction by a whole number.

Example Find $\frac{3}{4}$ of 8.

Step 1: Because you are finding $\frac{3}{4}$ of 8, divide 8 into 4 equal parts.

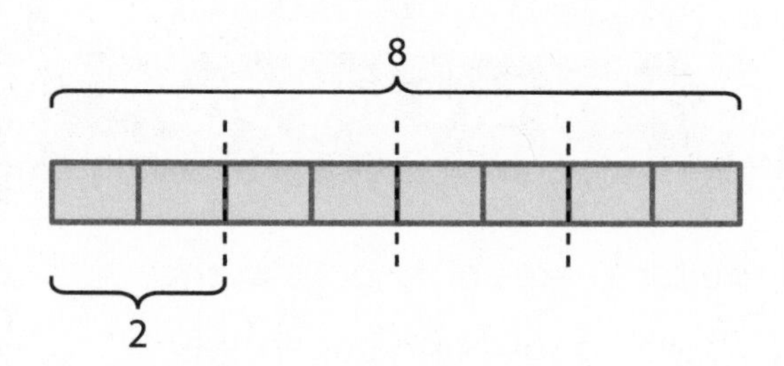

Each of the four equal parts is 2.

Step 2: Because you are finding $\frac{3}{4}$ of 8, take 3 of the parts.

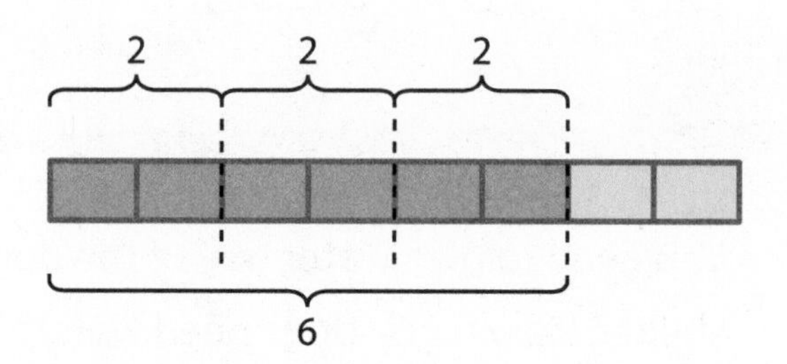

So, $\frac{3}{4}$ of $8 = \frac{3}{4} \times 8 =$ ______.

Example Find $\frac{5}{8} \times 4$.

Step 1: Because you are finding $\frac{5}{8}$ of 4, divide 4 into 8 equal parts.

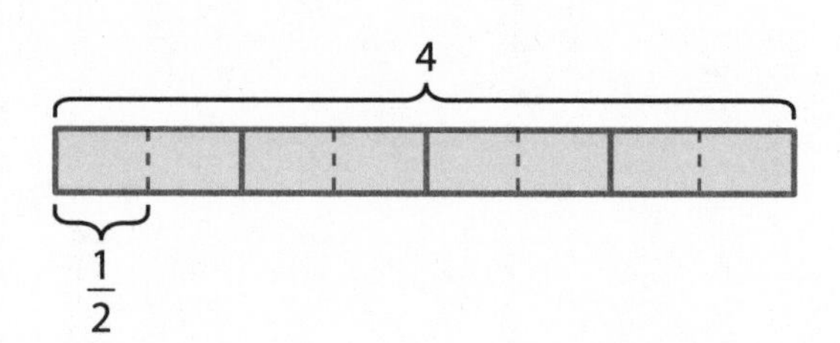

Each of the eight equal parts is $\frac{1}{2}$.

Step 2: Because you are finding $\frac{5}{8}$ of 4, take 5 of the parts.

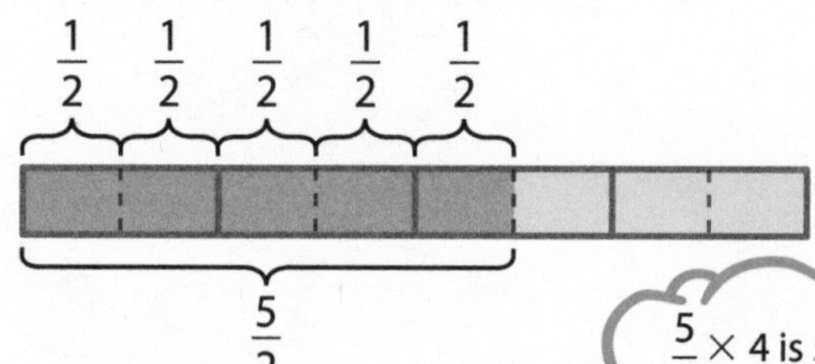

So, $\frac{5}{8} \times 4 =$ ______.

$\frac{5}{8} \times 4$ is 5 parts when you divide 4 into 8 parts.

Show and Grow *I can do it!*

1. Find $\frac{2}{5}$ of 10.

2. Find $\frac{7}{12} \times 6$.

Name ______________________________

Apply and Grow: Practice

Multiply. Use a model to help.

3. $\frac{5}{6}$ of 12

4. $\frac{2}{3} \times 9$

5. $\frac{1}{5} \times 10$

6. $\frac{3}{5}$ of 5

7. $\frac{1}{6}$ of 3

8. $\frac{3}{8} \times 4$

9. You have 25 beads. You use $\frac{2}{5}$ of the beads to make a bracelet. How many beads do you use?

10. **Writing** Write and solve a real-life problem for the expression.

$$\frac{3}{4} \times 20$$

11. **YOU BE THE TEACHER** Descartes finds $\frac{2}{3} \times 6$. Is he correct? Explain.

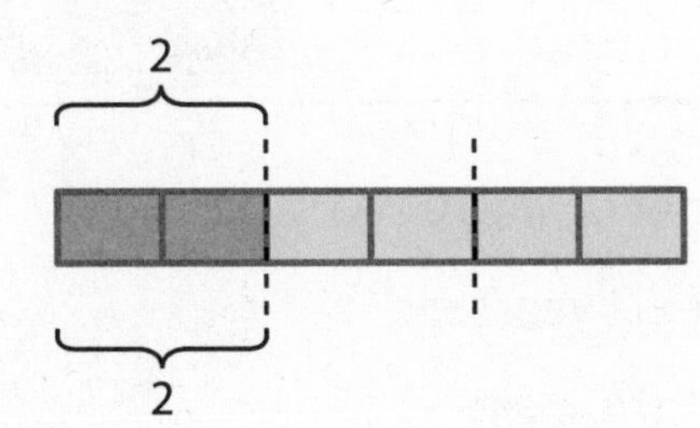

I divided 6 into 3 equal parts. Then I took 2 parts. So, the product is $\frac{2}{6}$, or $\frac{1}{3}$.

Think and Grow: Modeling Real Life

Example A recipe calls for 2 cups of rice. You only have $\frac{3}{4}$ of that amount. How much more rice do you need?

Find the number of cups of rice that you have by finding $\frac{3}{4}$ of 2.

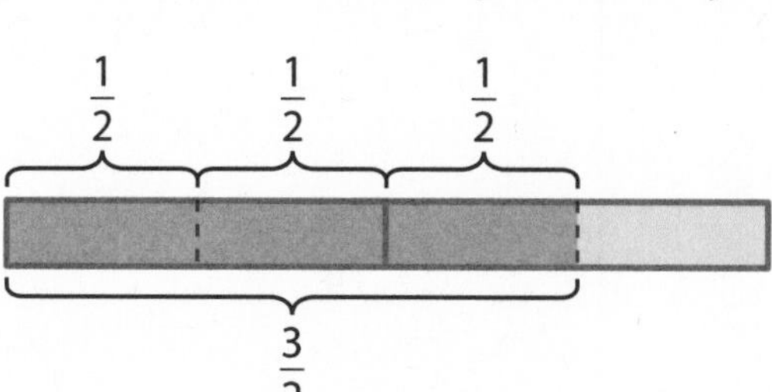

You have _____ cups of rice.

Subtract the amount of rice you have from the amount of rice you need.

$2 - \_\_\_\_\_ = \_\_\_\_\_$

So, you need _____ cup more of rice.

Show and Grow *I can think deeper!*

12. You have 12 tokens. You use $\frac{3}{4}$ of them to play a pinball game. How many tokens do you have left?

13. A male lion sleeps $\frac{5}{6}$ of each day. How many hours does the lion sleep in 1 week?

14. **DIG DEEPER!** In a class of 20 students, $\frac{1}{10}$ of the students are 10 years old, $\frac{4}{5}$ of the students are 11 years old, and the rest are 12 years old. How many more 11-year-olds than 12-year-olds are in the class?

Name ______________________________

Homework & Practice 9.2

Learning Target: Multiply fractions by whole numbers.

Example Find $\frac{3}{10}$ of 5.

Step 1: Because you are finding $\frac{3}{10}$ of 5, divide 5 into 10 equal parts.

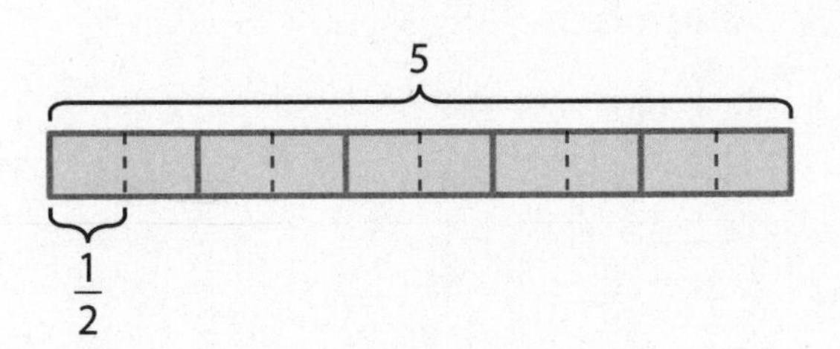

Each of the ten equal parts is $\frac{1}{2}$.

Step 2: Because you are finding $\frac{3}{10}$ of 5, take 3 of the parts.

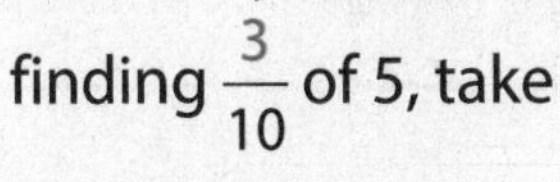
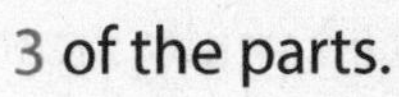
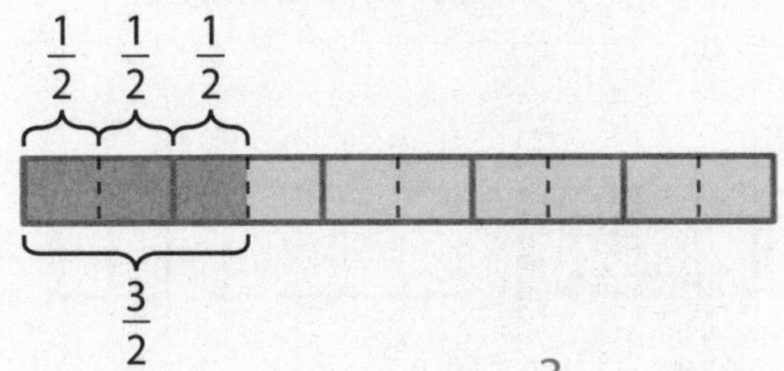

So, $\frac{3}{10}$ of 5 = $\frac{3}{10} \times 5$ = $\frac{3}{2}$.

Multiply. Use a model to help.

1. $\frac{2}{3} \times 6$

2. $\frac{3}{5}$ of 10

3. $\frac{1}{2}$ of 4

4. $\frac{1}{4} \times 12$

5. $\frac{5}{6} \times 3$

6. $\frac{3}{4}$ of 2

7. You have 27 foam balls. You use $\frac{1}{3}$ of the balls for a model. How many balls do you use?

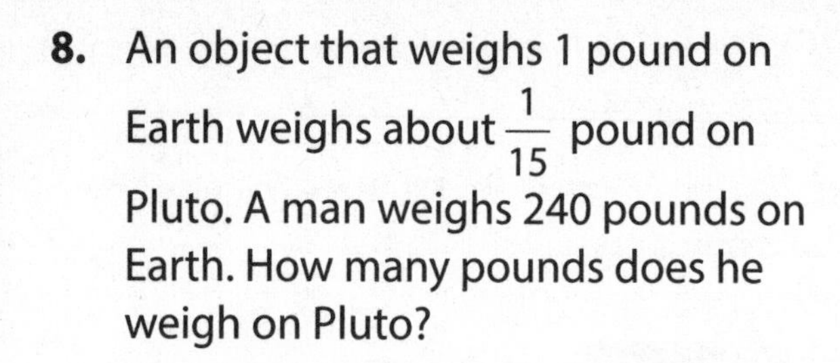

8. An object that weighs 1 pound on Earth weighs about $\frac{1}{15}$ pound on Pluto. A man weighs 240 pounds on Earth. How many pounds does he weigh on Pluto?

9. **MP Structure** Write a multiplication equation represented by the model.

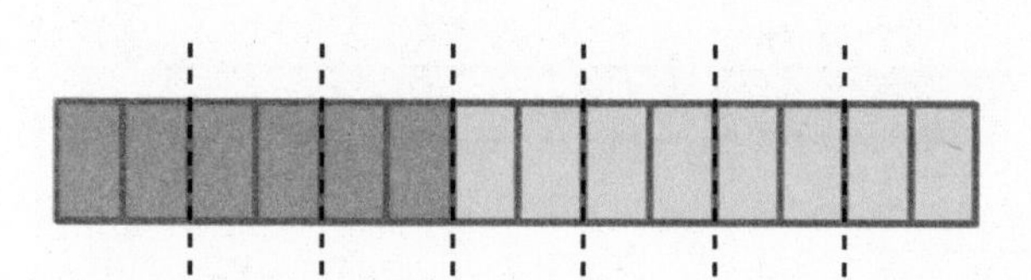

10. **DIG DEEPER!** Find each missing number.

$\frac{1}{3} \times \square = 4$ $\qquad$ $\frac{2}{5} \times 15 = \square$

$\frac{1}{\square} \times 10 = 5$ $\qquad$ $\frac{\square}{6} \times 18 = 15$

11. **Modeling Real Life** You have 28 craft sticks. You use $\frac{4}{7}$ of them for a project. How many craft sticks do you have left?

12. **Modeling Real Life** A mother otter spends $\frac{1}{3}$ of each day feeding her baby. How many hours does the mother otter spend feeding her baby in 1 week?

Review & Refresh

Estimate the quotient.

13. $5{,}692 \div 5$

14. $309 \div 12$

15. $2{,}987 \div 53$

Name ______________________________

Multiply Fractions and Whole Numbers

9.3

Learning Target: Multiply fractions and whole numbers.

Success Criteria:
- I can use a rule to multiply a whole number by a fraction.
- I can use a rule to multiply a fraction by a whole number.

Explore and Grow

Use models to help you complete the table. What do you notice about each expression and its product?

Expression	Product
$2 \times \frac{7}{10}$	
$5 \times \frac{2}{3}$	
$\frac{3}{4} \times 4$	
$\frac{5}{8} \times 2$	

Construct Arguments Explain how to multiply fractions and whole numbers without using models.

Think and Grow: Multiply Fractions and Whole Numbers

Key Idea You can find the product of a fraction and a whole number by multiplying the numerator and the whole number. Then write the result over the denominator.

Example Find $2 \times \frac{5}{6}$.

Multiply the numerator and the whole number.

$$2 \times \frac{5}{6} = \frac{\square \times \square}{6}$$

$$= \frac{\square}{6}, \text{ or } \frac{\square}{\square}$$

$\frac{10}{6}$

$\frac{1}{6}$	$\frac{1}{6}$	$\frac{1}{6}$	$\frac{1}{6}$	$\frac{1}{6}$	$\frac{1}{6}$
$\frac{1}{6}$	$\frac{1}{6}$	$\frac{1}{6}$	$\frac{1}{6}$	$\frac{1}{6}$	$\frac{1}{6}$

$2 \times \frac{5}{6}$

Example Find $\frac{5}{6} \times 2$.

Multiply the numerator and the whole number.

$$\frac{5}{6} \times 2 = \frac{\square \times \square}{6}$$

$$= \frac{\square}{6}, \text{ or } \frac{\square}{\square}$$

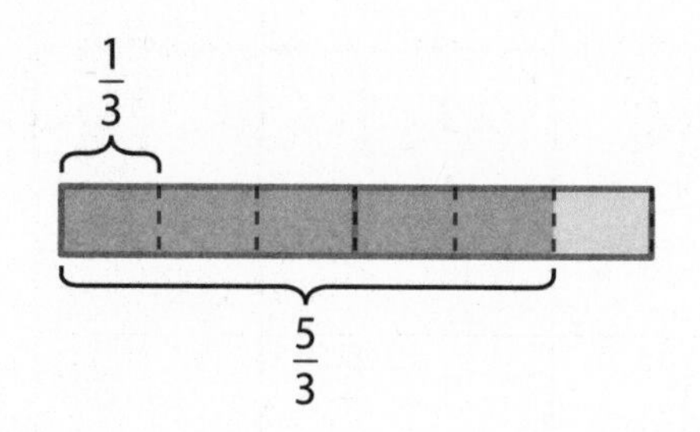

Divide 2 into 6 parts and take 5 of the parts.

Show and Grow *I can do it!*

Multiply.

1. $3 \times \frac{5}{8} = $ _____

2. $6 \times \frac{4}{9} = $ _____

3. $\frac{2}{5} \times 15 = $ _____

Name ____________________

Apply and Grow: Practice

Multiply.

4. $\frac{3}{5} \times 2 =$ ______

5. $5 \times \frac{2}{9} =$ ______

6. $\frac{5}{6} \times 4 =$ ______

7. $8 \times \frac{3}{10} =$ ______

8. $\frac{1}{5} \times 7 =$ ______

9. $9 \times \frac{5}{12} =$ ______

10. $15 \times \frac{5}{8} =$ ______

11. $\frac{3}{4} \times 20 =$ ______

12. $\frac{7}{9} \times 5 =$ ______

13. One-tenth of the 50 states in the United States of America have a mockingbird as their state bird. How many states have a mockingbird as their state bird?

14. **Writing** Explain why $9 \times \frac{2}{3}$ is equivalent to $\frac{2}{3} \times 9$.

15. MP **Reasoning** Without calculating, determine which product is greater. Explain.

$\frac{1}{8} \times 24$ $\frac{7}{8} \times 24$

Think and Grow: Modeling Real Life

Example Newton buys 27 songs. Two-thirds of them are classical songs. Descartes buys 16 songs. Seven-eighths of them are classical songs. Who buys more classical songs? How many more?

Multiply $\frac{2}{3}$ by 27 to find the number of classical songs Newton buys. Multiply $\frac{7}{8}$ by 16 to find the number Descartes buys.

Newton: $\frac{2}{3} \times 27 = \frac{\square \times \square}{3} = \frac{\square}{3} = \_\_\_\_$

Descartes: $\frac{7}{8} \times 16 = \frac{\square \times \square}{8} = \frac{\square}{8} = \_\_\_\_$

So, ________ buys more classical songs.

Subtract the products to find how many more.

$18 - 14 = \_\_\_\_$

________ buys ____ more classical songs than ________.

Show and Grow *I can think deeper!*

16. You take 48 pictures on a walking tour. Five-twelfths of them are of buildings. Your friend takes 45 pictures. Six-fifteenths of them are of buildings. Who takes more pictures of buildings? How many more?

17. You have 72 rocks in your rock collection. Five-eighths of them are sedimentary, one-sixth of them are igneous, and the rest are metamorphic. How many of your rocks are metamorphic?

18. **DIG DEEPER!** Each day, you spend $\frac{3}{4}$ hour reading and $\frac{1}{2}$ hour writing in a journal. How many total hours do you spend reading and writing in 1 week? Describe two ways to solve the problem.

Name ____________________

Homework & Practice 9.3

Learning Target: Multiply fractions and whole numbers.

Example Find $\frac{7}{8} \times 2$.

Multiply the numerator and the whole number.

$$\frac{7}{8} \times 2 = \frac{\boxed{7} \times \boxed{2}}{8}$$

$$= \frac{\boxed{14}}{8}, \text{ or } \frac{\boxed{7}}{\boxed{4}}$$

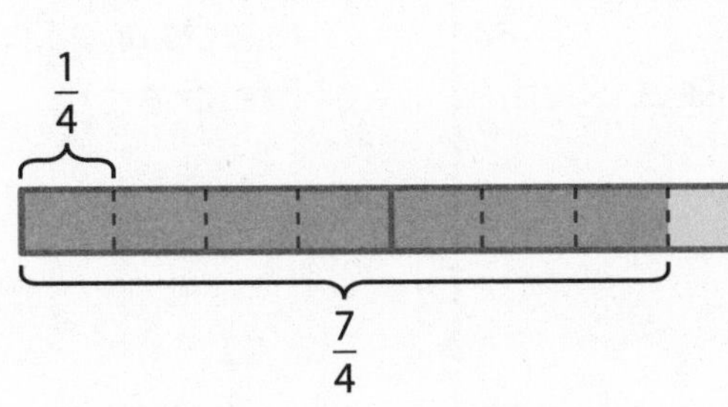

$\frac{1}{4}$
$\frac{7}{4}$

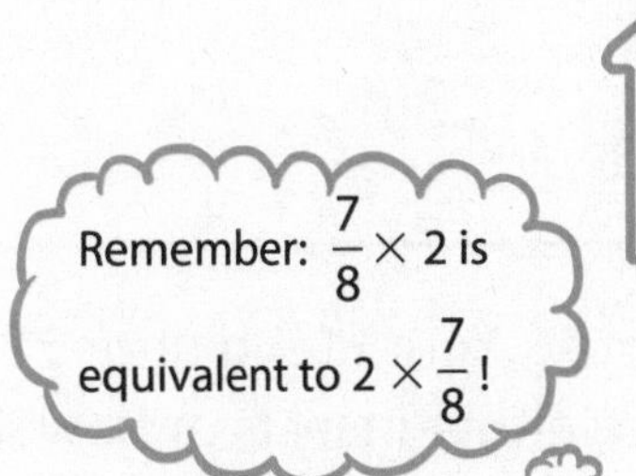

Remember: $\frac{7}{8} \times 2$ is equivalent to $2 \times \frac{7}{8}$!

Multiply.

1. $\frac{5}{6} \times 3 =$ ______

2. $\frac{2}{3} \times 6 =$ ______

3. $7 \times \frac{1}{8} =$ ______

4. $2 \times \frac{1}{2} =$ ______

5. $\frac{4}{5} \times 9 =$ ______

6. $4 \times \frac{5}{12} =$ ______

7. $\frac{1}{4} \times 24 =$ ______

8. $16 \times \frac{3}{8} =$ ______

9. $\frac{7}{10} \times 25 =$ ______

10. You spend $\frac{3}{4}$ hour jumping rope every week for 8 weeks. How many hours do you jump rope altogether?

11. MP **Logic** Your friend finds 25 items that are either insects or flowers. She says that $\frac{1}{6}$ of the items are insects. Can this be true? Explain.

12. **Open-Ended** Write two different pairs of fractions that could represent the insects and flowers your friend finds in Exercise 11.

13. **Modeling Real Life** Newton bakes 56 treats. Five-eighths of them contain peanut butter. Descartes bakes 120 treats. Five-sixths of them contain peanut butter. Who bakes more peanut butter treats? How many more?

14. **Modeling Real Life** Your class conducts an egg-dropping experiment with 60 eggs. Three-fifths of the eggs break open, one-sixth of the eggs crack, and the rest do not break at all. How many of the eggs do *not* crack or break open?

Review & Refresh

Add.

15. $5\frac{5}{8} + 6\frac{3}{4} =$ ______

16. $1\frac{5}{6} + 8\frac{1}{12} =$ ______

17. $3\frac{1}{2} + \frac{3}{5} + 2\frac{7}{10} =$ ______

Name ______________________________

Use Models to Multiply Fractions 9.4

Learning Target: Use models to multiply a fraction by a fraction.

Success Criteria:

- I can divide a whole into equal parts.
- I can divide a unit fraction into equal parts.
- I can use a model to find the product of two fractions.

Explore and Grow

Fold a sheet of paper in half. Shade $\frac{1}{4}$ of either half. What fraction of the entire sheet of paper did you shade? Draw a model to support your answer.

Reasoning What multiplication expression does your model represent? Explain your reasoning.

Think and Grow: Use Models to Multiply Fractions

You can use models to multiply a fraction by a fraction.

Example Find $\frac{1}{2} \times \frac{1}{3}$.

One Way: Use a tape diagram to find $\frac{1}{2}$ of $\frac{1}{3}$.

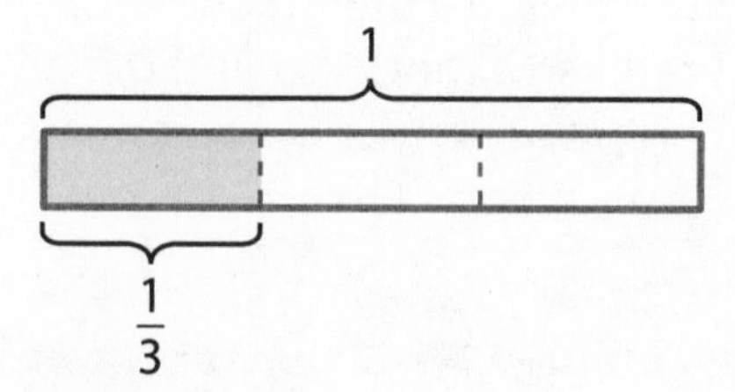

Step 1: Model $\frac{1}{3}$. Divide 1 whole into 3 equal parts.

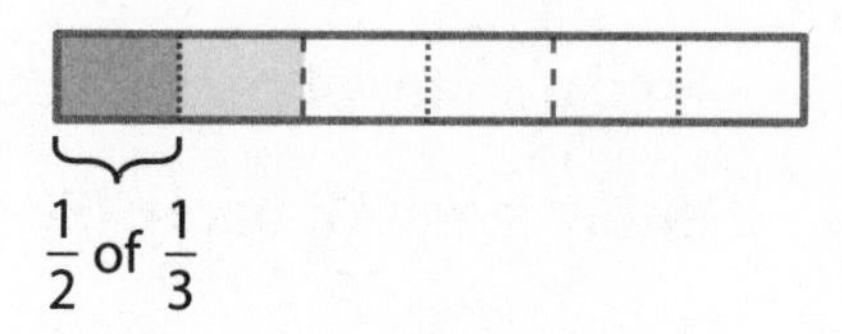

Step 2: To find $\frac{1}{2}$ of $\frac{1}{3}$, divide each $\frac{1}{3}$ into 2 equal parts.

Because the 6 parts make 1 whole, 1 part represents _____. So, $\frac{1}{2} \times \frac{1}{3} =$ _____.

Another Way: Use an area model with 2 rows and 3 columns to find $\frac{1}{2} \times \frac{1}{3}$.

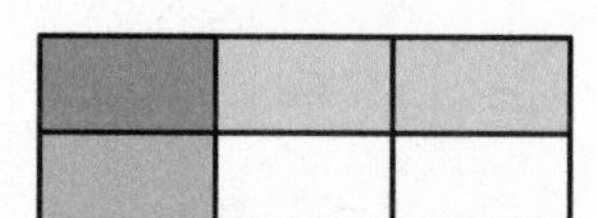

Step 1: Shade 1 of the 2 rows blue to represent $\frac{1}{2}$.

Step 2: Shade 1 of the 3 columns red to represent $\frac{1}{3}$.

The purple overlap shows the product.

_____ out of _____ parts is purple.

So, $\frac{1}{2} \times \frac{1}{3} =$ _____.

Show and Grow *I can do it!*

Multiply. Use a model to help.

1. $\frac{1}{3} \times \frac{1}{4} =$ _____

2. $\frac{2}{3} \times \frac{1}{2} =$ _____

Name ____________________

Apply and Grow: Practice

Multiply. Use a model to help.

3. $\frac{1}{2} \times \frac{1}{6} =$ ______

4. $\frac{1}{5} \times \frac{1}{8} =$ ______

5. $\frac{1}{4} \times \frac{1}{6} =$ ______

6. $\frac{2}{3} \times \frac{1}{3} =$ ______

Write a multiplication equation represented by the model.

7.

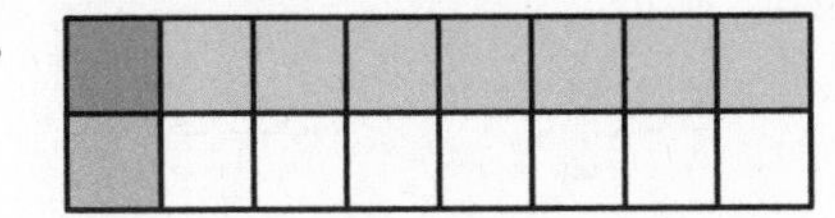

8.

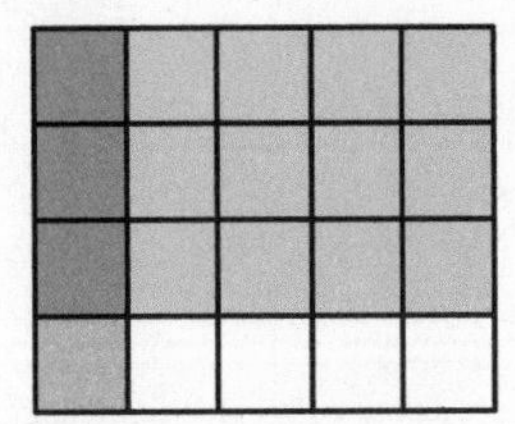

9. One-fifth of the students in your school have tried skating. Of those students, $\frac{1}{7}$ have tried ice skating. What fraction of students in your school have tried ice skating?

10. DIG DEEPER! Are both Newton and Descartes correct? Explain.

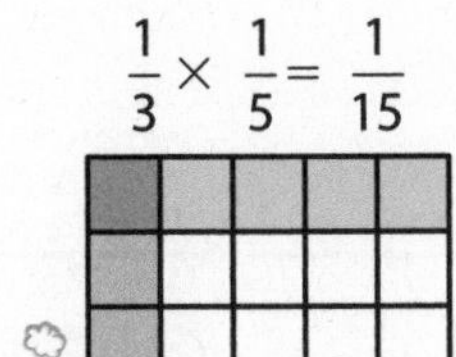

Think and Grow: Modeling Real Life

Example A recipe calls for $\frac{3}{4}$ teaspoon of cinnamon. You want to halve the recipe. What fraction of a teaspoon of cinnamon do you need?

Because you want to halve the recipe, multiply $\frac{1}{2}$ by $\frac{3}{4}$ to find how many teaspoons of cinnamon you need.

Use an area model with 2 rows and 4 columns to find $\frac{1}{2} \times \frac{3}{4}$.

Step 1: Shade 1 of the 2 rows blue to represent $\frac{1}{2}$.

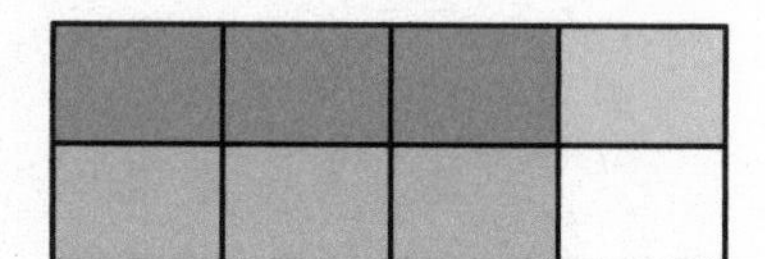

Step 2: Shade 3 of the 4 columns red to represent $\frac{3}{4}$.

The purple overlap shows the product.

_____ out of _____ parts are purple.

So, $\frac{1}{2} \times \frac{3}{4} =$ _____.

You need _____ teaspoon of cinnamon.

Show and Grow *I can think deeper!*

11. The mass of a mango is $\frac{2}{5}$ kilogram. The mass of a guava is $\frac{1}{4}$ as much as the mango. What is the mass of the guava?

12. A giant panda spends $\frac{2}{3}$ of 1 day eating and foraging. It spends $\frac{3}{4}$ of that time eating bamboo. What fraction of 1 day does the panda spend eating bamboo?

13. **DIG DEEPER!** You have a half-gallon carton of milk that you use only for cereal. You use the same amount each day for 5 days. There is $\frac{3}{8}$ of the carton left. How many cups of milk do you use each day? Explain.

Name ___________________________

Homework & Practice 9.4

Learning Target: Use models to multiply a fraction by a fraction.

Example Find $\frac{1}{2} \times \frac{1}{5}$.

One Way: Use a tape diagram to find $\frac{1}{2}$ of $\frac{1}{5}$.

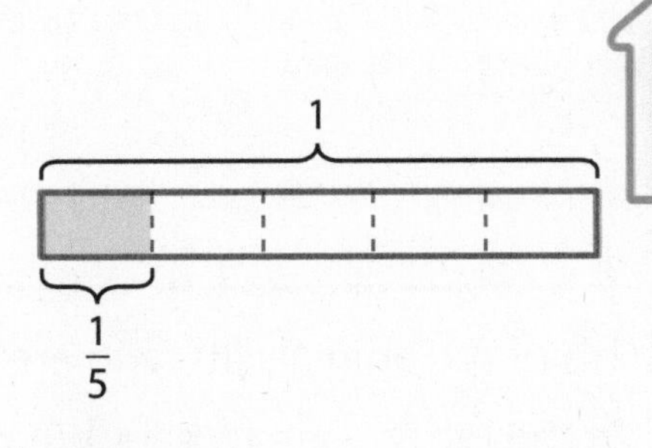

Step 1: Model $\frac{1}{5}$. Divide 1 whole into 5 equal parts.

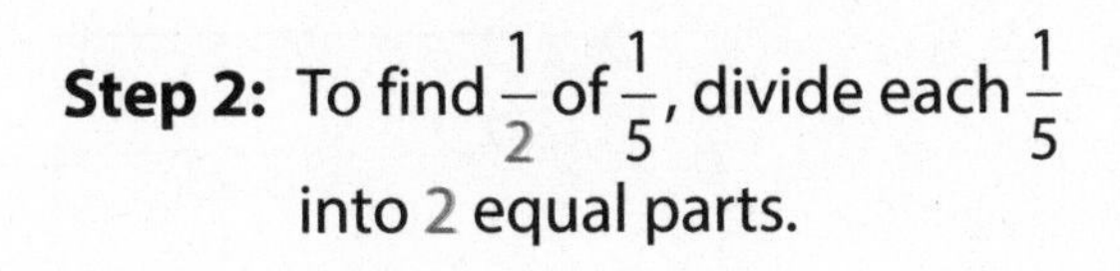

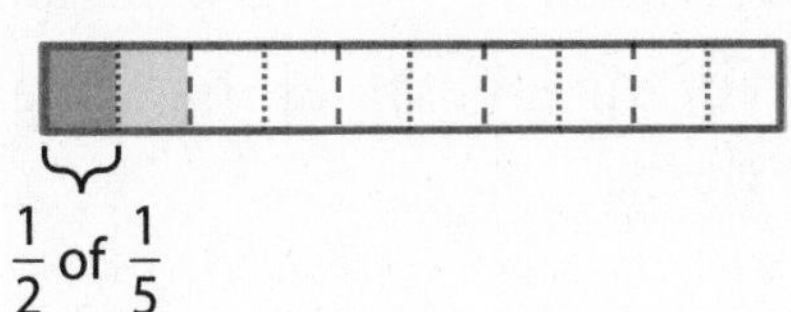

Step 2: To find $\frac{1}{2}$ of $\frac{1}{5}$, divide each $\frac{1}{5}$ into 2 equal parts.

Because the 10 parts make 1 whole, 1 part represents $\underline{\frac{1}{10}}$. So, $\frac{1}{2} \times \frac{1}{5} = \underline{\frac{1}{10}}$.

Another Way: Use an area model.

Shade 1 of the 2 rows blue to represent $\frac{1}{2}$. Shade 1 of the 5 columns red to represent $\frac{1}{5}$. The purple overlap shows the product.

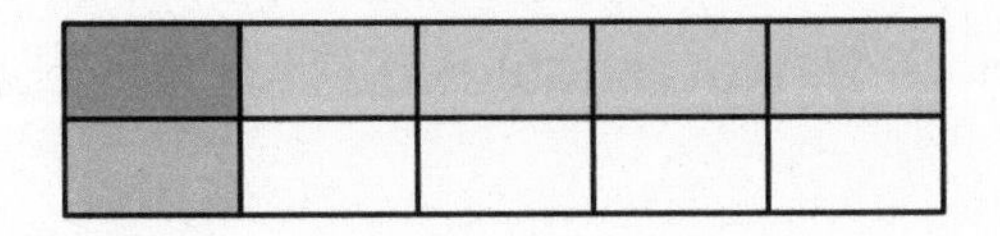

<u>1</u> out of <u>10</u> parts is purple.

So, $\frac{1}{2} \times \frac{1}{5} = \underline{\frac{1}{10}}$.

Multiply. Use a model to help.

1. $\frac{1}{3} \times \frac{1}{7} =$ ______

2. $\frac{1}{2} \times \frac{1}{9} =$ ______

3. $\frac{2}{5} \times \frac{1}{6} =$ ______

4. $\frac{3}{4} \times \frac{2}{7} =$ ______

Write a multiplication equation represented by the model.

5.

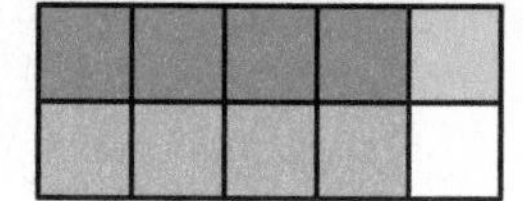

6.

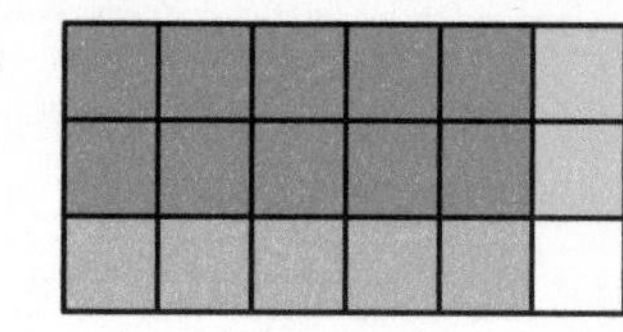

7. One sixth of the animals at a zoo are birds. Of the birds, $\frac{1}{3}$ are female. What fraction of the animals at the zoo are female birds?

8. **Writing** Write and solve a real-life problem for the expression.

$$\frac{2}{3} \times \frac{3}{7}$$

9. **Modeling Real Life** A Gouldian finch is $\frac{1}{2}$ the length of the sun conure. How long is the Gouldian finch?

Sun conure: $\frac{11}{12}$ ft

Gouldian finch

10. **DIG DEEPER!** A recipe calls for $\frac{2}{3}$ cup of chopped walnuts. You chop 4 walnuts and get $\frac{1}{4}$ of the amount you need. How much more of a cup of chopped walnuts do you need? All of your walnuts are the same size. How many more walnuts should you chop? Explain.

Review & Refresh

Subtract.

11. $5 - 1\frac{3}{4} =$ ____

12. $13\frac{1}{4} - 7\frac{5}{8} =$ ____

13. $12\frac{7}{10} - 5\frac{3}{10} - 1\frac{1}{5} =$ ____

Name ______________________________

Multiply Fractions 9.5

Learning Target: Multiply a fraction by a fraction.

Success Criteria:

- I can multiply the numerators of two fractions.
- I can multiply the denominators of two fractions.
- I can use a rule to find the product of two fractions.

Explore and Grow

Use models to help you complete the table. What do you notice about each expression and its product?

Expression	Product
$\frac{1}{2} \times \frac{1}{5}$	
$\frac{3}{4} \times \frac{1}{2}$	
$\frac{1}{2} \times \frac{3}{5}$	
$\frac{2}{3} \times \frac{2}{3}$	

MP **Construct Arguments** Explain how to multiply two fractions without using a model.

Think and Grow: Multiply Fractions

Key Idea You can find the product of a fraction and a fraction by multiplying the numerators and multiplying the denominators.

Example Find $\frac{1}{2} \times \frac{3}{2}$.

Multiply the numerators and multiply the denominators.

$\frac{1}{2} \times \frac{3}{2} = \frac{\square \times \square}{\square \times \square}$

$= \frac{\square}{\square}$

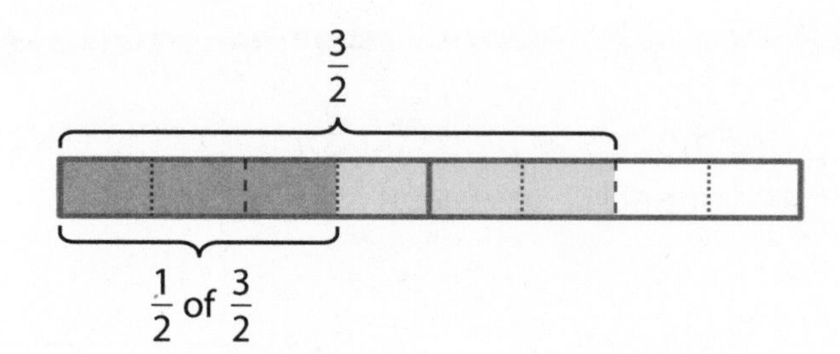

Example Find $\frac{5}{6} \times \frac{3}{5}$.

Multiply the numerators and multiply the denominators.

$\frac{5}{6} \times \frac{3}{5} = \frac{\square \times \square}{\square \times \square}$

$= \frac{\square}{\square}$, or $\frac{\square}{\square}$

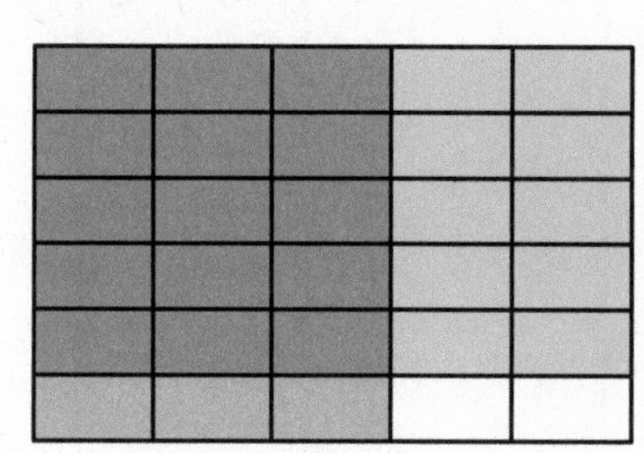

15 out of 30 parts are shaded twice.

Show and Grow *I can do it!*

Multiply.

1. $\frac{1}{2} \times \frac{4}{3} =$ _____

2. $\frac{2}{5} \times \frac{2}{3} =$ _____

3. $\frac{3}{4} \times \frac{5}{8} =$ _____

Name ______________________________

Apply and Grow: Practice

Multiply.

4. $\frac{1}{4} \times \frac{1}{4} =$ ______

5. $\frac{5}{6} \times \frac{7}{10} =$ ______

6. $\frac{6}{9} \times \frac{8}{2} =$ ______

7. $\frac{21}{100} \times \frac{3}{5} =$ ______

8. $\frac{1}{12} \times \frac{9}{4} =$ ______

9. $\frac{4}{7} \times \frac{8}{8} =$ ______

Evaluate.

10. $\left(\frac{1}{2} \times \frac{7}{8}\right) \times 2 =$ ______

11. $\left(\frac{7}{6} - \frac{5}{6}\right) \times \frac{2}{3} =$ ______

12. $\frac{9}{10} \times \left(\frac{4}{9} + \frac{1}{3}\right) =$ ______

13. At a school, $\frac{3}{4}$ of the students play a sport. Of the students that play a sport, $\frac{1}{5}$ play baseball. What fraction of the students at the school play baseball?

14. **MP Reasoning** Descartes says he can find the product of a whole number and a fraction the same way he finds the product of two fractions. Explain why his reasoning makes sense.

15. **Writing** Explain how multiplying fractions is different than adding and subtracting fractions.

Think and Grow: Modeling Real Life

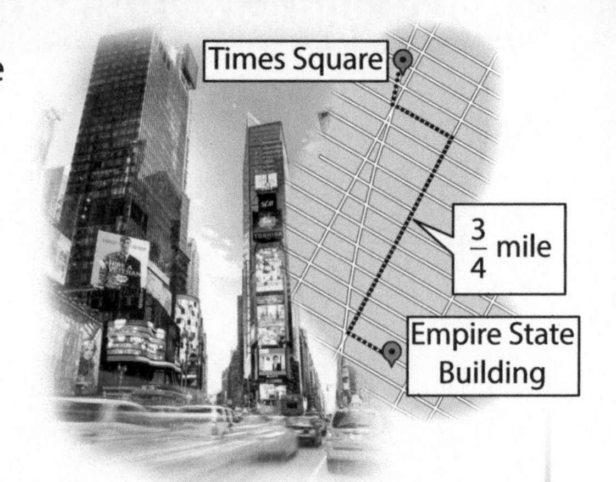

Example A tourist is walking from the Empire State Building to Times Square. She is $\frac{2}{3}$ of the way there. What fraction of a mile does she have left to walk?

Find the distance she has walked. Because she has walked $\frac{2}{3}$ of $\frac{3}{4}$ mile, multiply $\frac{2}{3}$ by $\frac{3}{4}$.

$$\frac{2}{3} \times \frac{3}{4} = \frac{\square \times \square}{\square \times \square}$$

$$= \frac{\square}{\square}, \text{ or } \frac{\square}{\square}$$

The tourist has walked ______ mile.

Subtract the product from $\frac{3}{4}$ mile to find how many miles she has left to walk.

$$\frac{3}{4} - \frac{1}{2} = \frac{3}{4} - \frac{\square}{4}$$

Rewrite $\frac{1}{2}$ as $\frac{1 \times 2}{2 \times 2} = \frac{2}{4}$.

$$= \frac{\square}{4}$$

So, the tourist has ______ mile left to walk.

Show and Grow *I can think deeper!*

16. At a zoo, $\frac{3}{5}$ of the animals are mammals. Of the mammals, $\frac{5}{12}$ are primates. What fraction of the animals at the zoo are *not* primates?

17. You have an album of 216 trading cards. One page contains $\frac{1}{24}$ of the cards. On that page, $\frac{2}{3}$ of the cards are epic. You only have one page with any epic cards. How many epic cards do you have?

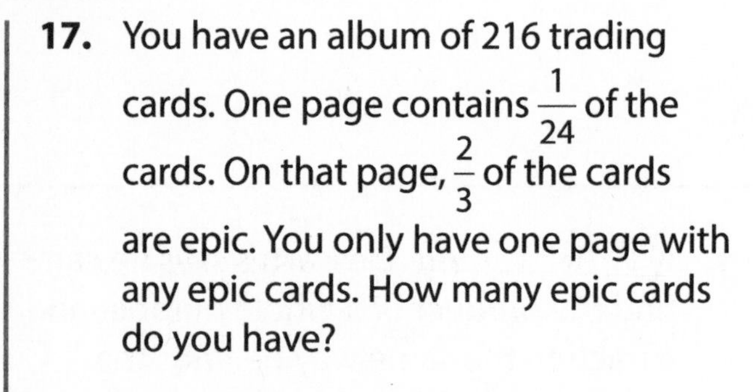

18. **DIG DEEPER!** In a class, $\frac{2}{5}$ of the students play basketball and $\frac{7}{10}$ play soccer. Of the students who play basketball, $\frac{2}{3}$ also play soccer. There are 30 students in the class. How many students play soccer but do *not* play basketball?

Name ______________________________

Homework & Practice 9.5

Learning Target: Multiply a fraction by a fraction.

Example Find $\frac{2}{3} \times \frac{7}{8}$.

Multiply the numerators and multiply the denominators.

$$\frac{2}{3} \times \frac{7}{8} = \frac{\boxed{2} \times \boxed{7}}{\boxed{3} \times \boxed{8}}$$

$$= \frac{\boxed{14}}{\boxed{24}}, \text{ or } \frac{\boxed{7}}{\boxed{12}}$$

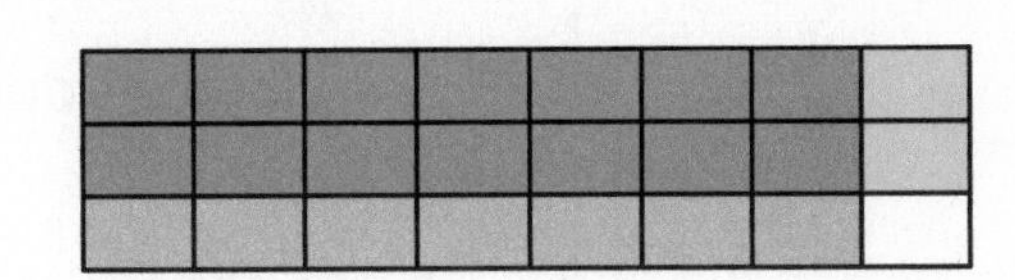

14 out of
24 parts are
shaded twice.

Multiply.

1. $\frac{1}{4} \times \frac{1}{5} =$ ______

2. $\frac{2}{7} \times \frac{1}{2} =$ ______

3. $\frac{9}{10} \times \frac{2}{3} =$ ______

4. $\frac{5}{8} \times \frac{5}{6} =$ ______

5. $\frac{9}{7} \times \frac{3}{4} =$ ______

6. $\frac{11}{100} \times \frac{2}{5} =$ ______

7. $\frac{7}{20} \times \frac{6}{2} =$ ______

8. $\frac{15}{16} \times \frac{1}{3} =$ ______

9. $\frac{5}{12} \times \frac{3}{10} =$ ______

Evaluate.

10. $3 \times \left(\frac{2}{3} \times \frac{1}{8}\right) =$ ______

11. $\left(\frac{1}{3} + \frac{1}{3}\right) \times \frac{4}{5} =$ ______

12. $\frac{6}{7} \times \left(\frac{3}{4} - \frac{5}{12}\right) =$ ______

13. A pancake recipe calls for $\frac{1}{3}$ cup of water. You want to halve the recipe. What fraction of a cup of water do you need?

14. MP **Number Sense** Which is greater, $\frac{3}{4} \times \frac{1}{5}$ or $\frac{3}{4} \times \frac{1}{8}$? Explain.

15. MP **Reasoning** Is $\frac{17}{24} \times \frac{7}{8}$ equal to $\frac{17}{8} \times \frac{7}{24}$? Explain.

16. MP **Number Sense** In which equations does $k = \frac{5}{6}$?

$\frac{1}{2} \times \frac{5}{3} = k$

$\frac{1}{10} \times k = \frac{1}{12}$

$\frac{1}{3} \times k = \frac{2}{9}$

$\frac{5}{4} \times \frac{2}{3} = k$

17. **Modeling Real Life** At a town hall meeting, $\frac{37}{50}$ of the members are present. Of those who are present, $\frac{1}{2}$ vote in favor of a new park. What fraction of the members do *not* vote in favor of the new park?

18. **Modeling Real Life** There are 50 U.S. states. Seven twenty-fifths of the states share a land border with Canada or Mexico. Of those states, $\frac{2}{7}$ share a border with Mexico. How many states share a border with Canada?

Review & Refresh

Evaluate. Check whether your answer is reasonable.

19. $15.67 + 4 + 6.5 =$ _____

20. $20.7 - 9.54 + 25.81 =$ _____

Name ________________________________

Find Areas of Rectangles 9.6

Learning Target: Find areas of rectangles.

Success Criteria:

- I can find the area of a rectangle with unit fraction side lengths.
- I can find the number of rectangles with unit fraction side lengths it takes to fill a rectangle.
- I can find the area of a rectangle with fractional side lengths.

Explore and Grow

Draw and cut out a rectangle that has any two of the side lengths below.

$\frac{1}{2}$ ft $\qquad$ $\frac{1}{3}$ ft $\qquad$ $\frac{1}{4}$ ft

Use several copies of your rectangle to create a unit square. What is the area (in square feet) of each small rectangle? Explain your reasoning.

Reasoning How can you use a rectangle with unit fraction side lengths to find the area of the rectangle below? Explain your reasoning.

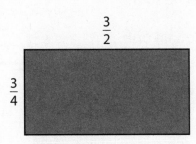

Think and Grow: Find Areas of Rectangles

One way to find the area of a rectangle with fractional side lengths is to fill it with smaller rectangles.

Example Find the area of the rectangle.

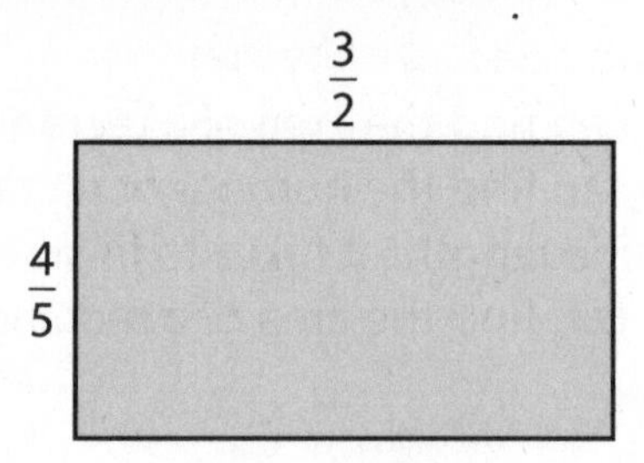

Step 1: Use the denominators of the side lengths to find a smaller rectangle with unit fraction side lengths.

$\frac{1}{5}$ $\frac{1}{2}$

Step 2: Find the area of a rectangle with side lengths of $\frac{1}{5}$ and $\frac{1}{2}$.

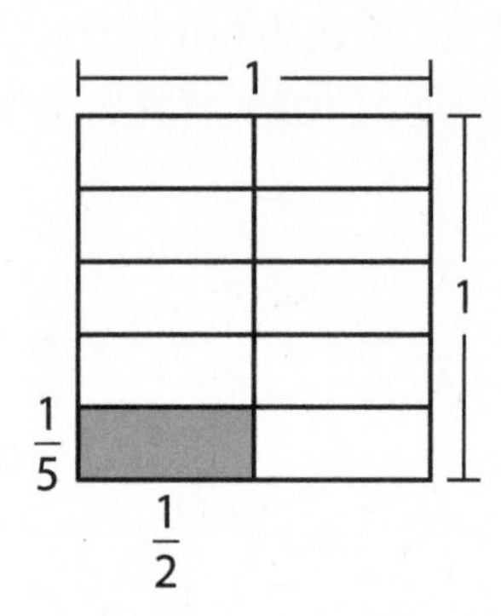

There are $5 \times 2 =$ _____ of these rectangles in a unit square, so the area of each is _____ square unit.

Step 3: Find the area of the large rectangle.

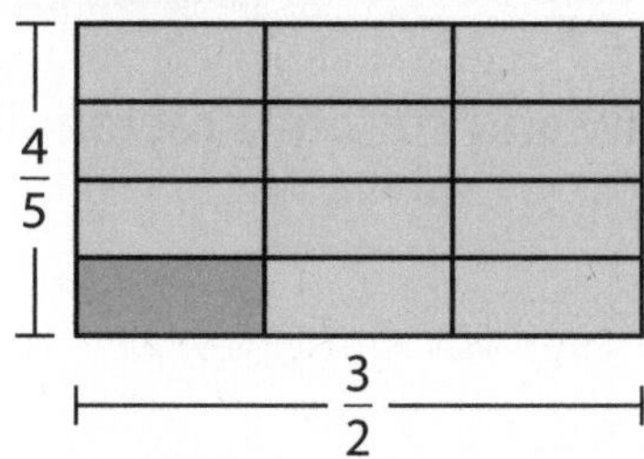

It takes $4 \times 3 =$ _____ of the smaller rectangles to fill the large rectangle. So, the area of the large rectangle is

$12 \times \frac{1}{10} =$ _____, or _____ square units.

This example shows why you can multiply the length and the width to find the area of a rectangle with fractional side lengths.

$\frac{1}{5} \times \frac{1}{2} = \frac{1}{10}$ and $\frac{4}{5} \times \frac{3}{2} = \frac{12}{10}$

Show and Grow *I can do it!*

1. Find the area of the shaded region.

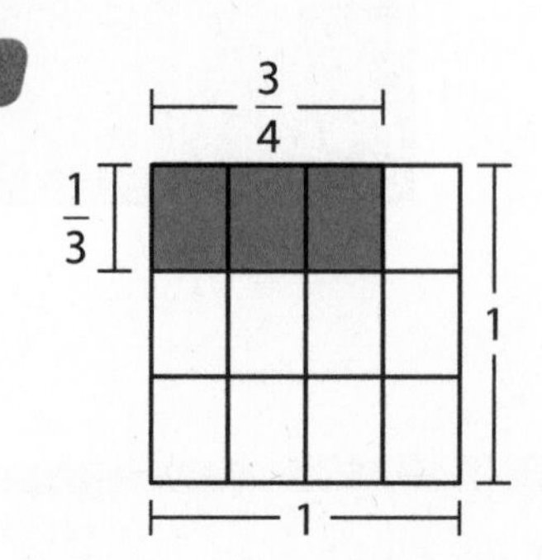

Name ______________________________

Apply and Grow: Practice

Find the area of the shaded region.

2.

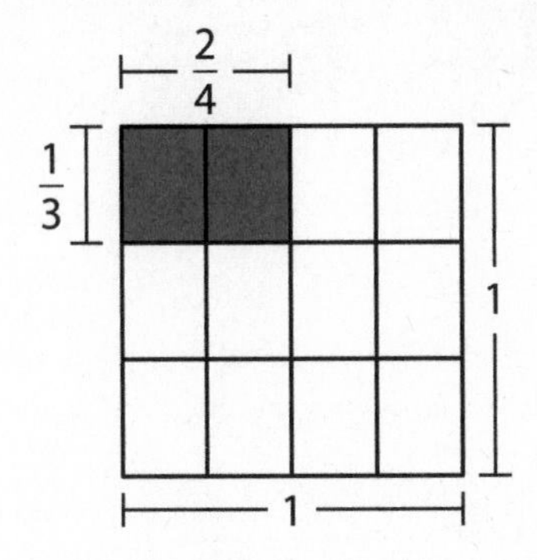

3.

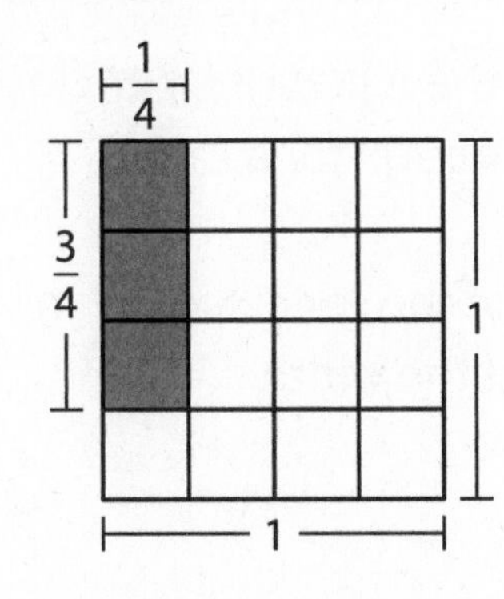

Use rectangles with unit fraction side lengths to find the area of the rectangle.

4. $\frac{2}{5}$

$\frac{3}{2}$

5.

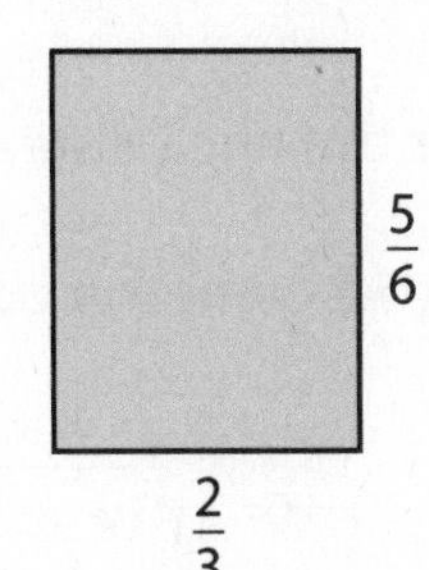

6. Find the area of a rectangle with side lengths of $\frac{5}{8}$ and $\frac{4}{3}$.

7. Find the area of a rectangle with side lengths of $\frac{7}{9}$ and $\frac{1}{2}$.

8. MP **Reasoning** Can you find the area of a rectangle with fractional side lengths the same way you find the area of a rectangle with whole number side lengths? Explain.

9. **YOU BE THE TEACHER** Your friend says she can find the area of a square given only one fractional side length. Is your friend correct? Explain.

Think and Grow: Modeling Real Life

Example A zoo needs an outdoor enclosure with an area of at least $\frac{3}{10}$ square kilometer for a camel. Is the rectangular enclosure shown large enough for a camel?

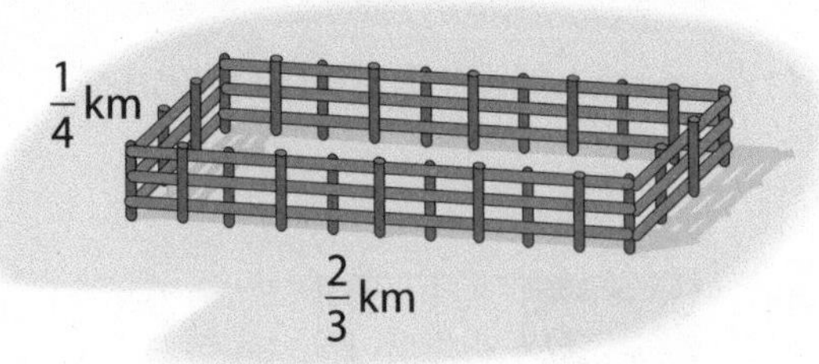

Find the area of the enclosure by multiplying the length and the width.

$\frac{1}{4} \times \frac{2}{3} = \frac{1 \times 2}{4 \times 3}$

$= \frac{\square}{\square}$, or $\frac{\square}{\square}$ square kilometer

To determine whether the enclosure is large enough, compare $\frac{1}{6}$ and $\frac{3}{10}$.

Write the fractions with a common denominator.

$\frac{1}{6} = \frac{1 \times 10}{6 \times 10} = \frac{\square}{60}$ $\frac{3}{10} = \frac{3 \times 6}{10 \times 6} = \frac{\square}{60}$

$\frac{\square}{60} \bigcirc \frac{\square}{60}$

So, the enclosure ____________ large enough for a camel.

Show and Grow *I can think deeper!*

10. The area of a square dog kennel is 4 square yards. Will the square mat fit in the kennel?

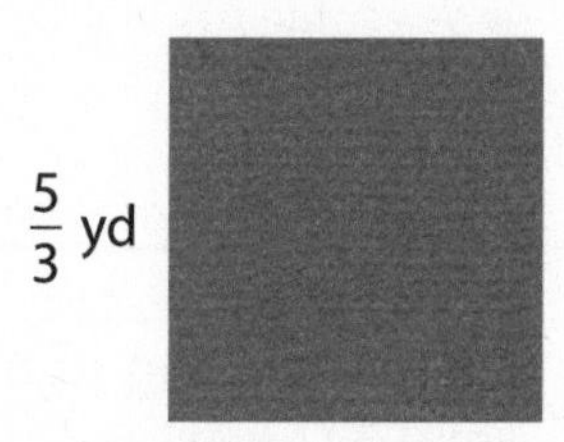

11. **DIG DEEPER!** The side lengths of each chalk art square are $\frac{11}{4}$ meters. The side lengths of the square zone around each chalk square are an additional $\frac{3}{4}$ meter. How many square meters of concrete are used to create the chalk walk shown?

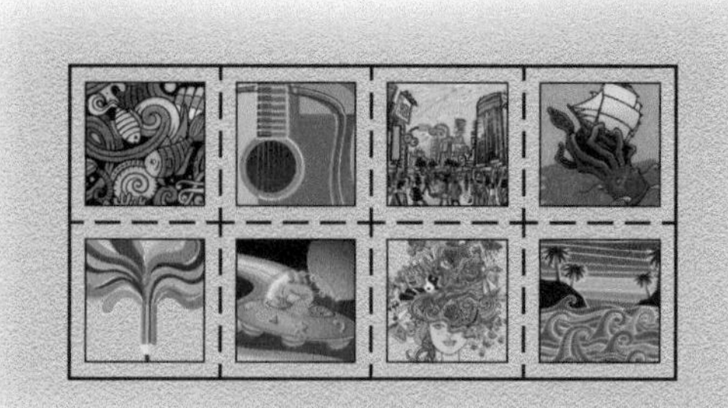

Name ____________________

Homework & Practice 9.6

Learning Target: Find areas of rectangles.

Example Find the area of the rectangle.

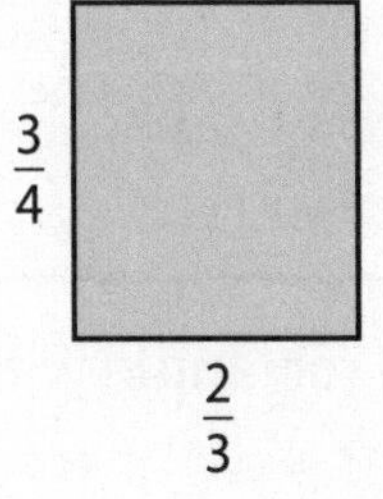

Step 1: Use the denominators of the side lengths to find a smaller rectangle with unit fraction side lengths.

$\frac{1}{4}$ $\frac{1}{3}$

Step 2: Find the area of a rectangle with side lengths of $\frac{1}{4}$ and $\frac{1}{3}$.

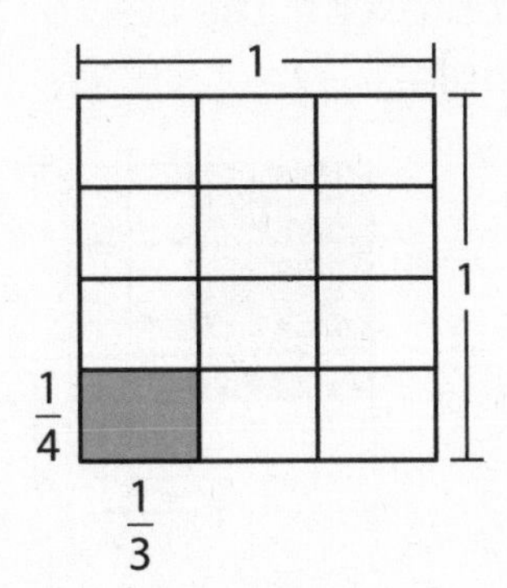

There are $4 \times 3 =$ 12 of these rectangles in a unit square, so the area of each is $\frac{1}{12}$ square unit.

Step 3: Find the area of the large rectangle.

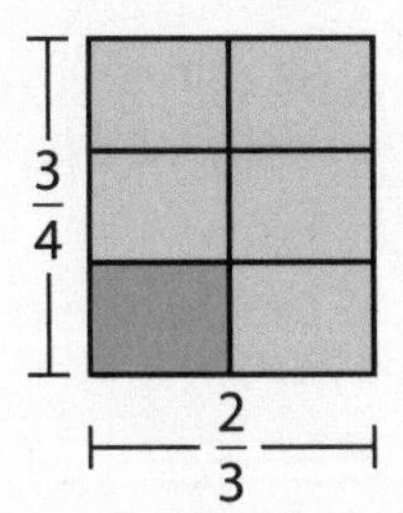

It takes $3 \times 2 =$ 6 of the smaller rectangles to fill the large rectangle. So, the area of the large rectangle is

$6 \times \frac{1}{12} = \frac{6}{12}$,

or $\frac{1}{2}$ square unit.

Find the area of the shaded region.

1.

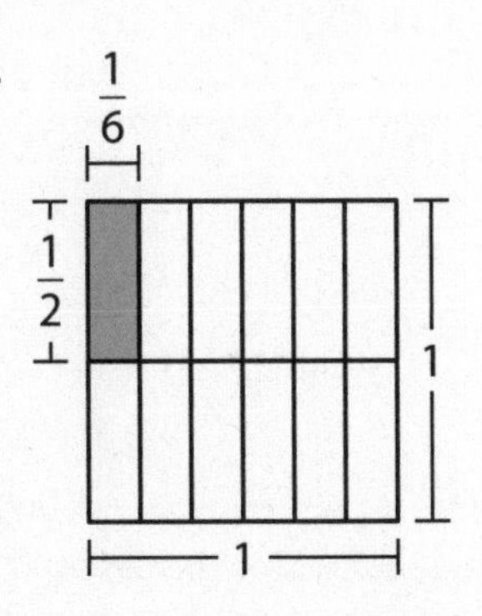

2.

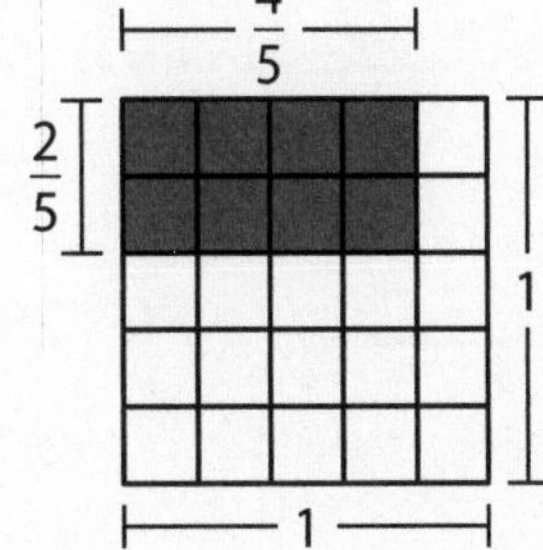

Use rectangles with unit fraction side lengths to find the area of the rectangle.

3.

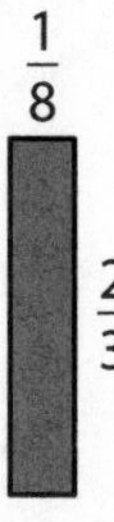

4.

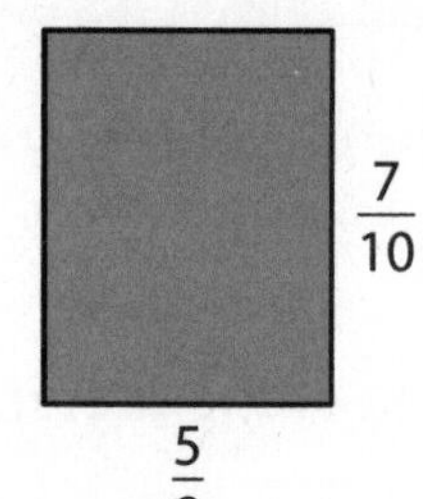

5. Find the area of a rectangle with side lengths of $\frac{3}{4}$ and $\frac{5}{12}$.

6. Find the area of a square with a side length of $\frac{9}{16}$.

7. Open-Ended The area of a rectangle is $\frac{16}{24}$. What are possible side lengths of the rectangle?

8. MP **Structure** Write an expression that represents the area of the shaded rectangle.

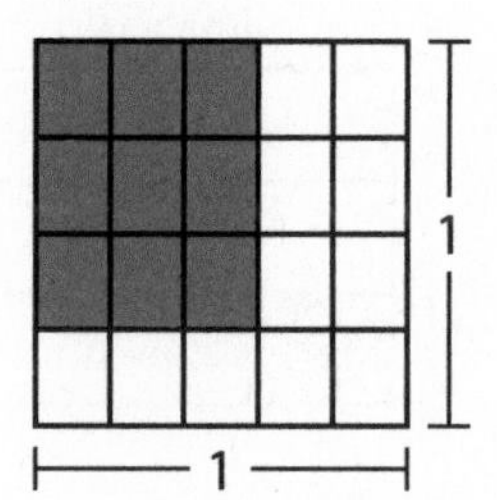

9. Modeling Real Life The area of a square table is 9 square feet. Will the board game fit on the table?

Review & Refresh

Divide. Then check your answer.

10. $365 \div 14 =$ ______

11. $282 \div 27 =$ ______

12. $601 \div 72 =$ ______

Name ___________________________________

Multiply Mixed Numbers 9.7

Learning Target: Multiply a mixed number by a mixed number.

Success Criteria:

- I can use a model to find the product of two mixed numbers.
- I can rewrite mixed numbers as improper fractions to find their products.
- I can find the product of two mixed numbers.

Explore and Grow

Find the area of the rectangle. Explain how you found your answer.

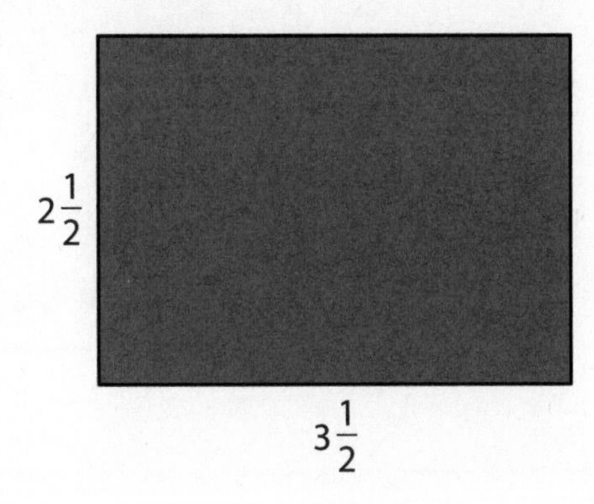

Structure Find the area using a different method. Explain how you found your answer.

Think and Grow: Multiply Mixed Numbers

You can use a model to find the product of two mixed numbers. You can also write the mixed numbers as improper fractions and then multiply.

Example Find $1\frac{1}{2} \times 2\frac{3}{4}$.

One Way: Use an area model.

Step 1: Write each mixed number as a sum. $1\frac{1}{2} = 1 + \frac{1}{2}$ $2\frac{3}{4} = 2 + \frac{3}{4}$

Step 2: Draw an area model that represents the product of the sums.

Step 3: Find the sum of the areas of the sections.

$(2 + 1) + \left(\frac{3}{4} + \frac{3}{8}\right) = 3 + \frac{\square}{\square}$

$= \square\frac{\square}{\square}$

	2	$\frac{3}{4}$
1	$1 \times 2 = 2$	$1 \times \frac{3}{4} = \frac{3}{4}$
$\frac{1}{2}$	$\frac{1}{2} \times 2 = 1$	$\frac{1}{2} \times \frac{3}{4} = \frac{3}{8}$

So, $1\frac{1}{2} \times 2\frac{3}{4} =$ _____.

Another Way: Write each mixed number as an improper fraction, then multiply.

$1\frac{1}{2} \times 2\frac{3}{4} = \frac{3}{2} \times \frac{11}{4}$

$= \frac{\square \times \square}{\square \times \square}$

$= \frac{\square}{\square}$, or $\square\frac{\square}{\square}$

Multiply the numerators and multiply the denominators.

Show and Grow *I can do it!*

Multiply.

1. $2\frac{1}{2} \times 1\frac{1}{2} =$ _____

2. $3\frac{1}{4} \times 2\frac{2}{3} =$ _____

Name ______________________________

Apply and Grow: Practice

Multiply.

3. $1\frac{3}{4} \times 2\frac{1}{6} =$ ______

4. $4\frac{1}{3} \times 1\frac{5}{6} =$ ______

5. $3\frac{2}{5} \times 1\frac{9}{10} =$ ______

6. $2\frac{3}{8} \times 3\frac{1}{2} =$ ______

Evaluate.

7. $5\frac{1}{4} \times \frac{2}{5} \times 6\frac{1}{12} =$ ______

8. $3\frac{2}{3} \times \left(10\frac{7}{8} - 2\frac{1}{4}\right) =$ ______

9. **YOU BE THE TEACHER** Your friend uses the model to find $4\frac{1}{2} \times 3\frac{2}{3}$. Is your friend correct? Explain.

$4\frac{1}{2}$

$3\frac{2}{3}$

$3 \times 4 = 12$

$\frac{2}{3} \times \frac{1}{2} = \frac{1}{3}$

$12 + \frac{1}{3} = 12\frac{1}{3}$

10. MP **Logic** Find the missing numbers.

$\square\frac{1}{4} \times 5\frac{1}{\square} = 6\frac{7}{8}$

Think and Grow: Modeling Real Life

Example A construction crew is paving 15 miles of a highway. The crew paves $4\frac{2}{10}$ miles each month. Does the crew finish paving the highway in $3\frac{1}{2}$ months?

Find the length of the highway the crew paves by multiplying the number of months by the number of miles they pave each month. Write each mixed number as an improper fraction, then multiply.

$3\frac{1}{2} \times 4\frac{2}{10} = \frac{7}{2} \times \frac{42}{10}$

$= \frac{\square \times \square}{\square \times \square}$

$= \frac{\square}{\square}$, or $\square\frac{\square}{\square}$

So, the crew paves ______ miles.

Estimate ________

Reasonable? ________ is close to ________. ✓

Compare the length the crew paves to the amount that needs to be paved.

The crew ________________ finish paving the highway in $3\frac{1}{2}$ months.

Show and Grow *I can think deeper!*

11. You have 3 cups of strawberries. You want to make $1\frac{1}{2}$ batches of the recipe. Do you have enough strawberries?

Strawberry Smoothie

$1\frac{3}{4}$ cups of strawberries

$\frac{3}{4}$ cup of yogurt

1 teaspoon lemon juice

2 tablespoons honey

1 cup ice

12. On Monday, you roller-skate $6\frac{1}{4}$ miles. On Tuesday, you skate $1\frac{2}{5}$ times as far as you did on Monday. How many total miles do you roller-skate on Monday and Tuesday combined?

13. **DIG DEEPER!** An artist paints a rectangular mural. The mural is $4\frac{1}{3}$ feet wide. The length is $2\frac{1}{4}$ times the width. What is the area of the mural?

Name ______________________________

Homework & Practice 9.7

Learning Target: Multiply a mixed number by a mixed number.

Example Find $1\frac{2}{3} \times 3\frac{4}{5}$.

One Way: Use an area model.

Step 1: Write each mixed number as a sum. $1\frac{2}{3} = 1 + \frac{2}{3}$ $3\frac{4}{5} = 3 + \frac{4}{5}$

Step 2: Draw an area model that represents the product of the sums.

Step 3: Find the sum of the areas of the sections.

$(3 + 2) + \left(\frac{4}{5} + \frac{8}{15}\right) = 5 + \frac{\boxed{20}}{\boxed{15}}$

$= \boxed{6}\frac{\boxed{1}}{\boxed{3}}$

	3	$\frac{4}{5}$
1	$1 \times 3 = 3$	$1 \times \frac{4}{5} = \frac{4}{5}$
$\frac{2}{3}$	$\frac{2}{3} \times 3 = 2$	$\frac{2}{3} \times \frac{4}{5} = \frac{8}{15}$

So, $1\frac{2}{3} \times 3\frac{4}{5} = \underline{6\frac{1}{3}}$.

Another Way: Write each mixed number as an improper fraction, then multiply.

$1\frac{2}{3} \times 3\frac{4}{5} = \frac{5}{3} \times \frac{19}{5}$

$= \frac{\boxed{5} \times \boxed{19}}{\boxed{3} \times \boxed{5}}$

$= \frac{\boxed{95}}{\boxed{15}}$, or $\boxed{6}\frac{\boxed{1}}{\boxed{3}}$

Multiply.

1. $1\frac{1}{2} \times 1\frac{1}{8} =$ ______

2. $1\frac{5}{6} \times 2\frac{1}{4} =$ ______

Evaluate.

3. $2\frac{3}{8} \times 2\frac{3}{4} =$ _____

4. $4\frac{1}{6} \times 3\frac{2}{7} =$ _____

5. $2\frac{1}{3} \times 3\frac{9}{10} \times 5\frac{1}{5} =$ _____

6. $\left(1\frac{7}{8} + 4\frac{4}{5}\right) \times 2\frac{1}{12} =$ _____

7. MP **Structure** Find the missing numbers.

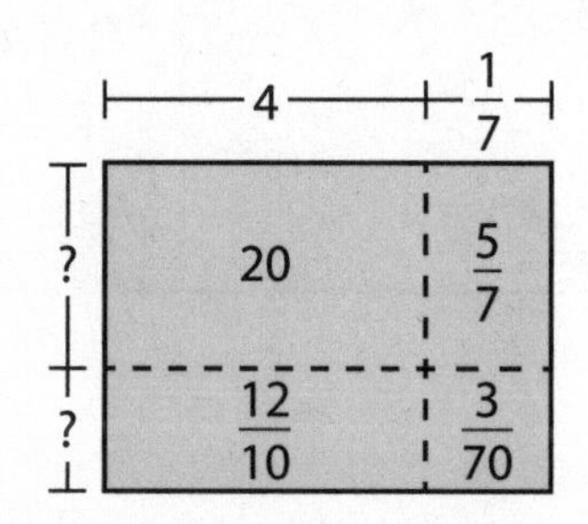

8. **YOU BE THE TEACHER** Your friend finds $1\frac{11}{12} \times 2\frac{3}{8}$. Is your friend correct? Explain.

$$1\frac{11}{12} \times 2\frac{3}{8} = \frac{23}{12} \times \frac{19}{8} = \frac{437}{96}, \text{ or } 4\frac{53}{96}$$

9. **Modeling Real Life** Your friend earns $7\frac{1}{2}$ dollars each hour. Will she earn enough money to buy a $35 toy after working $4\frac{3}{4}$ hours?

10. **Modeling Real Life** One class collects $8\frac{1}{4}$ pounds of recyclable materials. Another class collects $1\frac{1}{2}$ times as many pounds as the first class. How many pounds of recyclable materials do the two classes collect altogether?

Review & Refresh

Find the product.

11. $6 \times 5.7 =$ _____

12. $0.84 \times 9 =$ _____

Name ____________________

Compare Factors and Products 9.8

Learning Target: Compare a product to each of its factors.

Success Criteria:

- I can determine whether a number is less than, greater than, or equal to 1.
- I can compare a product to each of its factors.
- I can explain why a product is less than, greater than, or equal to each of its factors.

Explore and Grow

Without calculating, order the rectangles by area from least to greatest. Explain your reasoning.

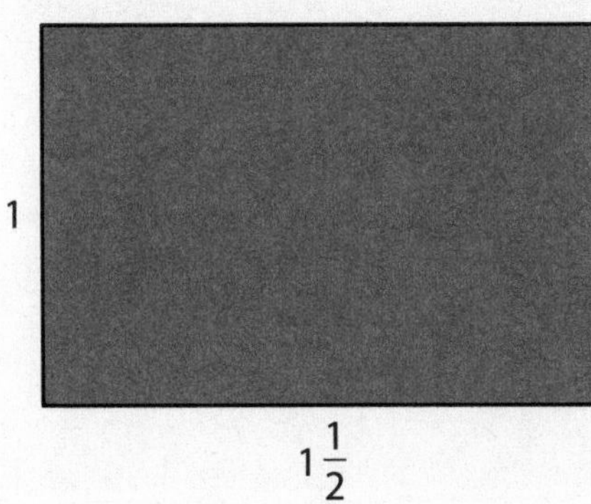

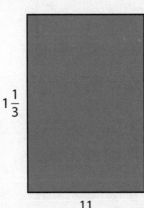

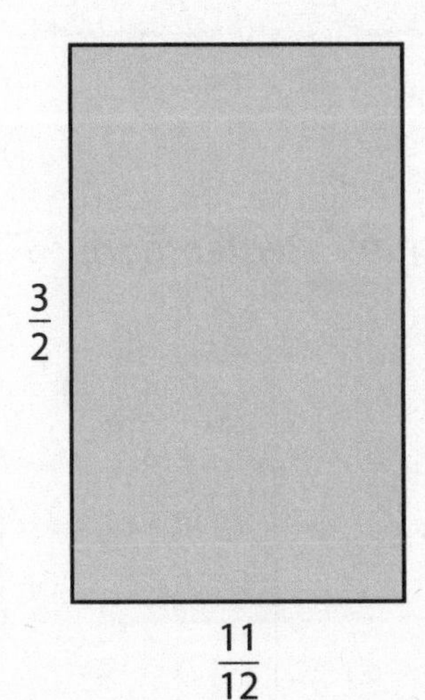

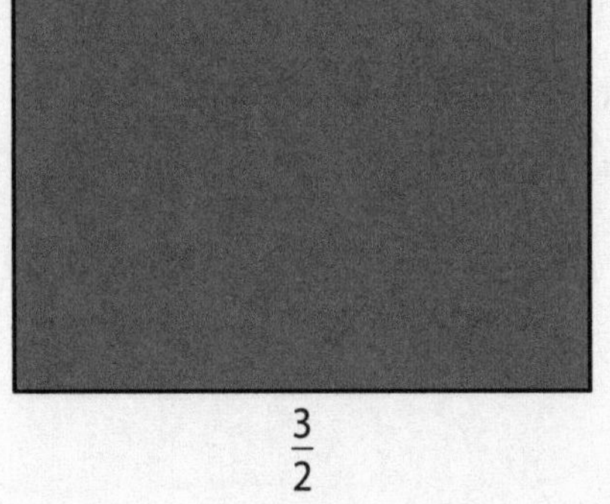

Construct Arguments Explain your strategy to your partner. Compare your strategies.

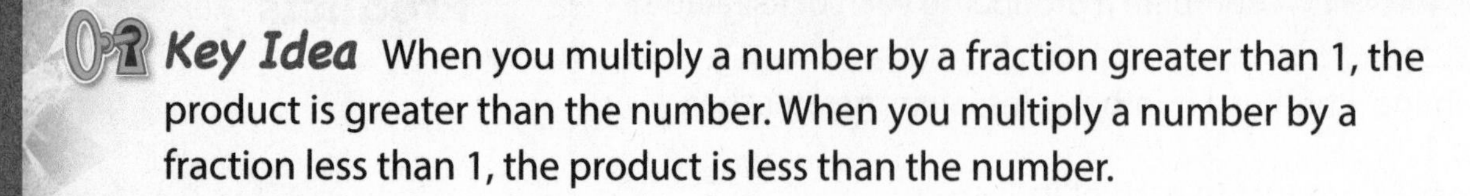

Think and Grow: Compare Factors and Products

Key Idea When you multiply a number by a fraction greater than 1, the product is greater than the number. When you multiply a number by a fraction less than 1, the product is less than the number.

Example Without calculating, tell whether the product $3\frac{1}{8} \times \frac{5}{6}$ is *less than*, *greater than*, or *equal to* each of its factors.

Because $3\frac{1}{8}$ ◯ 1, the product $3\frac{1}{8} \times \frac{5}{6}$ is ________ $\frac{5}{6}$.

Because $\frac{5}{6}$ ◯ 1, the product $3\frac{1}{8} \times \frac{5}{6}$ is ________ $3\frac{1}{8}$.

Example Without calculating, tell whether the product $\frac{5}{3} \times \frac{2}{2}$ is *less than*, *greater than*, or *equal to* each of its factors.

Because $\frac{5}{3}$ ◯ 1, the product $\frac{5}{3} \times \frac{2}{2}$ is ________ $\frac{2}{2}$.

Because $\frac{2}{2}$ ◯ 1, the product $\frac{5}{3} \times \frac{2}{2}$ is ________ $\frac{5}{3}$.

$\frac{5}{3} \times \frac{2}{2} = \frac{5 \times 2}{3 \times 2} = \frac{10}{6}$ shows you that $\frac{5}{3}$ and $\frac{10}{6}$ are equivalent. You can think of this as multiplying by 1.

$\frac{5}{3} \times \frac{2}{2} = \frac{5}{3} \times 1$

Show and Grow *I can do it!*

Without calculating, tell whether the product is *less than*, *greater than*, or *equal to* each of its factors.

1. $8 \times \frac{3}{10}$

2. $\frac{4}{4} \times 5\frac{2}{3}$

3. $\frac{4}{3} \times \frac{1}{6}$

Name ______________________________

Apply and Grow: Practice

Without calculating, tell whether the product is *less than*, *greater than*, or *equal to* each of its factors.

4. $\frac{1}{4} \times \frac{1}{12}$

5. $3\frac{4}{5} \times 6\frac{7}{8}$

6. $\frac{1}{6} \times \frac{10}{10}$

7. $\frac{2}{3} \times 5$

8. $\frac{7}{10} \times 4\frac{8}{9}$

9. $\frac{9}{2} \times 1\frac{3}{4}$

Without calculating, order the products from least to greatest.

10. $\frac{5}{6} \times 1\frac{8}{9}$ $\quad \frac{5}{6} \times \frac{1}{3}$ $\quad \frac{5}{6} \times \frac{7}{7}$

11. $\frac{1}{6} \times \frac{1}{4}$ $\quad \frac{1}{10} \times \frac{1}{4}$ $\quad 5\frac{7}{10} \times \frac{1}{4}$

12. **YOU BE THE TEACHER** Your friend says that $\frac{1}{2} \times 8$ is half as much as 8. Is your friend correct? Explain.

13. **DIG DEEPER!** Without calculating, tell whether the product is *less than*, *greater than*, or *equal to* $3\frac{3}{4}$. Explain.

$$\left(\frac{1}{2} \times 3\frac{3}{4}\right) \times \frac{2}{7}$$

Think and Grow: Modeling Real Life

Shot put is a track and field throwing event that uses a heavy metal ball called a shot.

Example Men's shot put competitions use a shot with a mass of $7\frac{1}{4}$ kilograms. The mass of a bowling ball is $\frac{7}{8}$ as much as the mass of the shot. Is the mass of the bowling ball less than, greater than, or equal to the mass of the shot?

Write an expression to represent the mass of the bowling ball.

The mass of the bowling ball is $\frac{7}{8}$ as much as the mass of the shot, so multiply the mass of the shot by $\frac{7}{8}$.

$7\frac{1}{4} \times \frac{7}{8}$

You can also say that the mass of the shot is $\frac{8}{7}$ times as much as the mass of the bowling ball.

Use the factor $\frac{7}{8}$ to compare the masses.

Because $\frac{7}{8}$ ◯ 1, the product $7\frac{1}{4} \times \frac{7}{8}$ is ______________ $7\frac{1}{4}$.

So, the mass of the bowling ball is ____________ the mass of the shot.

Show and Grow *I can think deeper!*

14. You practice playing the keyboard for $3\frac{1}{2}$ hours. Your friend practices playing the keyboard for $\frac{5}{4}$ as many hours as you. Does your friend practice for fewer hours, more hours, or the same number of hours as you?

15. The original price of a telescope is \$99. The sale price is $\frac{4}{5}$ of the original price. An astrologist buys the telescope at its sale price and uses a half-off coupon. What fraction of the original price does the astrologist pay for the telescope?

16. **DIG DEEPER!** The Abraj Al-Bait Clock Tower is $\frac{6}{10}$ kilometer tall. Zifeng Tower is $\frac{3}{4}$ as tall as the clock tower. Is Zifeng Tower shorter than, taller than, or the same height as the Abraj Al-Bait Clock Tower? What is the height of each tower in meters?

Name ______________________________

Homework & Practice 9.8

Learning Target: Compare a product to each of its factors.

Example Without calculating, tell whether the product $\frac{1}{3} \times \frac{6}{7}$ is *less than, greater than,* or *equal to* each of its factors.

Because $\frac{1}{3}$ ⓧ 1, the product $\frac{1}{3} \times \frac{6}{7}$ is less than $\frac{6}{7}$.

Because $\frac{6}{7}$ ⓧ 1, the product $\frac{1}{3} \times \frac{6}{7}$ is less than $\frac{1}{3}$.

Without calculating, tell whether the product is *less than, greater than,* or *equal to* each of its factors.

1. $1\frac{3}{4} \times 6$

2. $\frac{5}{12} \times \frac{1}{6}$

3. $\frac{2}{7} \times \frac{5}{5}$

4. $3\frac{4}{5} \times 2\frac{9}{10}$

5. $8 \times \frac{2}{3}$

6. $1\frac{7}{8} \times \frac{1}{4}$

Without calculating, order the products from least to greatest.

7. $\frac{1}{3} \times \frac{4}{5}$ $\quad \frac{1}{3} \times \frac{6}{6}$ $\quad \frac{1}{3} \times 8\frac{2}{9}$

8. $2\frac{1}{2} \times \frac{3}{5}$ $\quad 4 \times \frac{3}{5}$ $\quad \frac{1}{12} \times \frac{3}{5}$

9. MP **Logic** Without calculating, use >, <, or = to make the statement true. Explain.

$$3\frac{1}{3} \times \frac{1}{6} \bigcirc 3\frac{1}{2}$$

10. MP **Logic** Without calculating, determine which number makes the statement true.

_____ $\times 1\frac{7}{8}$ is greater than $1\frac{7}{8}$.

1 $\quad$ $\frac{1}{2}$ $\quad$ $1\frac{4}{5}$

11. MP **Reasoning** Why does multiplying by a fraction greater than one result in a product greater than the original number?

12. **Modeling Real Life** You snowboard $1\frac{7}{8}$ miles. Your friend snowboards $3\frac{2}{3}$ times as far as you. Does your friend snowboard fewer miles, more miles, or the same number of miles as you?

13. **Modeling Real Life** A pet owner has three dogs. The youngest dog weighs $\frac{1}{4}$ as much as the second oldest dog. The oldest dog weighs $1\frac{1}{4}$ as much as the second oldest. The second oldest weighs 20 pounds. Which dog weighs the most? the least?

Review & Refresh

Compare.

14. 40.5 $\bigcirc$ 40.13

15. 13.90 $\bigcirc$ 13.9

16. 32.006 $\bigcirc$ 32.06

Name ______________________________

Performance Task 9

1. You see a rock formation at a national park. The formation has layers that formed millions of years ago when particles settled in water and became rock. You make a model of the rock formation using $\frac{3}{16}$-inch foam sheets.

   a. The three types of sedimentary rocks are limestone, sandstone, and shale. Use the number of foam sheets to find the height of each sedimentary rock layer.

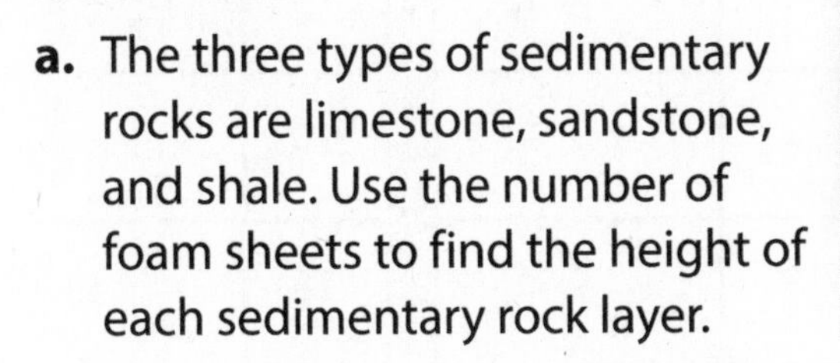

Type	Number of Foam Sheets	Height of Model (inches)
Topsoil		
Limestone	12	
Sandstone	8	
Shale	20	
Granite		3

   b. What is the combined height of the sedimentary rock layers?

   c. Will you use more foam sheets for the granite layers or the shale layers? Explain.

   d. The height of the topsoil layer is $1\frac{1}{4}$ times the height of the sandstone layer. How many foam sheets do you use in the topsoil layer?

   e. On your model, 1 inch represents 40 feet. What is the actual height of the rock formation?

   f. Why do you think the rock formation has layers?

Fraction Connection: Multiplication

Directions:

1. Players take turns rolling three dice.
2. On your turn, evaluate the expression indicated by your roll and cover the answer.
3. The first player to get four in a row, horizontally, vertically, or diagonally, wins!

Roll	Evaluate
3	$4 \times \frac{3}{8}$
4	$\frac{1}{2} \times \frac{7}{9}$
5	$\frac{3}{5} \times 9$
6	$2\frac{1}{4} \times 1\frac{2}{3}$
7	$7 \times \frac{2}{5}$
8	$\frac{7}{8} \times \frac{2}{3}$
9	$\frac{5}{6} \times 8$
10	$3\frac{1}{2} \times 2\frac{3}{4}$

Roll	Evaluate
11	$6 \times \frac{3}{7}$
12	$\frac{4}{5} \times \frac{3}{4}$
13	$\frac{7}{12} \times 4$
14	$1\frac{2}{7} \times 2\frac{2}{5}$
15	$5 \times \frac{1}{4}$
16	$\frac{9}{10} \times \frac{4}{5}$
17	$\frac{2}{9} \times 3$
18	$2\frac{3}{4} \times 3\frac{2}{3}$

$6\frac{2}{3}$	$2\frac{4}{7}$	$\frac{18}{25}$	$3\frac{3}{4}$
$3\frac{3}{35}$	$1\frac{1}{2}$	$9\frac{5}{8}$	$\frac{2}{3}$
$2\frac{4}{5}$	$\frac{3}{5}$	$2\frac{1}{3}$	$\frac{7}{18}$
$5\frac{2}{5}$	$10\frac{1}{12}$	$\frac{7}{12}$	$1\frac{1}{4}$

Name ______________________________

Chapter 9 Practice

9.1 Multiply Whole Numbers by Fractions

Multiply.

1. $5 \times \frac{1}{2} =$ _____

2. $2 \times \frac{7}{10} =$ _____

3. $9 \times \frac{5}{8} =$ _____

4. $6 \times \frac{71}{100} =$ _____

5. $4 \times \frac{8}{5} =$ _____

6. $7 \times \frac{5}{3} =$ _____

9.2 Use Models to Multiply Fractions by Whole Numbers

Multiply. Use a model to help.

7. $\frac{2}{5}$ of 20

8. $\frac{1}{6} \times 12$

9. $\frac{1}{3} \times 6$

10. $\frac{5}{16}$ of 8

11. Modeling Real Life You have 24 apples. You use $\frac{1}{4}$ of them to make a single serving of applesauce. How many apples do you *not* use?

9.3 Multiply Fractions and Whole Numbers

Multiply.

12. $\frac{3}{5} \times 15 =$ ______

13. $\frac{9}{10} \times 30 =$ ______

14. $48 \times \frac{3}{4} =$ ______

15. $11 \times \frac{5}{9} =$ ______

16. $\frac{1}{6} \times 19 =$ ______

17. $7 \times \frac{13}{50} =$ ______

9.4 Use Models to Multiply Fractions

Multiply. Use a model to help.

18. $\frac{1}{2} \times \frac{1}{10} =$ ______

19. $\frac{1}{5} \times \frac{1}{9} =$ ______

20. $\frac{1}{6} \times \frac{1}{7} =$ ______

21. $\frac{1}{3} \times \frac{1}{8} =$ ______

22. $\frac{2}{5} \times \frac{1}{3} =$ ______

23. $\frac{2}{3} \times \frac{3}{5} =$ ______

9.5 Multiply Fractions

Evaluate.

24. $\frac{1}{3} \times \frac{1}{8} =$ _____

25. $\frac{5}{6} \times \frac{1}{4} =$ _____

26. $\frac{7}{2} \times \frac{2}{5} =$ _____

27. $\frac{9}{10} \times \frac{3}{7} =$ _____

28. $\frac{4}{5} \times \frac{13}{100} =$ _____

29. $\frac{11}{25} \times \frac{3}{4} =$ _____

30. $9 \times \left(\frac{2}{9} \times \frac{1}{2}\right) =$ _____

31. $\left(\frac{1}{10} + \frac{7}{10}\right) \times \frac{3}{4} =$ _____

32. $\frac{4}{7} \times \left(\frac{5}{8} - \frac{1}{4}\right) =$ _____

9.6 Find Areas of Rectangles

Use rectangles with unit fraction side lengths to find the area of the rectangle.

33.

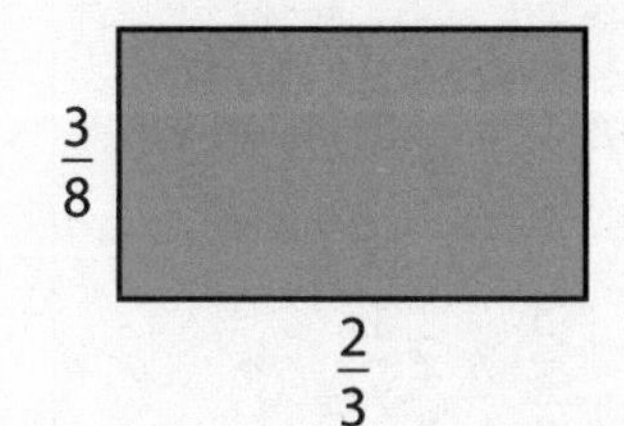

$\frac{3}{8}$

$\frac{2}{3}$

34.

$\frac{7}{12}$

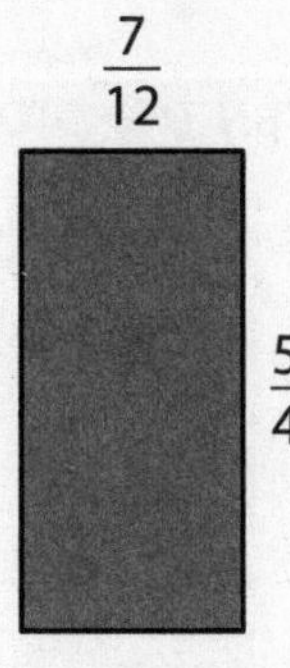

$\frac{5}{4}$

35. Find the area of a rectangle with side lengths of $\frac{2}{9}$ and $\frac{1}{10}$.

36. Find the area of a square with a side length of $\frac{3}{4}$.

9.7 Multiply Mixed Numbers

Multiply.

37. $1\frac{1}{4} \times 2\frac{1}{4} =$ _____

38. $3\frac{4}{5} \times 1\frac{1}{2} =$ _____

39. $5\frac{1}{3} \times 4\frac{7}{8} =$ _____

40. $4\frac{5}{6} \times 2\frac{9}{10} \times \frac{1}{8} =$ _____

41. MP **Logic** Find the missing numbers.

$$4\frac{1}{\square} \times 2\frac{1}{5} = 9\frac{\square}{10}$$

9.8 Compare Factors and Products

Without calculating, tell whether the product is *less than*, *greater than*, or *equal to* each of its factors.

42. $1\frac{1}{2} \times 3$

43. $\frac{1}{8} \times \frac{1}{6}$

44. $\frac{5}{5} \times 2\frac{4}{7}$

45. $\frac{3}{4} \times 2\frac{11}{12}$

46. $3\frac{1}{9} \times 2\frac{7}{8}$

47. $\frac{9}{8} \times \frac{5}{2}$

10 Divide Fractions

Chapter Learning Target:
Understand dividing fractions.

Chapter Success Criteria:

- I can identify fractions as division.
- I can explain mixed numbers as quotients.
- I can divide fractions.
- I can solve a problem using fractions.

- What are some tasks you can program a robot to complete?
- Your robot takes $\frac{1}{2}$ minute to complete each task in a robotics competition. When you know the total time, how can you find the number of tasks your robot completes?

Name ______________________________

10 Vocabulary

Review Words
addition
division
inverse operations
multiplication
proper fraction
subtraction

Organize It

Use the review words to complete the graphic organizer. Then evaluate the expression.

Operations that "undo" each other

$2 + 7 = 9 \longleftrightarrow 9 - 7 = 2$

$4 \times 3 = 12 \longleftrightarrow 12 \div 3 = 4$

Define It

What am I?

A fraction less than 1

$48 \div 8 = A$ $810 \div 9 = N$ $20 \div 4 = F$ $18 \div 9 = O$

$15 \div 5 = T$ $240 \div 3 = C$ $80 \div 4 = I$ $70 \div 7 = R$

$42 \div 6 = P$ $300 \div 5 = E$

7	10	2	7	60	10

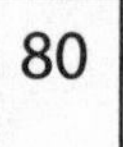
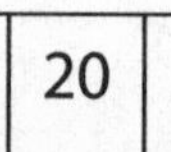

5	10	6	80	3	20	2	90

Name ___________________________________

Interpret Fractions as Division 10.1

Learning Target: Understand how fractions relate to division.

Success Criteria:

- I can use a model to divide two whole numbers that have a fraction as the quotient.
- I can use an equation to divide two whole numbers that have a fraction as the quotient.
- I can interpret a fraction as division.

Explore and Grow

You share 4 sheets of construction paper equally among 8 people. Write a division expression that represents the situation.

What fraction of a sheet of paper does each person get? Use a model to support your answer.

Structure How can you check your answer using multiplication?

Think and Grow: Divide Whole Numbers

You can use models to divide whole numbers that have a fraction as the quotient.

Example Find $2 \div 3$.

One Way: Use a tape diagram. Show 2 wholes. Divide each whole into 3 equal parts.

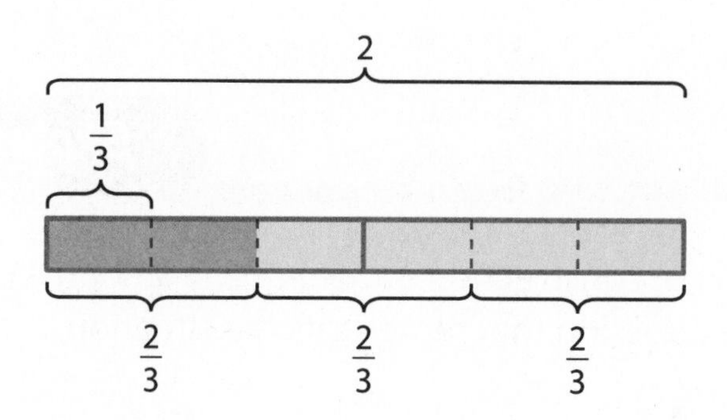

When you divide 2 wholes into 3 equal parts,

each part is $\frac{\square}{\square}$ of a whole. So, $2 \div 3 = \frac{\square}{\square}$.

Another Way: Use an area model. Show 2 wholes. Divide each whole into 3 equal parts. Then separate the parts into 3 equal groups.

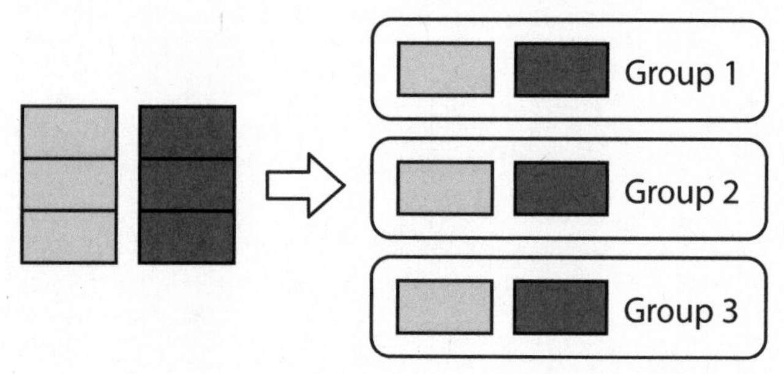

There are 2 wholes. Each group gets $\frac{\square}{\square}$ of each whole.

So, $2 \div 3 = 2 \times \frac{1}{3} = \frac{\square}{\square}$.

Show and Grow *I can do it!*

Divide. Use a model to help.

1. $2 \div 4 =$ ______

2. $1 \div 3 =$ ______

Name ______________________

Apply and Grow: Practice

Divide. Use a model to help.

3. $1 \div 8 =$ _____

4. $1 \div 4 =$ _____

5. $2 \div 6 =$ _____

6. $2 \div 5 =$ _____

7. $3 \div 7 =$ _____

8. $5 \div 6 =$ _____

9. How many 6s are in 1?

10. How many 10s are in 9?

11. MP **Number Sense** For which equations does $k = 8$?

$3 \div k = \frac{3}{8}$

$k \div 9 = \frac{8}{9}$

$2 \div 16 = k$

$8 \div k = \frac{8}{15}$

12. **Writing** Write and solve a real-life problem for $7 \div 12$.

Think and Grow: Modeling Real Life

Example Three fruit bars are shared equally among 4 friends. What fraction of a fruit bar does each friend get?

Divide 3 by 4 to find what fraction of a fruit bar each friend gets.

Use an area model to find $3 \div 4$. Show 3 whole fruit bars. Divide each fruit bar into 4 equal parts. Then separate the parts into 4 equal groups.

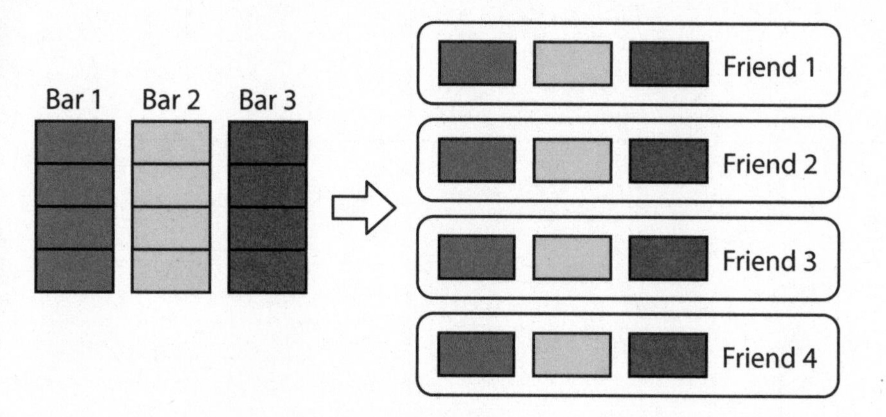

You can interpret a fraction as division of the numerator by the denominator.
$\frac{a}{b} = a \div b$

There are 3 whole fruit bars. Each friend gets $\frac{\square}{\square}$ of each fruit bar.

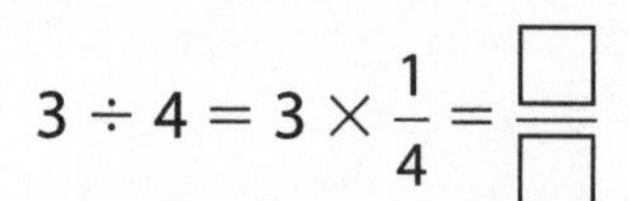

$3 \div 4 = 3 \times \frac{1}{4} = \frac{\square}{\square}$

So, each friend gets $\frac{\square}{\square}$ of a fruit bar.

Show and Grow *I can think deeper!*

13. Four circular lemon slices are shared equally among 8 glasses of water. What fraction of a lemon slice does each glass get?

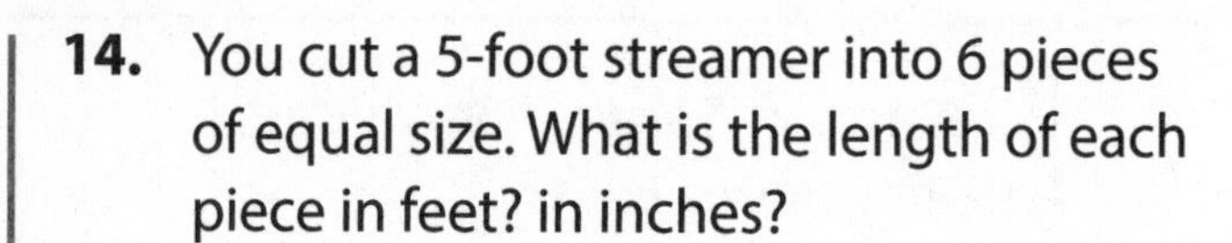

14. You cut a 5-foot streamer into 6 pieces of equal size. What is the length of each piece in feet? in inches?

15. **DIG DEEPER!** A fruit drink is made using $\frac{7}{4}$ quarts of orange juice and $\frac{5}{4}$ quarts of pineapple juice. The drink is shared equally among 12 guests. What fraction of a quart does each guest get?

Name ______________________

Homework & Practice 10.1

Learning Target: Understand how fractions relate to division.

Example Find $3 \div 5$.

Use an area model. Show 3 wholes. Divide each whole into 5 equal parts.

Then separate the parts into 5 equal groups.

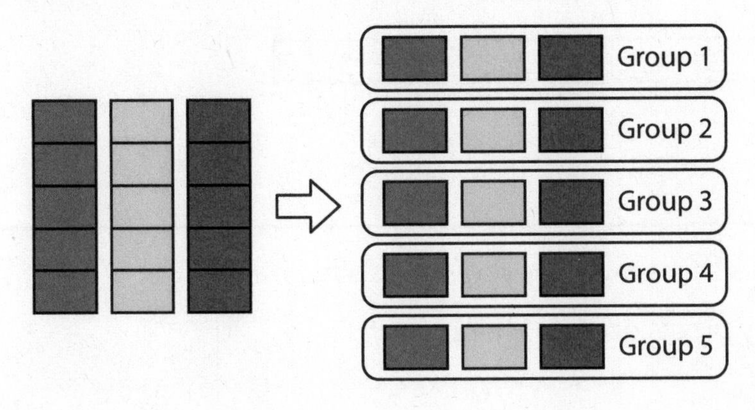

Group 1
Group 2
Group 3
Group 4
Group 5

You can interpret a fraction as division of the numerator by the denominator.
$\frac{a}{b} = a \div b$

There are 3 wholes. Each group gets $\frac{1}{5}$ of each whole.

So, $3 \div 5 = 3 \times \frac{1}{5} = \frac{3}{5}$.

Divide. Use a model to help.

1. $1 \div 6 =$ ______

2. $1 \div 7 =$ ______

3. $1 \div 5 =$ ______

4. $3 \div 4 =$ ______

5. $6 \div 7 =$ ______

6. $5 \div 9 =$ ______

7. **YOU BE THE TEACHER** Your friend says $\frac{5}{12}$ is equivalent to $12 \div 5$. Is your friend correct? Explain.

8. **Writing** Explain how fractions and division are related.

9. **MP Structure** Write a division equation represented by the model.

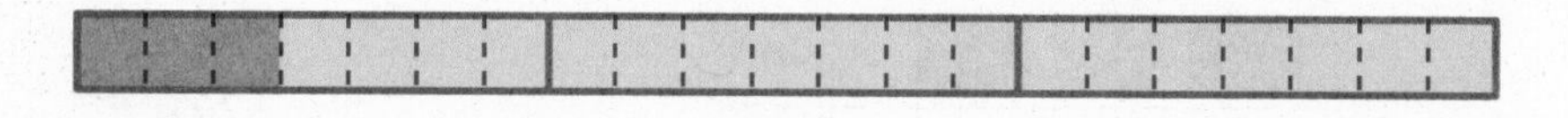

10. **MP Number Sense** Eight friends share multiple vegetable pizzas, and each gets $\frac{3}{8}$ of a pizza. How many pizzas do they share?

11. **Modeling Real Life** Seven friends each run an equal part of a 5-kilometer relay race. What fraction of a kilometer does each friend complete?

12. **Modeling Real Life** A group of friends equally share 3 bags of pretzels. Each friend gets $\frac{3}{5}$ of a bag of pretzels. How many friends are in the group?

Review & Refresh

Multiply.

13. $9 \times \frac{2}{3} =$ _____

14. $5 \times \frac{7}{10} =$ _____

15. $3 \times \frac{5}{12} =$ _____

Name ________________________________

Mixed Numbers as Quotients 10.2

Learning Target: Understand how mixed numbers relate to division.

Success Criteria:

- I can use a model to divide two whole numbers that have a mixed number as the quotient.
- I can use an equation to divide two whole numbers that have a mixed number as the quotient.
- I can write and solve a real-life problem involving division of whole numbers.

Explore and Grow

You share 6 sheets of construction paper equally among 4 people. Write a division expression that represents the situation.

How much paper does each person get? Use a model to support your answer.

Precision Does each person get less than or more than 1 sheet of paper? Use the dividend and divisor to explain why your answer makes sense.

Think and Grow: Divide Whole Numbers

You can use models to divide whole numbers that have a mixed number as the quotient.

Example Find $3 \div 2$.

One Way: Use a tape diagram. Show 3 wholes. Divide each whole into 2 equal parts.

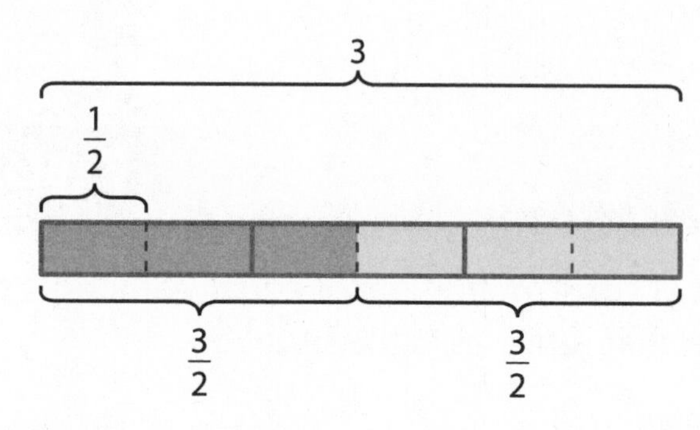

When you divide 3 wholes into 2 equal parts,

each part is $\frac{\square}{\square}$ of a whole. So, $3 \div 2 = \frac{\square}{\square}$ or $\square\frac{\square}{\square}$.

Another Way: Use an area model. Show 3 wholes. Divide each whole into 2 equal parts. Then separate the parts into 2 equal groups.

Group 1

Group 2

There are 3 wholes. Each group gets $\frac{\square}{\square}$ of each whole.

So, $3 \div 2 = 3 \times \frac{1}{2} = \frac{\square}{\square}$, or $\square\frac{\square}{\square}$.

Show and Grow *I can do it!*

Divide. Use a model to help.

1. $5 \div 3 =$ _____

2. $7 \div 2 =$ _____

Name ______________________________

Apply and Grow: Practice

Divide. Use a model to help.

3. $12 \div 7 =$ _____

4. $25 \div 20 =$ _____

5. $15 \div 4 =$ _____

6. $6 \div 13 =$ _____

7. $16 \div 8 =$ _____

8. $92 \div 50 =$ _____

9. How many 3s are in 7?

10. How many 6s are in 21?

11. **YOU BE THE TEACHER** Your friend says that $\frac{35}{6}$ is equivalent to $35 \div 6$. Is your friend correct? Explain.

12. **Writing** Write and solve a real-life problem for $24 \div 5$.

Think and Grow: Modeling Real Life

Example You share 7 bales of hay equally among 3 horse stalls. How many whole bales are in each stall? What fractional amount of a bale is in each stall?

Divide 7 by 3 to find how many bales of hay are in each stall. Use an area model to help.

$7 \div 3 = \frac{\square}{\square}$

$= \square\frac{\square}{\square}$

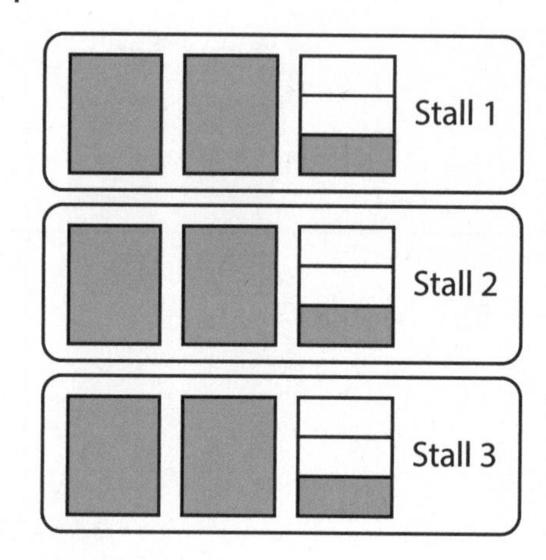

Three stalls each get 2 bales, which is 6 bales total. Each stall gets $\frac{1}{3}$ of the remaining bale.

Interpret each part of the quotient.

Whole number part: ______ There are ______ whole bales in each stall.

Fractional part: ______ There is ______ of a bale in each stall.

So, there are ______ whole bales and ______ of a bale in each stall.

Show and Grow *I can think deeper!*

13. Six muffins are shared equally among 4 friends. How many whole muffins does each friend get? What fractional amount of a muffin does each friend get?

14. A cyclist bikes 44 miles in 5 days. She bikes the same distance each day. Does she bike more than $8\frac{1}{2}$ miles each day? Explain.

15. **DIG DEEPER!** At Table A, 4 students share 7 packs of clay equally. At Table B, 5 students share 8 packs of clay equally. At which table does each student get a greater amount of clay? Explain.

Name ______________________________

Homework & Practice 10.2

Learning Target: Understand how mixed numbers relate to division.

Example Find 6 ÷ 4.

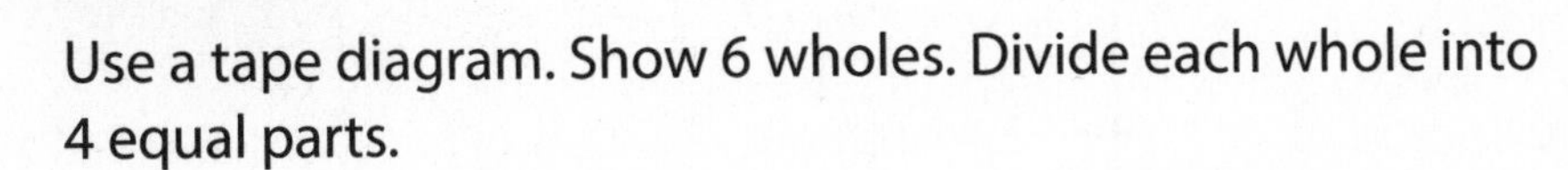

Use a tape diagram. Show 6 wholes. Divide each whole into 4 equal parts.

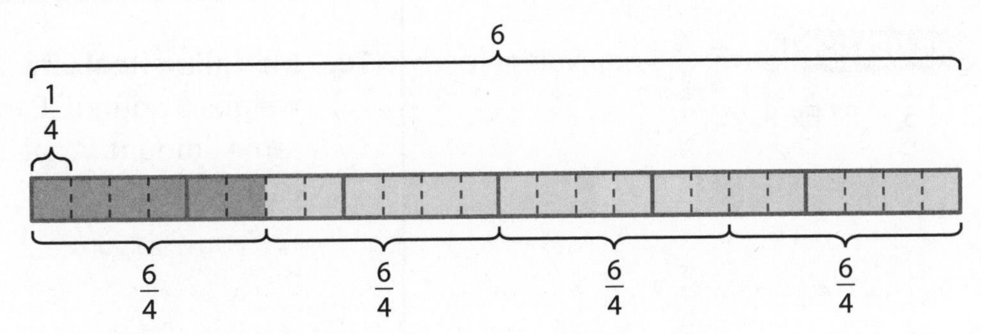

When you divide 6 wholes into 4 equal parts, each part is $\frac{6}{4}$ of a whole.

So, $6 \div 4 = \frac{6}{4}$, or $1\frac{1}{2}$.

Divide. Use a model to help.

1. 5 ÷ 2 = _____

2. 10 ÷ 7 = _____

3. 3 ÷ 9 = _____

4. 11 ÷ 4 = _____

5. 13 ÷ 6 = _____

6. 45 ÷ 8 = _____

7. MP **Number Sense** Between which two whole numbers is the quotient of 74 and 9?

8. MP **Reasoning** Three friends want to share 22 baseball cards. For this situation, why does the quotient 7 R1 make more sense than the quotient $7\frac{1}{3}$?

9. DIG DEEPER! Is $\frac{2}{5} \times 3$ equivalent to $2 \times 3 \div 5$? Explain.

10. **Modeling Real Life** A bag of 4 balls weighs 6 pounds. Each ball weighs the same amount. What is the weight of each ball?

11. **Modeling Real Life** Zookeepers order 600 pounds of bamboo for the pandas. The bamboo lasts 7 days. How many whole pounds of bamboo do the pandas eat each day? What fractional amount of a pound do the pandas eat each day?

12. **Modeling Real Life** A plumber has 20 feet of piping. He cuts the piping into 6 equal pieces. Is each piece greater than, less than, or equal to $3\frac{1}{2}$ feet?

Review & Refresh

Add.

13. $\frac{2}{9} + \frac{2}{3} =$ _____

14. $\frac{1}{10} + \frac{3}{4} =$ _____

15. $\frac{3}{5} + \frac{5}{6} + \frac{1}{5} =$ _____

Name ______________________________

Divide Whole Numbers by Unit Fractions

10.3

Learning Target: Divide whole numbers by unit fractions.

Success Criteria:

- I can use a model to divide a whole number by a unit fraction.
- I can use an equation to divide a whole number by a unit fraction.
- I can write and solve a real-life problem involving division of a whole number and a unit fraction.

Explore and Grow

Write a real-life problem that can be represented by $6 \div \frac{1}{2}$.

What is the solution to the problem? Use a model to support your answer.

Structure How can you check your answer using multiplication?

Think and Grow: Divide Whole Numbers by Unit Fractions

You can use models to divide whole numbers by unit fractions.

Example Find $4 \div \frac{1}{3}$.

One Way: Use a tape diagram to find how many $\frac{1}{3}$s are in 4. There are 4 wholes.

Divide each whole into 3 equal parts. Each part is $\frac{1}{3}$.

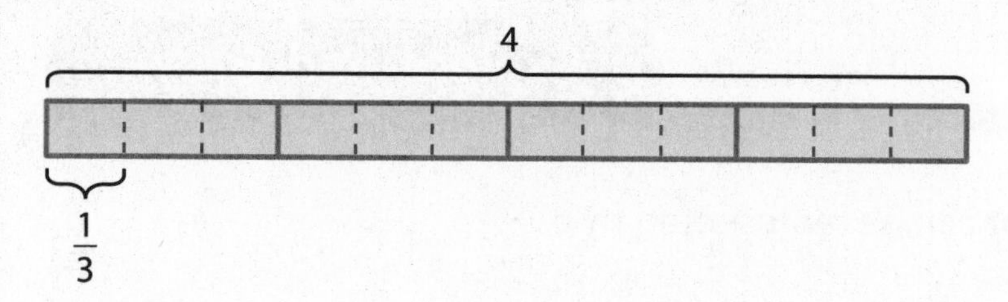
4
$\frac{1}{3}$

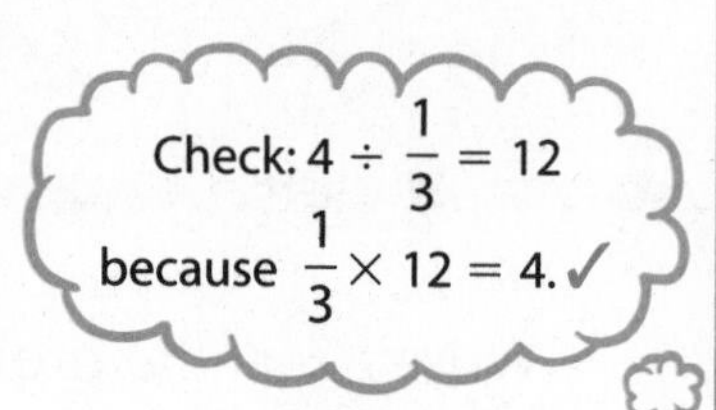
Check: $4 \div \frac{1}{3} = 12$
because $\frac{1}{3} \times 12 = 4$. ✓

Because there are 3 one-thirds in 1 whole, there are

$4 \times$ _____ $=$ _____ one-thirds in 4 wholes.

So, $4 \div \frac{1}{3} =$ _____.

Another Way: Use an area model to find how many $\frac{1}{3}$s are in 4.

There are 4 wholes. Divide each whole into 3 equal parts. Each part is $\frac{1}{3}$.

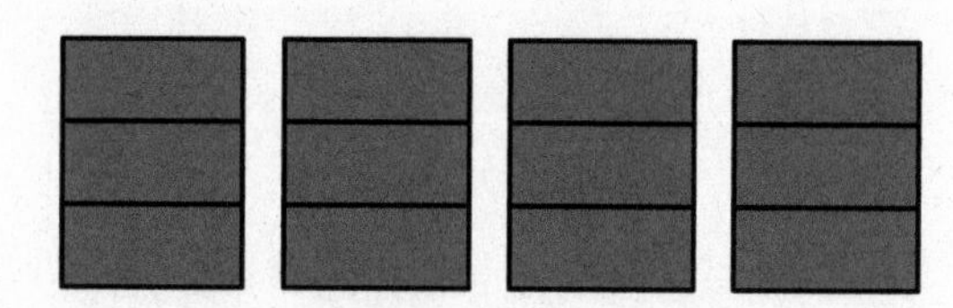

Because there are 3 one-thirds in 1 whole, there are

$4 \times$ _____ $=$ _____ one-thirds in 4 wholes.

So, $4 \div \frac{1}{3} =$ _____.

Show and Grow *I can do it!*

Divide. Use a model to help.

1. $3 \div \frac{1}{2} =$ _____

2. $2 \div \frac{1}{5} =$ _____

Name ______________________________

Apply and Grow: Practice

Divide. Use a model to help.

3. $1 \div \frac{1}{3} = $ _____

4. $3 \div \frac{1}{5} = $ _____

5. $5 \div \frac{1}{3} = $ _____

6. $4 \div \frac{1}{4} = $ _____

7. $7 \div \frac{1}{2} = $ _____

8. $2 \div \frac{1}{7} = $ _____

9. How many $\frac{1}{4}$s are in 5?

10. How many $\frac{1}{6}$s are in 2?

11. **YOU BE THE TEACHER** Newton finds $6 \div \frac{1}{3}$. Is he correct? Explain.

12. **Writing** Write and solve a real-life problem for $4 \div \frac{1}{2}$.

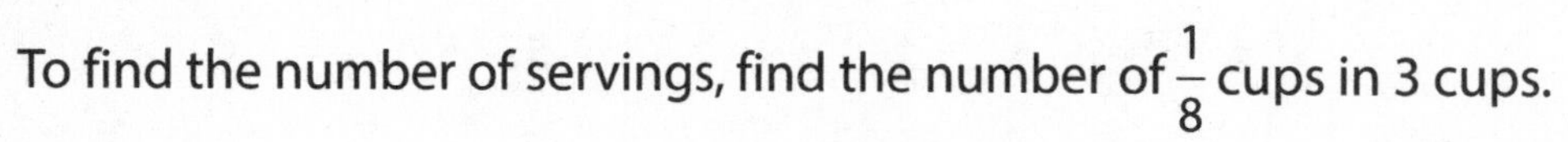

Think and Grow: Modeling Real Life

Example A chef makes 3 cups of salsa. A serving of salsa is $\frac{1}{8}$ cup. How many servings does the chef make?

To find the number of servings, find the number of $\frac{1}{8}$ cups in 3 cups.

Use an area model to find $3 \div \frac{1}{8}$. Divide each cup into 8 equal parts.

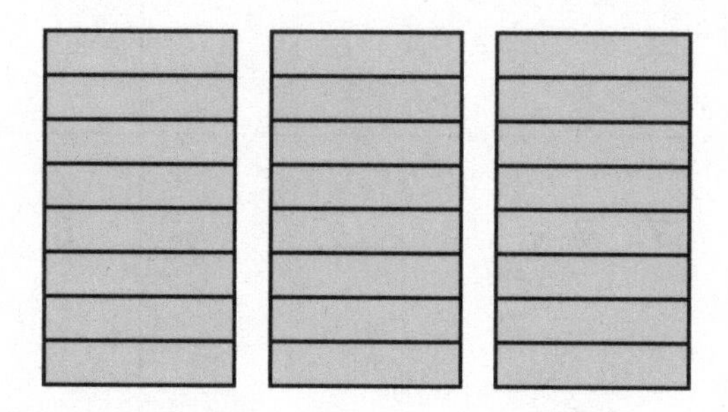

Notice that dividing a whole number by a unit fraction is the same as multiplying the whole number by the demoninator.
$a \div \frac{1}{b} = a \times b$

Because there are 8 one-eighth-cup servings in 1 cup,

there are $3 \times$ _____ $=$ _____ one-eighth-cup servings in 3 cups.

So, the chef makes _____ $\frac{1}{8}$-cup servings.

Show and Grow *I can think deeper!*

13. A litter of kittens weighs a total of 2 pounds. Each newborn kitten weighs $\frac{1}{4}$ pound. How many kittens are in the litter?

14. You put signs on a walking trail that is 7 miles long. You put a sign at the start and at the end of the trail. You also put a sign every $\frac{1}{10}$ mile. How many signs do you put on the trail?

15. **DIG DEEPER!** You have 2 boards that are each 8 feet long. You cut $\frac{1}{2}$-foot pieces to make square picture frames. How many picture frames can you make?

Name ______________________________

Homework & Practice 10.3

Learning Target: Divide whole numbers by unit fractions.

Example Find $6 \div \frac{1}{4}$.

One Way: Use a tape diagram to find how many $\frac{1}{4}$s are in 6. There are 6 wholes. Divide each whole into 4 equal parts. Each part is $\frac{1}{4}$.

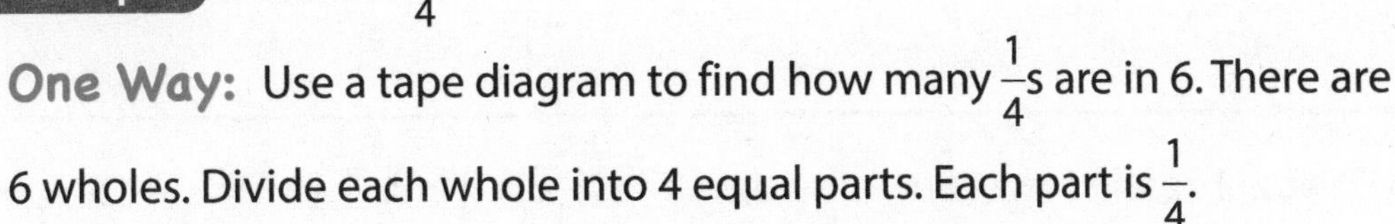

$\frac{1}{4}$

Because there are 4 one-fourths in 1 whole, there are

$6 \times$ __4__ $=$ __24__ one-fourths in 6 wholes.

So, $6 \div \frac{1}{4} =$ __24__.

Another Way: Use an area model to find how many $\frac{1}{4}$s are in 6.

There are 6 wholes. Divide each whole into 4 equal parts.

Each part is $\frac{1}{4}$.

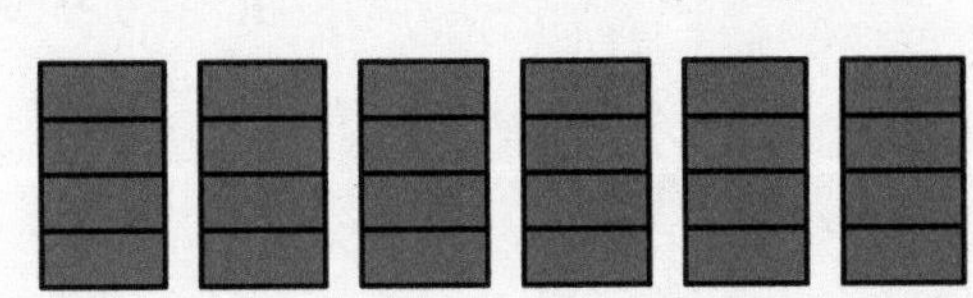

Because there are 4 one-fourths in 1 whole, there are

$6 \times$ __4__ $=$ __24__ one-fourths in 6 wholes.

So, $6 \div \frac{1}{4} =$ __24__.

Divide. Use a model to help.

1. $1 \div \frac{1}{9} =$ ______

2. $2 \div \frac{1}{3} =$ ______

3. $5 \div \frac{1}{2} =$ ______

Divide. Use a model to help.

4. $9 \div \frac{1}{4} =$ ______

5. $7 \div \frac{1}{3} =$ ______

6. $8 \div \frac{1}{5} =$ ______

7. MP **Number Sense** Explain how you can check your answer for Exercise 6.

8. **YOU BE THE TEACHER** Descartes finds $5 \div \frac{1}{4}$. Is he correct? Explain.

$$5 \div \frac{1}{4} = \frac{20}{4} \div \frac{1}{4} = 20$$

9. **Modeling Real Life** You need $\frac{1}{2}$ pound of clay to make a pinch pot. How many pinch pots can you make with 12 pounds of clay?

10. **Modeling Real Life** Your art teacher has 5 yards of yellow string and 4 yards of green string. She cuts both colors of string into $\frac{1}{3}$-yard pieces to hang student artwork. How many pieces of student artwork can she hang?

Review & Refresh

Multiply.

11. $\frac{2}{5} \times \frac{3}{4} =$ ______

12. $\frac{1}{8} \times \frac{5}{8} =$ ______

13. $\frac{4}{9} \times \frac{2}{7} =$ ______

Name ____________________

Divide Unit Fractions by Whole Numbers 10.4

Learning Target: Divide unit fractions by whole numbers.

Success Criteria:

- I can use a model to divide a unit fraction by a whole number.
- I can use an equation to divide a unit fraction by a whole number.
- I can write and solve a real-life problem involving division of a unit fraction and a whole number.

Explore and Grow

Write a real-life problem that can be represented by $\frac{1}{2} \div 3$.

What is the solution to the problem? Use a model to support your answer.

MP **Precision** Is the answer greater than or less than 1? Explain.

Think and Grow: Divide Unit Fractions by Whole Numbers

You can use models to divide unit fractions by whole numbers.

Example Find $\frac{1}{3} \div 4$.

One Way: Use a tape diagram. Divide $\frac{1}{3}$ into 4 equal parts.

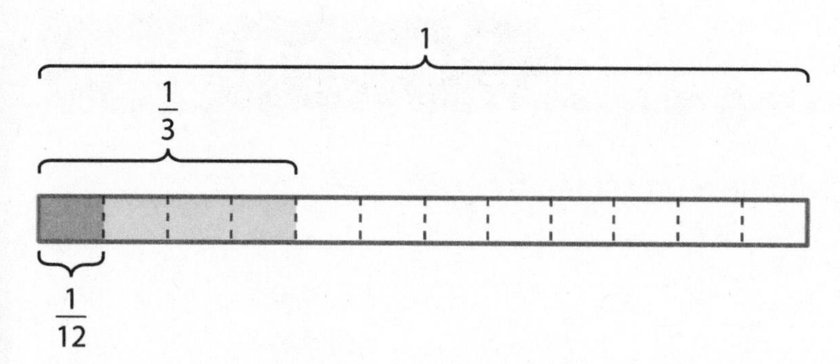

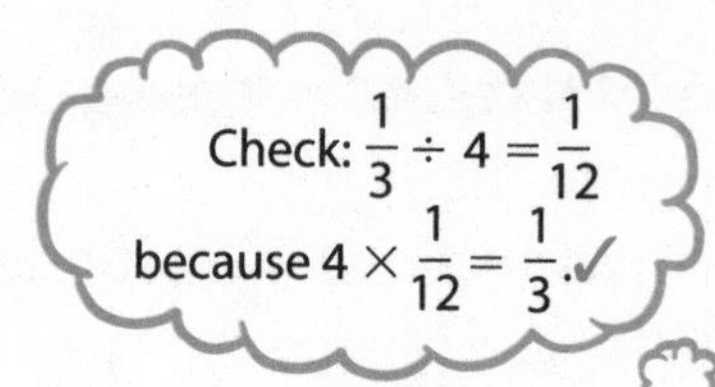

Check: $\frac{1}{3} \div 4 = \frac{1}{12}$
because $4 \times \frac{1}{12} = \frac{1}{3}$. ✓

Each of the 4 equal parts of $\frac{1}{3}$ represents _____ of the whole.

So, $\frac{1}{3} \div 4 =$ _____.

Another Way: Use an area model to represent 1 whole.
Divide $\frac{1}{3}$ into 4 equal parts.

$\frac{1}{3}$ {

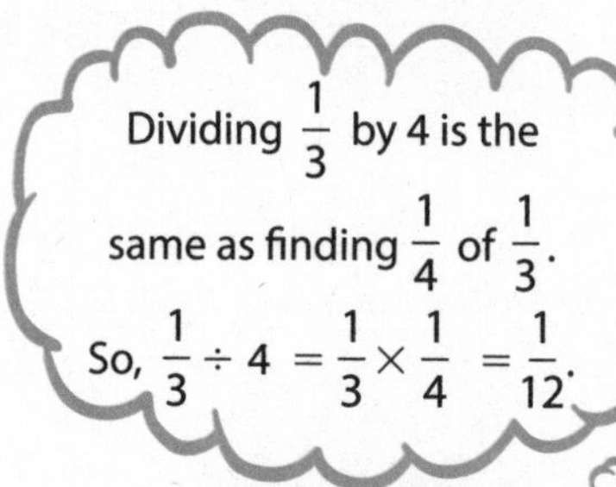

Dividing $\frac{1}{3}$ by 4 is the
same as finding $\frac{1}{4}$ of $\frac{1}{3}$.
So, $\frac{1}{3} \div 4 = \frac{1}{3} \times \frac{1}{4} = \frac{1}{12}$.

Each of the 4 equal parts of $\frac{1}{3}$ represents _____ of the whole.

So, $\frac{1}{3} \div 4 =$ _____.

Show and Grow *I can do it!*

Divide. Use a model to help.

1. $\frac{1}{4} \div 2 =$ _____

2. $\frac{1}{2} \div 5 =$ _____

Name ______

Apply and Grow: Practice

Divide. Use a model to help.

3. $\frac{1}{5} \div 3 =$ ______

4. $\frac{1}{6} \div 2 =$ ______

5. $\frac{1}{3} \div 5 =$ ______

6. $\frac{1}{5} \div 4 =$ ______

7. $\frac{1}{3} \div 3 =$ ______

8. $\frac{1}{8} \div 2 =$ ______

9. How many 6s are in $\frac{1}{2}$?

10. How many 2s are in $\frac{1}{3}$?

11. **Writing** Write and solve a real-life problem for $\frac{1}{2} \div 7$.

12. MP **Reasoning** Complete the statements.

Dividing $\frac{1}{4}$ by 7 is the same as finding ☐ of ☐.

So, $\frac{1}{4} \div 7 =$ ☐ $\times$ ☐ $=$ ☐.

Think and Grow: Modeling Real Life

Example You melt $\frac{1}{4}$ quart of soap. You pour the soap into 4 of the same-sized molds. What fraction of a quart of soap does each mold hold?

You are dividing $\frac{1}{4}$ quart into 4 equal parts, so you need to find $\frac{1}{4} \div 4$.

Use an area model to find $\frac{1}{4} \div 4$. Divide $\frac{1}{4}$ quart into 4 equal parts.

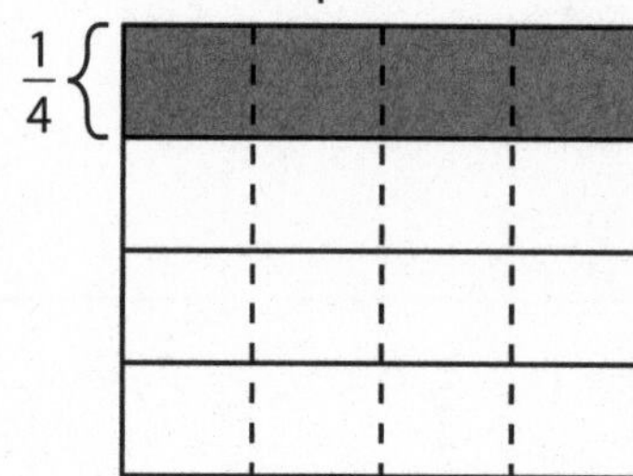

Dividing by a number b is the same as multiplying by $\frac{1}{b}$.

So, $\frac{1}{a} \div b = \frac{1}{a} \times \frac{1}{b} = \frac{1}{a \times b}$.

Each of the 4 equal parts of $\frac{1}{4}$ represents ______ of the whole.

So, each mold holds ______ quart of soap.

Show and Grow I can think deeper!

13. You buy $\frac{1}{2}$ pound of grapes. You equally divide the grapes into 2 bags. What fraction of a pound of grapes do you put into each bag?

14. You have $\frac{1}{8}$ cup of red sand, $\frac{1}{4}$ cup of blue sand, and $\frac{1}{2}$ cup of white sand. You equally divide the sand into 3 containers. What fraction of a cup of sand do you pour into each container?

15. **DIG DEEPER!** You, your friend, and your cousin share $\frac{1}{2}$ of a vegetable pizza and $\frac{1}{4}$ of a cheese pizza. The pizzas are the same size. What fraction of a pizza do you get in all?

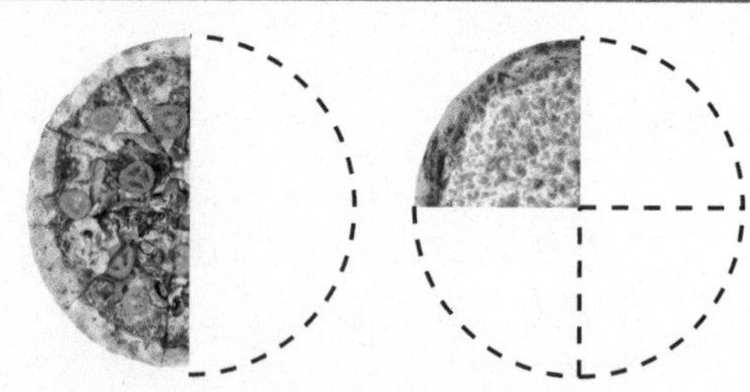

Name ___________________________

Learning Target: Divide unit fractions by whole numbers.

Example Find $\frac{1}{2} \div 5$.

One Way: Use a tape diagram. Divide $\frac{1}{2}$ into 5 equal parts.

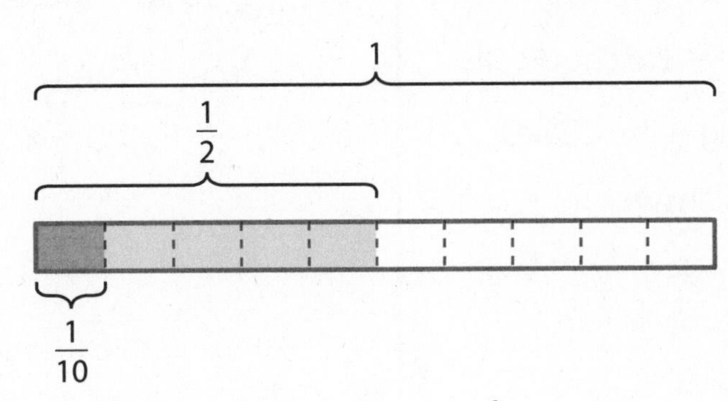

Each of the 5 equal parts of $\frac{1}{2}$ represents $\underline{\frac{1}{10}}$ of the whole.

So, $\frac{1}{2} \div 5 = \underline{\frac{1}{10}}$.

Dividing by a number b is the same as multiplying by $\frac{1}{b}$.

So, $\frac{1}{a} \div b = \frac{1}{a} \times \frac{1}{b} = \frac{1}{a \times b}$.

Another Way: Use an area model to represent 1 whole. Divide $\frac{1}{2}$ into 5 equal parts.

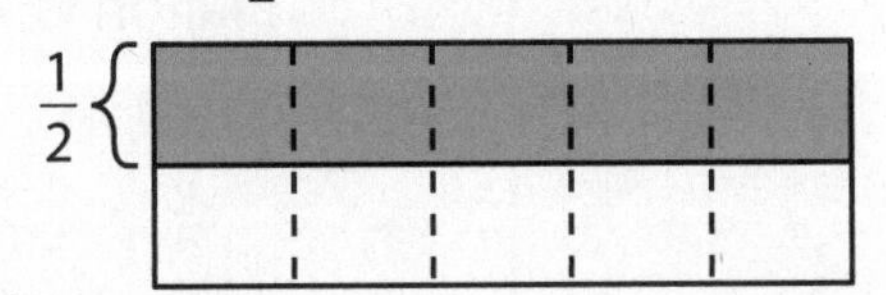

Each of the 5 equal parts of $\frac{1}{2}$ represents $\underline{\frac{1}{10}}$ of the whole.

So, $\frac{1}{2} \div 5 = \underline{\frac{1}{10}}$.

Divide. Use a model to help.

1. $\frac{1}{3} \div 4 =$ ______

2. $\frac{1}{6} \div 3 =$ ______

3. $\frac{1}{4} \div 5 =$ ______

Divide. Use a model to help.

4. $\frac{1}{5} \div 9 =$ _____

5. $\frac{1}{8} \div 6 =$ _____

6. $\frac{1}{7} \div 4 =$ _____

7. **YOU BE THE TEACHER** Your friend divides $\frac{1}{3}$ by 7 to get $\frac{1}{21}$. He checks his answer by multiplying $\frac{1}{21} \times \frac{1}{3}$. Does your friend check his answer correctly? Explain.

8. **MP Logic** Find the missing numbers.

$$\frac{1}{4} \div \square = \frac{1}{4} \times \frac{\square}{6} = \frac{1}{\square \times 6}$$

9. **Modeling Real Life** You win tickets that you can exchange for prizes. You exchange $\frac{1}{5}$ of your tickets and then divide them equally among 3 prizes. What fraction of your tickets do you spend on each prize?

10. **DIG DEEPER!** You have $\frac{1}{8}$ gallon of melted crayon wax. You pour the wax equally into 8 different molds to make new crayons. What fraction of a cup of melted wax is in each mold? Think: 1 gallon is 16 cups.

Review & Refresh

Find the quotient.

11. $0.9 \div 0.1 =$ _____

12. $38.6 \div 100 =$ _____

13. $2.57 \div 0.01 =$ _____

Name ______________________________

Problem Solving: Fraction Division 10.5

Learning Target: Solve multi-step word problems involving division with fractions.

Success Criteria:

- I can understand a problem.
- I can make a plan to solve.
- I can solve a problem using an equation.

Explore and Grow

You want to make a $\frac{1}{3}$ batch of the recipe. How you can use division to find the amount of each ingredient you need?

Zucchini Fries

- 4 zucchini
- $\frac{1}{3}$ cup cheese
- $\frac{1}{2}$ tablespoon spices
- 2 tablespoons olive oil

Reasoning Without calculating, explain how you can tell whether you need more than or less than 1 tablespoon of olive oil.

Think and Grow: Problem Solving: Fraction Division

Example You have 4 cups of yellow paint and 3 cups of blue paint. How many batches of green paint can you make?

Green Paint

$\frac{1}{2}$ cup yellow paint

$\frac{1}{3}$ cup blue paint

Understand the Problem

What do you know?

- You have 4 cups of yellow paint and 3 cups of blue paint.
- One batch of green paint is made of $\frac{1}{2}$ cup of yellow and $\frac{1}{3}$ cup of blue.

What do you need to find?

- You need to find how many batches of green paint you can make.

Make a Plan

How will you solve?

- Find how many batches are possible from yellow, and how many from blue.
- Choose the lesser number of batches.

Solve

Number of batches possible = Amount of yellow paint ÷ Amount needed for one batch

Let y represent the number of batches you can make from the yellow.

$y = 4 \div \frac{1}{2}$

$= 4 \times$ _____

$=$ _____

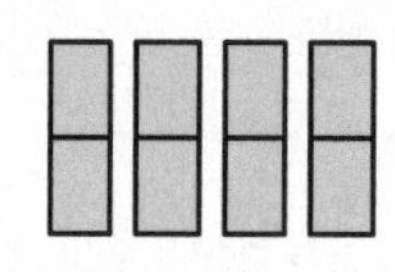

Number of batches possible = Amount of blue paint ÷ Amount needed for one batch

Let b represent the number of batches you can make from the blue.

$b = 3 \div \frac{1}{3}$

$= 3 \times$ _____

$=$ _____

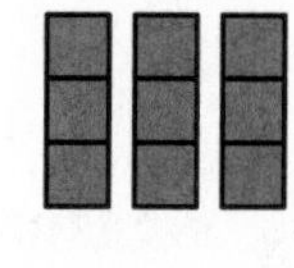

_____ $<$ _____

So, you can make _____ batches of green paint.

Show and Grow *I can do it!*

1. In the example, explain why you choose the fewer number of batches.

Name ____________________

Apply and Grow: Practice

Understand the problem. What do you know? What do you need to find? Explain.

2. A landowner donates 3 acres of land to a city. The mayor of the city uses 1 acre of the land for a playground and the rest of the land for community garden plots. Each garden plot is $\frac{1}{3}$ acre. How many plots are there?

3. Your friend uses $2\frac{1}{2}$ pounds of dried fruit and $\frac{1}{2}$ pound of raisins to make trail mix. How many $\frac{1}{2}$-pound packages can he make?

Understand the problem. Then make a plan. How will you solve? Explain.

4. A craftsman uses $\frac{3}{4}$ gallon of paint to paint 4 identical dressers. He uses the same amount of paint on each dresser. How much paint does he use to paint 7 of the same dressers?

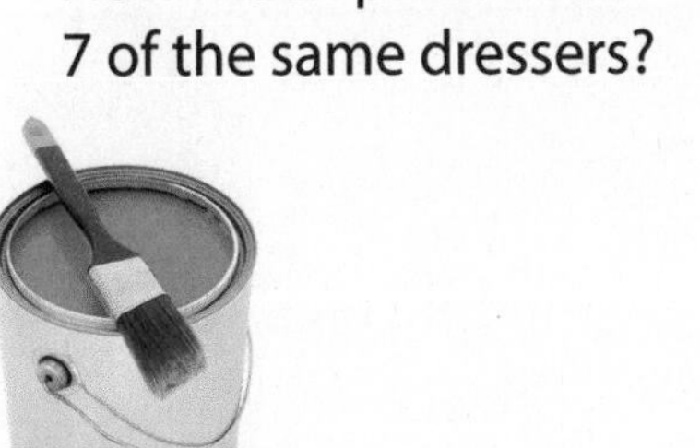

5. An airplane travels 125 miles in $\frac{1}{4}$ hour. It travels the same number of miles each hour. How many miles does the plane travel in 5 hours?

6. You make bows for gifts using $\frac{2}{3}$ yard of ribbon for each bow. You have 4 feet of red ribbon and 5 feet of green ribbon. How many bows can you make?

7. A landscaper buys 1 gallon of plant fertilizer. He uses $\frac{1}{5}$ of the fertilizer, and then divides the rest into 3 smaller bottles. How many gallons does he put into each bottle?

Think and Grow: Modeling Real Life

Example A sponsor donates \$0.10 to a charity for every $\frac{1}{4}$ kilometer of the triathlon an athlete completes. The athlete completes the entire triathlon. How much money does the sponsor donate?

Triathlon Components	
Sport	**Distance**
Swimming	1.9 km
Biking	90 km
Running	21.1 km

Think: What do you know? What do you need to find? How will you solve?

Write and solve an equation.

Add 1.9, 90, and 21.1 to find how many kilometers the athlete completes.

Divide the sum by $\frac{1}{4}$ to find how many $\frac{1}{4}$ kilometers the athlete completes.

Multiply the quotient by \$0.10 to find how much money the sponsor donates.

Total amount of money donated = (Swimming distance + Biking distance + Running distance) ÷ Distance counted for donation × Amount of money donated

Let m represent the total amount of money donated.

$$m = (1.9 + 90 + 21.1) \div \frac{1}{4} \times 0.10$$

$$= 113 \div \frac{1}{4} \times 0.10$$

$$= \_\_\_\_\_ \times 0.10$$

$$= \_\_\_\_\_$$

So, the sponsor donates \$_____ to the charity.

Show and Grow *I can think deeper!*

8. You earn \$5 for every $\frac{1}{2}$ hour you do yard work. How much money do you earn in 1 week?

Yard Work Completed in 1 Week	
Task	**Amount of Time**
Mowing lawns	$5\frac{1}{2}$ hours
Raking leaves	3 hours
Watering plants	$1\frac{1}{2}$ hours

Name ____________________

Homework & Practice 10.5

Learning Target: Solve multi-step word problems involving division with fractions.

Example You have 6 cups of pasta and 4 cups of pesto. You need $\frac{1}{2}$ cup of pasta and $\frac{1}{8}$ cup of pesto to make 1 serving of a recipe. How many servings of the recipe can you make?

Think: What do you know? What do you need to find? How will you solve?

Number of servings possible	=	Amount of pasta	÷	Amount needed for one serving

Let k represent the number of servings of the recipe you can make from the pasta.

$k = 6 \div \frac{1}{2}$

$= 6 \times \underline{\ 2\ }$

$= \underline{\ 12\ }$

Number of servings possible	=	Amount of pesto	÷	Amount needed for one serving

Let j represent the number of servings of the recipe you can make from the pesto.

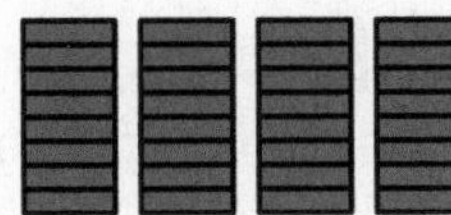

$j = 4 \div \frac{1}{8}$

$= 4 \times \underline{\ 8\ }$

$= \underline{\ 32\ }$

$\underline{\ 12\ } < \underline{\ 32\ }$

So, you can make $\underline{\ 12\ }$ servings of the recipe.

Understand the problem. Then make a plan. How will you solve? Explain.

1. A train travels 75 miles in $\frac{1}{2}$ hour. How many miles does the train travel in 8 hours?

2. You need $\frac{2}{3}$ yard of fabric to create a headband. You have 12 feet of blue fabric and 4 feet of yellow fabric. How many headbands can you make with all of the fabric?

3. An art teacher has 8 gallons of paint. Her class uses $\frac{3}{4}$ of the paint. The teacher divides the rest of the paint into 4 bottles. How much paint is in each bottle?

4. You mix $3\frac{1}{4}$ cups of frozen strawberries and $4\frac{1}{2}$ cups of frozen blueberries in a bowl. A smoothie requires $\frac{1}{2}$ cup of your berry mix. How many smoothies can you make?

5. **Modeling Real Life** A sponsor donates \$0.10 for every $\frac{1}{4}$ dollar donated at the locations shown. How much money does the sponsor donate?

Donation Collections

Location	Amount Collected
Pet store	\$25.25
Hardware store	\$12.50
Supermarket	\$63.25

6. **DIG DEEPER!** A nurse earns \$16 for every $\frac{1}{2}$ hour at work. How much money does she earn in 5 days?

Work Completed in 1 Day

Task	Amount of Time
Monitor patients	$6\frac{3}{4}$ hours
Do paperwork	1 hour
Research	$\frac{1}{4}$ hour

Review & Refresh

Find the quotient. Then check your answer.

7. $12\overline{)185.88}$

8. $24\overline{)74.4}$

9. $46\overline{)42.32}$

Name ______________________________

Performance Task 10

Your city has a robotics competition. Each team makes a robot that travels through a maze. The time each robot spends in the maze is used to find the team's score.

1. One-third of the students in your grade participate in the competition. The number of participating students is divided into 12 teams.

a. What fraction of the total number of students in your grade is on each team?

b. There are 3 students on each team. How many students are in your grade?

2. The maze for the competition is shown.

a. Write the length of the maze in feet.

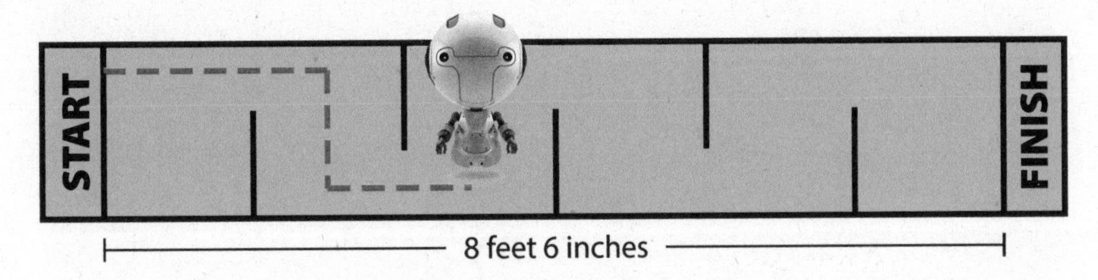

b. The length of the maze is divided into 6 equal sections. What is the length of each section of the maze?

3. Each team has 200 seconds to complete the maze. The rules require judges to use the expression $(200 - x) \div \frac{1}{5}$, where x is the total number of seconds, to find a team's total score.

a. Your robot completes the maze in 3 minutes 5 seconds. How many points does your team earn?

b. Do you think the team with the most points or the fewest points wins? Use an example to justify your answer.

Fraction Connection: Division

Directions:

1. Players take turns rolling three dice.
2. On your turn, evaluate the expression indicated by your roll and cover the answer.
3. The first player to get four in a row, horizontally, vertically, or diagonally, wins!

Roll	Evaluate
3	$3 \div 6$
4	$5 \div 2$
5	$4 \div \frac{1}{5}$
6	$\frac{1}{3} \div 2$
7	$2 \div 5$
8	$7 \div 3$
9	$6 \div \frac{1}{2}$
10	$\frac{1}{4} \div 3$

Roll	Evaluate
11	$4 \div 7$
12	$9 \div 4$
13	$2 \div \frac{1}{4}$
14	$\frac{1}{5} \div 4$
15	$5 \div 8$
16	$4 \div 2$
17	$8 \div \frac{1}{3}$
18	$\frac{1}{2} \div 5$

$2\frac{1}{2}$	$\frac{5}{8}$	$\frac{1}{12}$	2
$\frac{4}{7}$	$\frac{1}{2}$	24	$\frac{2}{5}$
8	12	$2\frac{1}{4}$	$\frac{1}{6}$
$\frac{1}{10}$	$2\frac{1}{3}$	20	$\frac{1}{20}$

Name ________________

Chapter Practice 10

10.1 Interpret Fractions as Division

Divide. Use a model to help.

1. $1 \div 2 =$ _____

2. $3 \div 10 =$ _____

3. $4 \div 7 =$ _____

4. $11 \div 15 =$ _____

5. $8 \div 9 =$ _____

6. $13 \div 20 =$ _____

7. **Modeling Real Life** Nine friends equally share 12 apples. What fraction of an apple does each friend get?

10.2 Mixed Numbers as Quotients

Divide. Use a model to help.

8. $8 \div 3 =$ _____

9. $6 \div 5 =$ _____

10. $10 \div 4 =$ _____

11. $20 \div 11 =$ _____

12. $25 \div 2 =$ _____

13. $64 \div 9 =$ _____

10.3 Divide Whole Numbers by Unit Fractions

Divide. Use a model to help.

14. $4 \div \frac{1}{2} =$ _____

15. $6 \div \frac{1}{5} =$ _____

16. $7 \div \frac{1}{4} =$ _____

17. $8 \div \frac{1}{3} =$ _____

18. $9 \div \frac{1}{2} =$ _____

19. $2 \div \frac{1}{10} =$ _____

10.4 Divide Unit Fractions by Whole Numbers

Divide. Use a model to help.

20. $\frac{1}{7} \div 2 =$ _____

21. $\frac{1}{2} \div 9 =$ _____

22. $\frac{1}{3} \div 7 =$ _____

23. $\frac{1}{6} \div 5 =$ _____

24. $\frac{1}{7} \div 3 =$ _____

25. $\frac{1}{8} \div 4 =$ _____

10.5 Problem Solving: Fraction Division

26. A mechanic buys 1 gallon of oil. She uses $\frac{1}{6}$ of the oil, and then divides the rest into 4 smaller bottles. How much does she put into each bottle?

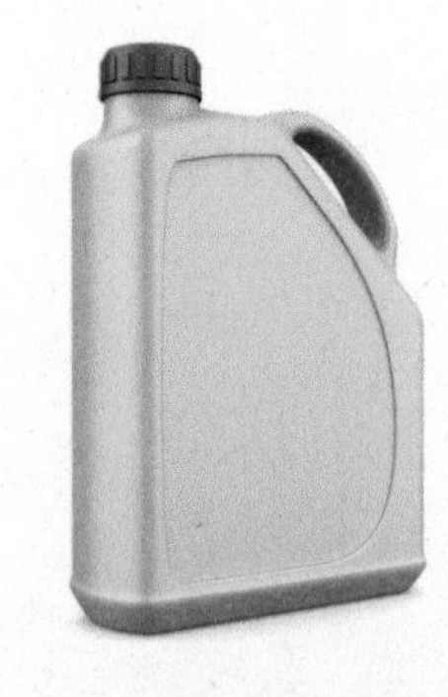

11 Convert and Display Units of Measure

Chapter Learning Target:
Understand measurement.

Chapter Success Criteria:
- I can identify length in metric units.
- I can describe mass and capacity in metric units.
- I can solve a problem using different ways to measure items.
- I can compare the values of two different forms of measurement.

- Passenger airliners fly people and cargo all over the world. What types of cargo do you think are flown on an airplane?
- An airliner can hold at most 30 tons of weight from passengers and cargo. Why might it be important for the pilot of an airliner to be able to convert between units of measure?

Name ___________________________

11 Vocabulary

Review Words
length
capacity
mass

Organize It

Use the review words to complete the graphic organizer.

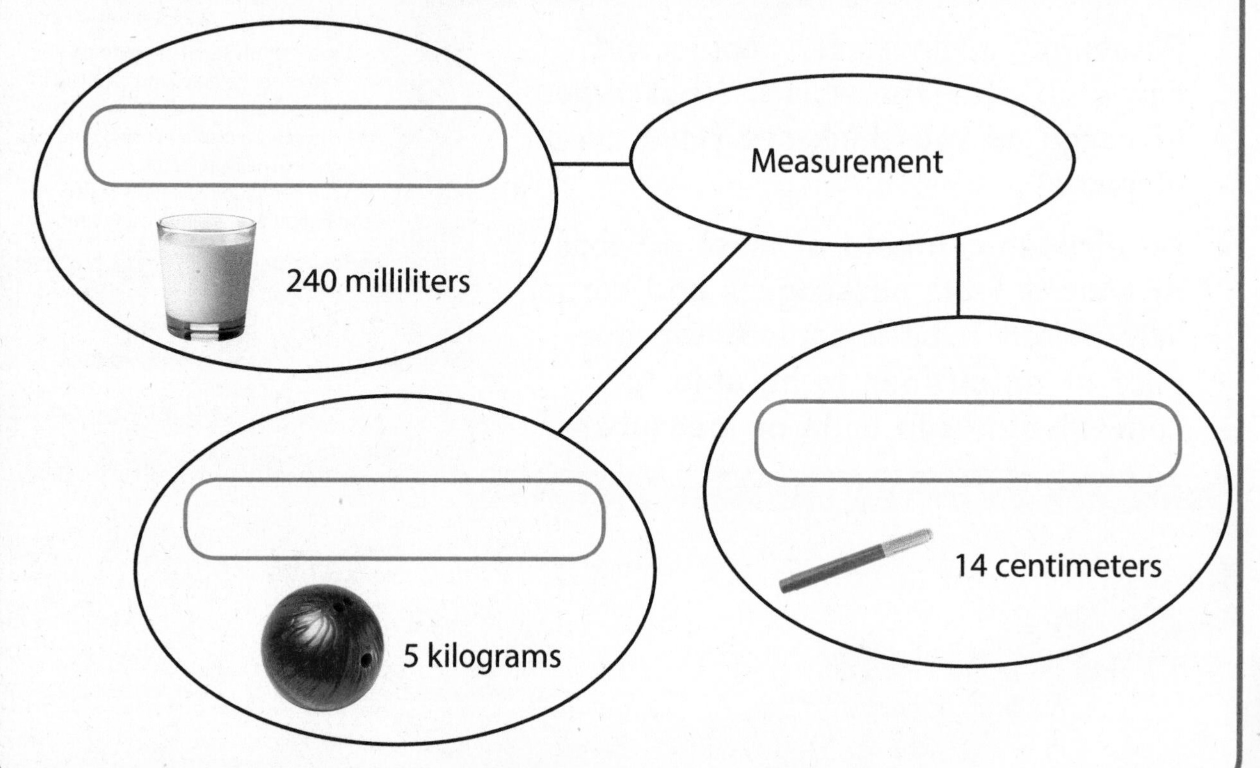

Define It

Use your vocabulary cards to complete each definition.

1. milligrams: A __________ unit used to measure __________

2. fluid ounces: A __________ unit used to measure __________

Chapter 11 Vocabulary Cards

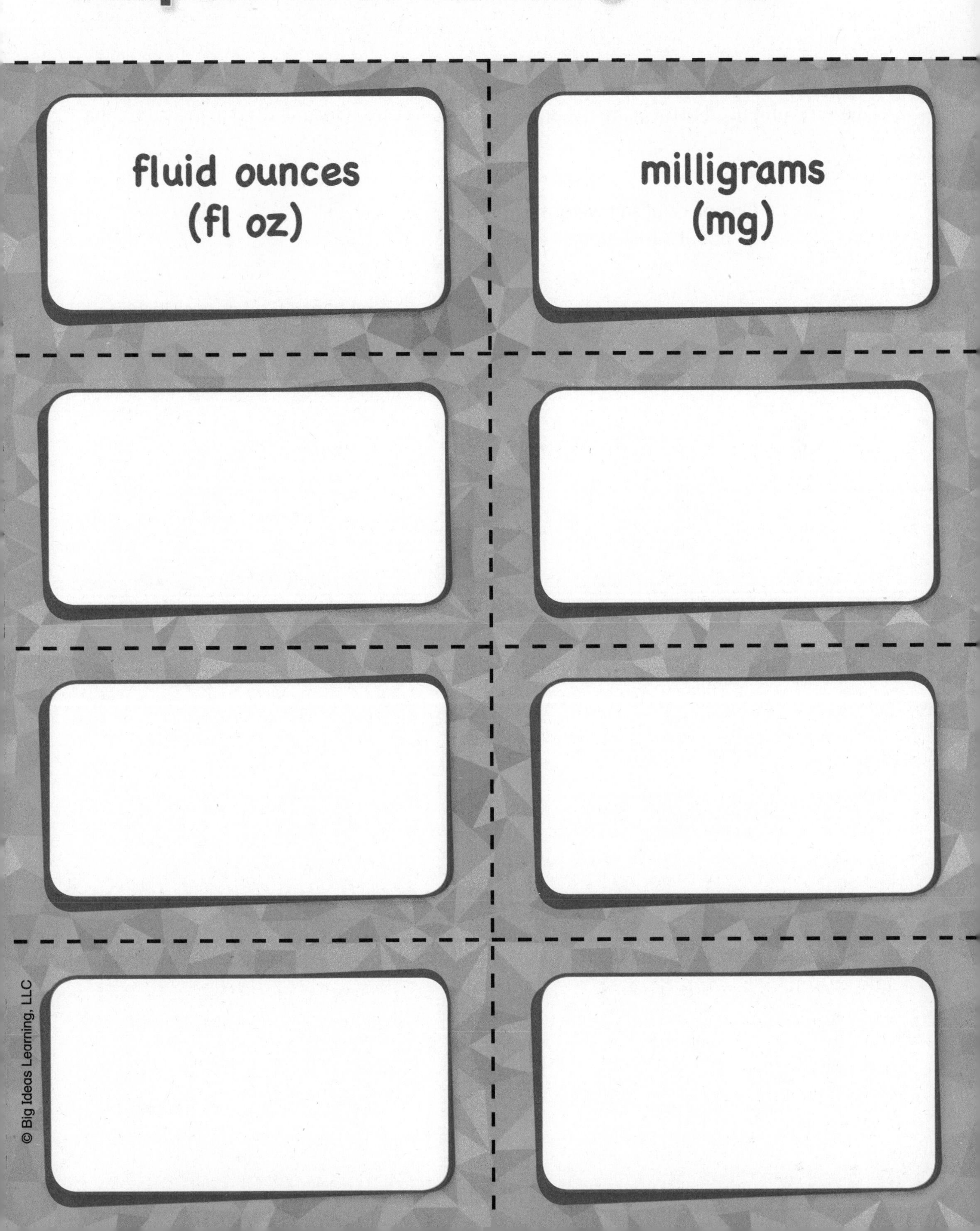

A metric unit used to measure mass

One piece of salt weighs about 1 milligram.

A customary unit used to measure capacity

There are 8 fluid ounces in 1 cup.

Name ____________________

Length in Metric Units 11.1

Learning Target: Write lengths using equivalent metric measures.

Success Criteria:

- I can compare the sizes of two metric units of length.
- I can write a metric length using a smaller metric unit.
- I can write a metric length using a larger metric unit.

Explore and Grow

Work with a partner. Find 3 objects in your classroom and use a meter stick to measure them. Record your measurements in the table.

Object	Length (meters)	Length (centimeters)	Length (millimeters)

1 centimeter is ________ times as long as 1 millimeter.

1 meter is ________ times as long as 1 centimeter.

1 meter is ________ times as long as 1 millimeter.

Structure How can you convert a metric length from a larger unit to a smaller unit? How can you convert a metric length from a smaller unit to a larger unit?

Think and Grow: Convert Metric Lengths

You can use powers of 10 to find equivalent measures in the metric system.

kilometer	hectometer	dekameter	meter	decimeter	centimeter	millimeter
10^3 m	10^2 m	10 m	1 m	0.1 m	0.01 m	0.001 m

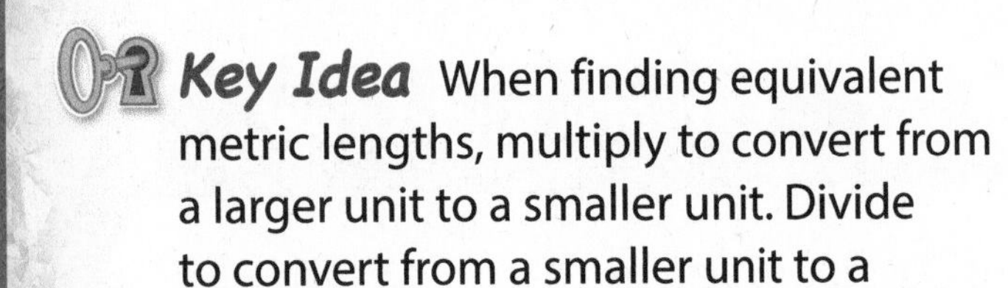

Key Idea When finding equivalent metric lengths, multiply to convert from a larger unit to a smaller unit. Divide to convert from a smaller unit to a larger unit.

Metric Units of Length
1 centimeter (cm) = 10 millimeters (mm)
1 meter (m) = 100 centimeters (cm)
1 kilometer (km) = 1,000 meters (m)

Example Convert 6 centimeters to millimeters.

There are ________ millimeters in 1 centimeter.

Because you are converting from a larger unit to a smaller unit, multiply.

6 × ________ = ________

So, 6 centimeters is ________ millimeters.

Example Convert 14,000 meters to kilometers.

There are ________ meters in 1 kilometer.

Because you are converting from a smaller unit to a larger unit, divide.

14,000 ÷ ________ = ________

So, 14,000 meters is ________ kilometers.

Show and Grow *I can do it!*

Convert the length.

1. 8.5 km = ________ m

2. 180 cm = ________ m

Name ___________________________________

Apply and Grow: Practice

Convert the length.

3. 150 m = _______ km

4. 90 cm = _______ mm

5. 0.03 m = _______ cm

6. 0.6 km = _______ cm

7. 800 mm = _______ m

8. 700 cm = _______ km

Compare.

9. 0.02 m ◯ 3 mm

10. 48,000 cm ◯ 0.48 km

11. 0.025 km ◯ 3,500 mm

12. The giant anteater has the longest tongue in relation to its body size of any mammal. Its tongue is about 0.6 meter long. How many centimeters long is its tongue?

13. MP **Number Sense** The length of an object can be written as *b* millimeters or *c* kilometers. Compare the values of *b* and *c*. Explain your reasoning.

14. **Writing** Why does the decimal point move to the left when converting from a smaller measure to a larger measure?

Think and Grow: Modeling Real Life

Mauna Kea, a volcano in Hawaii, rises about 4,200 meters above sea level.

Example The base of Mauna Kea extends about 5.76 kilometers below sea level. What is the total height of the volcano in meters?

Convert the distance below sea level to meters.

There are _______ meters in 1 kilometer.

5.76 × _______ = _______

So, the volcano extends _______ meters below sea level.

Add the distance below sea level and the distance above sea level.

$$\begin{array}{r} 4{,}200 \\ +\ \square \\ \hline \end{array}$$

So, the total height of the volcano is about _______ meters.

Show and Grow *I can think deeper!*

15. A pool is 3.65 meters deep. A diving board is 100 centimeters above the surface of the water. What is the distance from the diving board to the bottom of the pool in centimeters?

16. You hike 1.6 kilometers from a cabin to a lookout. You plan to hike the same way back. On your way back, you stop after 1,050 meters to look at a map. How many meters have you hiked so far? How many kilometers are you from the cabin?

17. **DIG DEEPER!** You can take one of two routes to school. Which route is longer? How much longer? Write your answer two different ways.

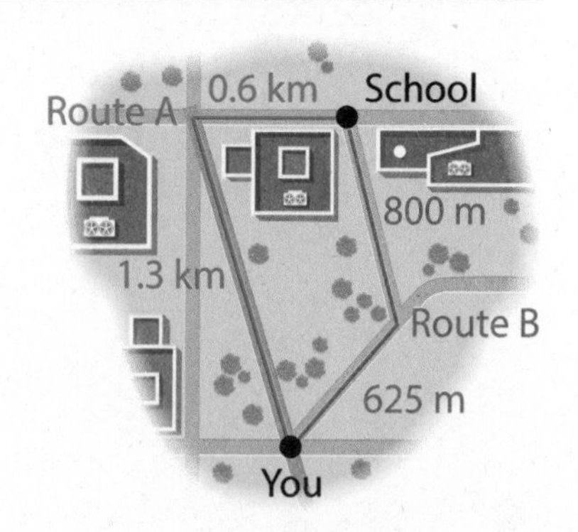

Name ______________________

Homework & Practice 11.1

Learning Target: Write lengths using equivalent metric measures.

Example Convert 0.02 kilometer to meters.

There are 1,000 meters in 1 kilometer.

Because you are converting from a larger unit to a smaller unit, multiply.

$0.02 \times$ 1,000 = 20

So, 0.02 kilometer is 20 meters.

Convert the length.

1. 0.8 cm = ________ mm

2. 7 m = ________ km

3. 6.4 km = ________ m

4. 1,300 mm = ________ cm

5. 91,000 cm = ________ km

6. 20,000 mm = ________ km

Compare.

7. 16 cm ◯ 1.6 m

8. 300 mm ◯ 0.3 m

9. 0.045 km ◯ 45 mm

10. Dolphins can hear sounds underwater that are 24 kilometers away. How many meters away can dolphins hear sounds underwater?

11. MP **Reasoning** How can you convert 7.8 meters to kilometers by moving the decimal point? Explain your reasoning.

12. **YOU BE THE TEACHER** Your friend divides by 100 to convert a length from meters to centimeters. Is your friend correct? Explain.

13. **Which One Doesn't Belong?** Which measurement does *not* belong with the other three?

0.5 km	500,000 mm
500 m	5,000 cm

14. **Modeling Real Life** A small chunk of ice called a *growler* breaks away from an iceburg. The growler sticks out of the water 840 millimeters and is 3.5 meters deep in the water. What is the total height of the growler in meters?

15. **DIG DEEPER!** A spaceship's route from Earth to the moon is 384,400 kilometers long. The spaceship travels 500,000 meters. How many kilometers does it have left to travel?

Review & Refresh

Write the fraction in simplest form.

16. $\frac{2}{8}$

17. $\frac{10}{100}$

18. $\frac{24}{16}$

Name ______________________________

Mass and Capacity in Metric Units 11.2

Learning Target: Write masses and capacities using equivalent metric measures.

Success Criteria:

- I can compare the sizes of two metric units of mass and capacity.
- I can write metric masses and capacities using smaller metric units.
- I can write metric masses and capacities using larger metric units.

Explore and Grow

Use a balance and weights to help you complete the statement.

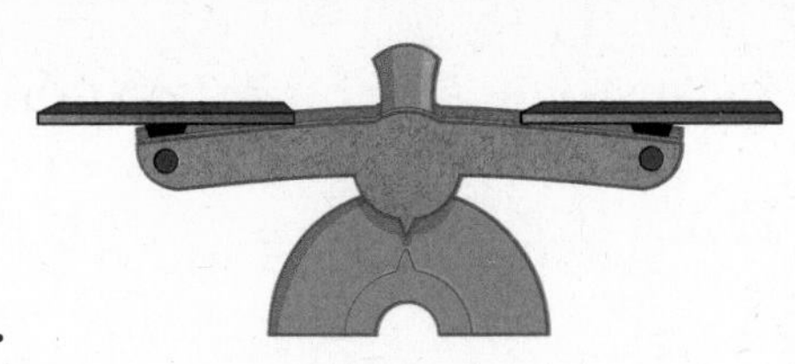

1 kilogram is __________ times as much as 1 gram.

Structure How can you convert kilograms to grams? How can you convert grams to kilograms?

Use a 1-liter beaker to help you complete the statement.

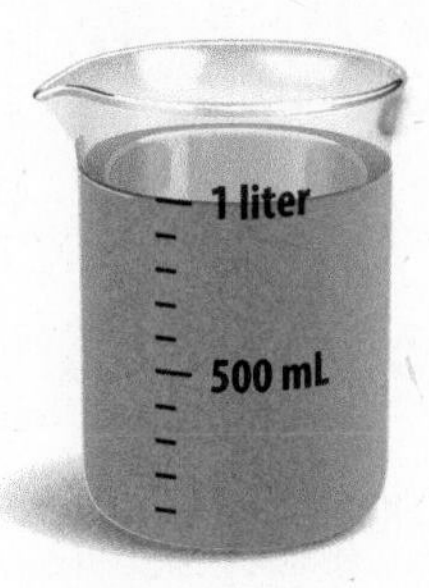

1 liter is __________ times as much as 1 milliliter.

Structure How can you convert liters to milliliters? How can you convert milliliters to liters?

Think and Grow: Convert Metric Measures

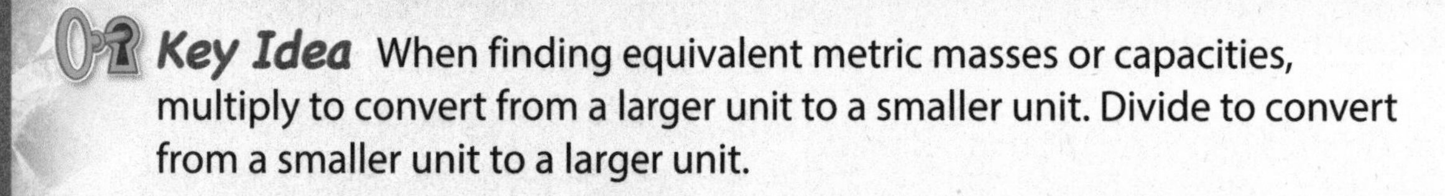

Key Idea When finding equivalent metric masses or capacities, multiply to convert from a larger unit to a smaller unit. Divide to convert from a smaller unit to a larger unit.

Metric Units of Mass
1 gram (g) = 1,000 **milligrams** (mg)
1 kilogram (kg) = 1,000 grams (g)

Metric Units of Capacity
1 liter (L) = 1,000 milliliters (mL)

Think, there will be more milligrams than grams. So, it makes sense to multiply.

Example Convert 12.4 grams to milligrams.

There are _____ milligrams in 1 gram.

Because you are converting from a larger unit to a smaller unit, multiply.

$12.4 \times$ _____ $=$ _____

So, 12.4 grams is _____ milligrams.

Example Convert 18,000 milliliters to liters.

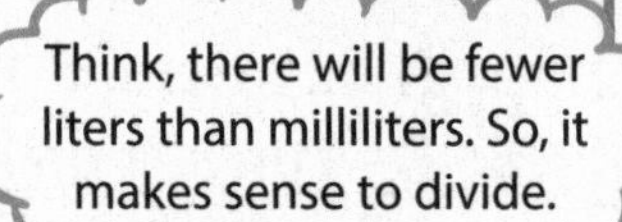

Think, there will be fewer liters than milliliters. So, it makes sense to divide.

There are _____ milliliters in 1 liter.

Because you are converting from a smaller unit to a larger unit, divide.

$18{,}000 \div$ _____ $=$ _____

So, 18,000 milliliters is _____ liters.

Show and Grow *I can do it!*

Convert the mass.

1. 8 kg = _____ g

2. 3,800 mg = _____ g

Convert the capacity.

3. 22,000 mL = _____ L

4. 4.6 L = _____ mL

Name ______

Apply and Grow: Practice

Convert the mass.

5. 5,000 g = ______ kg

6. 67 g = ______ mg

7. 0.2 kg = ______ mg

8. 30,000 mg = ______ kg

Convert the capacity.

9. 8 L = ______ mL

10. 70 mL = ______ L

11. 200 mL = ______ L

12. 0.4 L = ______ mL

13. What is the mass of the pumpkin in kilograms?

6,000 grams

14. **Which One Doesn't Belong?** Which one does *not* have the same capacity as the other three?

2,000 mL   2 L

2 mL 2×10^3 mL

15. DIG DEEPER! Order the masses from least to greatest. Explain how you converted the masses.

0.039 kg   14,000 mg   56 g   0.14 kg

Think and Grow: Modeling Real Life

Example You have a 5-kilogram bag of dog food. You give your dog 50 grams of food each day. How many days does the bag of food last?

Convert the mass of the bag to grams.

There are _____ grams in 1 kilogram.

$5 \times$ _____ $=$ _____

So, the bag contains _____ grams of dog food.

Divide the amount of dog food in the bag by the amount you give your dog each day.

_____ $\div 50 =$ _____

So, the bag lasts for _____ days.

Show and Grow *I can think deeper!*

16. You have 6 liters of juice to make frozen treats. You pour 30 milliliters of juice into each treat mold. How many treats can you make?

17. Your goal is to eat no more than 2.3 grams of sodium each day. You record the amounts of sodium you eat. How many more milligrams of sodium can you eat and not exceed your limit?

Meal	Sodium You Eat (milligrams)
Breakfast	210
Snack	250
Lunch	690

2 L

200 mL

18. **DIG DEEPER!** Which contains more juice, 3 of the bottles, or 32 of the juice boxes? How much more? Write your answer in milliliters.

Name ______________________________

Homework & Practice 11.2

Learning Target: Write masses and capacities using equivalent metric measures.

Example Convert 0.5 liter to milliliters.

There are 1,000 milliliters in 1 liter.

Because you are converting from a larger unit to a smaller unit, multiply.

$0.5 \times \underline{1,000} = \underline{500}$

So, 0.5 liter is 500 milliliters.

Convert the mass.

1. 9 g = _____ mg

2. 78 g = _____ kg

3. 260,000 mg = _____ kg

4. 0.148 kg = _____ mg

Convert the capacity.

5. 600 mL = _____ L

6. 3 L = _____ mL

7. 0.21 L = _____ mL

8. 35 mL = _____ L

9. There are 3.2 liters of iced tea in a pitcher. How many milliliters of iced tea are in the pitcher?

10. **YOU BE THE TEACHER** Your friend says that 0.04 kilogram is less than 4×10^5 milligrams. Is your friend correct? Explain.

11. **MP Number Sense** How does the meaning of each prefix relate to the metric units of mass and capacity?

Prefix	Definition
kilo-	one thousand
milli-	one thousandth

12. **Modeling Real Life** You have 9 kilograms of corn kernels. You put 450 grams of corn kernels in each bag. How many bags can you make?

13. **DIG DEEPER!** Your teacher has one of each of the beakers shown. You need to measure exactly 2 liters of liquid for an experiment. What are three different ways you can do this?

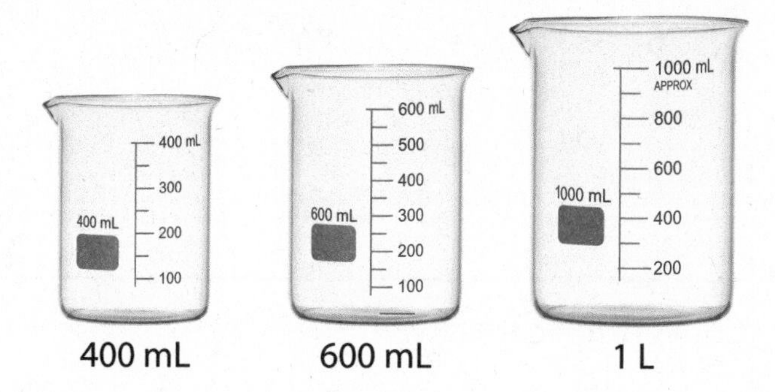

Review & Refresh

14. Newton rides to the store in a taxi. He owes the driver $12. He calculates the driver's tip by multiplying $12 by 0.15. How much money does he pay the driver, including the tip?

Name ______________________________

Length in Customary Units 11.3

Learning Target: Write lengths using equivalent customary measures.

Success Criteria:

- I can compare the sizes of two customary units of length.
- I can write a customary length using a smaller customary unit.
- I can write a customary length using a larger customary unit.

Explore and Grow

Work with a partner. Use a yard stick to draw 3 lines on a whiteboard that are 1 yard, 2 yards, and 3 yards in length. Then measure the lengths of the lines in feet and in inches. Record your measurements in the table.

Length (yards)	Length (feet)	Length (inches)
1		
2		
3		

1 foot is ______ times as long as 1 inch.

1 yard is ______ times as long as 1 foot.

1 yard is ______ times as long as 1 inch.

Structure How can you convert a customary length from a larger unit to a smaller unit? How can you convert a customary length from a smaller unit to a larger unit?

Think and Grow: Convert Customary Lengths

Key Idea When finding equivalent customary lengths, multiply to convert from a larger unit to a smaller unit. Divide to convert from a smaller unit to a larger unit.

Customary Units of Length
1 foot (ft) = 12 inches (in.)
1 yard (yd) = 3 feet (ft)
1 mile (mi) = 1,760 yards (yd)

Example Convert 3 miles to yards.

There are _____ yards in 1 mile.

Because you are converting from a larger unit to a smaller unit, multiply.

3 × _____ = _____

So, 3 miles is ________ yards.

Example Convert 42 inches to feet and inches.

The remainder shows the number of inches left over.

There are _____ inches in 1 foot.

Because you are converting from a smaller unit to a larger unit, divide.

42 ÷ _____ = _____ R _____

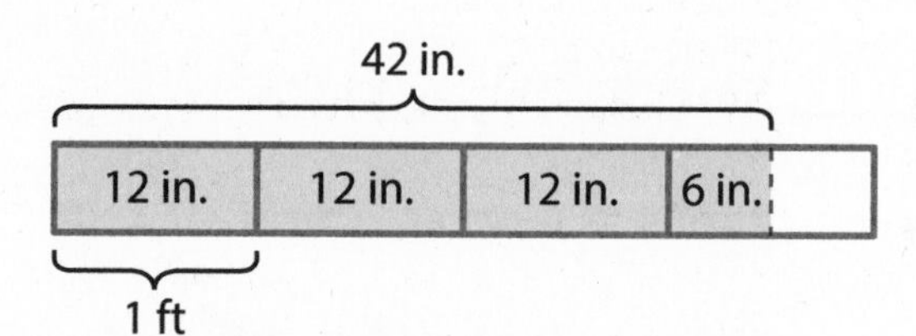

So, 42 inches is _____ feet _____ inches.

Show and Grow *I can do it!*

Convert the length.

1. $6\frac{1}{2}$ ft = _____ in.

2. 94 in. = _____ ft _____ in.

Name ________________________________

Apply and Grow: Practice

Convert the length.

3. $3\frac{3}{4}$ ft = _____ in.

4. 60 in. = _____ ft

5. 375 ft = _____ yd

6. $12\frac{2}{3}$ yd = _____ ft

7. 51 in. = _____ ft _____ in.

8. 5 yd = _____ in.

Compare.

9. $7\frac{1}{3}$ yd ◯ 22 ft

10. 54 in. ◯ 4 ft 8 in.

11. 216 in. ◯ 6 yd

12. A dugong is $8\frac{1}{3}$ feet long. How many inches long is the dugong?

The dugong lives in the Great Barrier Reef.

13. DIG DEEPER! Order the lengths from shortest to longest. Explain how you converted the lengths.

$5\frac{1}{2}$ feet | 5 feet 3 inches | $5\frac{2}{3}$ feet | $5\frac{3}{4}$ feet

Think and Grow: Modeling Real Life

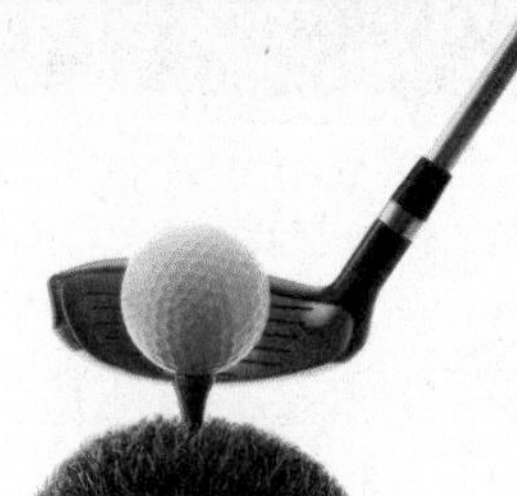

Example A golfer uses a device to determine that his golf ball is 265 feet from a hole. After his next shot, his ball is 15 feet short of the hole. How many yards did the golfer hit his ball?

Because the golfer's ball is 15 feet short of the hole after his next shot, subtract 15 from 265 to find how many feet the golfer hit his ball.

$265 - 15 =$ _____ feet

Convert the number of feet the golfer hit his ball to yards.

There are _____ feet in 1 yard.

Because you are converting from a smaller unit to a larger unit, divide.

$$\begin{array}{r} 83 \text{ R } \_\_\_ \\ 3\overline{)250} \\ - \square \\ \hline \square \\ - \square \\ \hline \square \end{array}$$

$\dfrac{\text{remainder}}{\text{divisor}} \rightarrow \dfrac{\square}{\square}$

To determine how many yards the golfer hit his ball, write the quotient and remainder as a mixed number.

_____ R _____ $= \square\dfrac{\square}{\square}$ yards So, the golfer hit his ball _______ yards.

Show and Grow *I can think deeper!*

14. A tree is 27 feet tall. After the tree is struck by lightning, it is 96 inches shorter. How many feet tall is the tree after it is struck by lightning?

---

15. **DIG DEEPER!** The rope ladder is $2\frac{1}{2}$ yards tall. Each knot is made using 16 inches of rope. How many feet of rope are used to make the ladder? Explain.

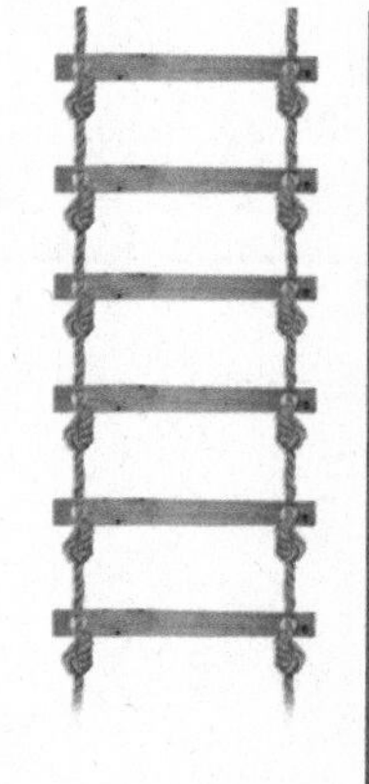

16. **DIG DEEPER!** The Chesapeake Bay Bridge is $4\frac{3}{10}$ miles long. Road work begins 1,200 yards from one end of the bridge and ends 2 miles from the other end of the bridge. How many yards long is the road work?

Name ______________________________

Homework & Practice 11.3

Learning Target: Write lengths using equivalent customary measures.

Example Convert $7\frac{3}{4}$ feet to inches.

There are __12__ inches in 1 foot.

Because you are converting from a larger unit to a smaller unit, multiply.

$7\frac{3}{4} \times \underline{12} = \frac{31}{4} \times \underline{12} = \underline{\frac{372}{4}}$

So, $7\frac{3}{4}$ feet is $\underline{\frac{372}{4}}$, or __93__ inches.

Convert the length.

1. $2\frac{1}{3}$ ft = ______ in.

2. 5 mi = ______ yd

3. $3\frac{1}{3}$ yd = ______ ft

4. 27 in. = ______ ft ______ in.

5. 108 in. = ______ yd

6. 34 in. = ______ ft

Compare.

7. $5\frac{3}{4}$ ft ◯ 65 in.

8. 19 in. ◯ 1 ft 5 in.

9. 2 mi ◯ 10,650 ft

10. A football player runs 93 yards. How many feet does he run?

11. MP **Precision** Write whether you would use *multiplication* or *division* for each conversion.

yards to feet: ______

miles to inches: ______

feet to miles: ______

inches to feet: ______

miles to yards: ______

12. MP **Reasoning** Match each measurement with the best customary unit of measure.

height of a jump	inches
length of a crayon	feet
length of a river	yards
length of a football field	miles

13. **Modeling Real Life** How long is the velociraptor in yards?

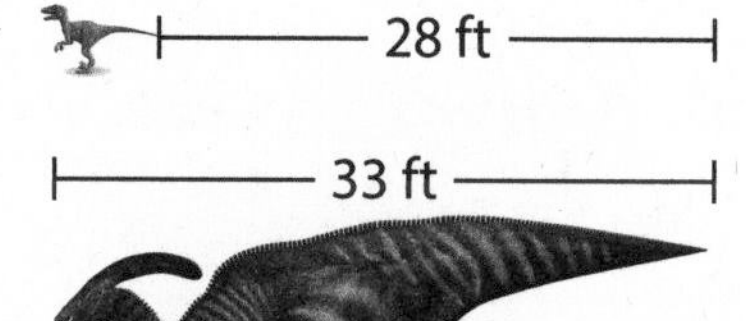

14. **DIG DEEPER!** You wrap a cube-shaped box with ribbon as shown. The ribbon is wrapped around all of the faces of the cube. You use 9 inches of ribbon for the bow. How many inches of ribbon do you use altogether? Explain.

Review & Refresh

Find the quotient.

15. $3{,}200 \div 40 =$ ______

16. $5{,}400 \div 9 =$ ______

17. $600 \div 20 =$ ______

Name ______________________________

Weight in Customary Units 11.4

Learning Target: Write weights using equivalent customary measures.

Success Criteria:
- I can compare the sizes of two customary units of weight.
- I can write a customary weight using a smaller customary unit.
- I can write a customary weight using a larger customary unit.

Explore and Grow

Work with a partner. Use the number line to help you complete each statement.

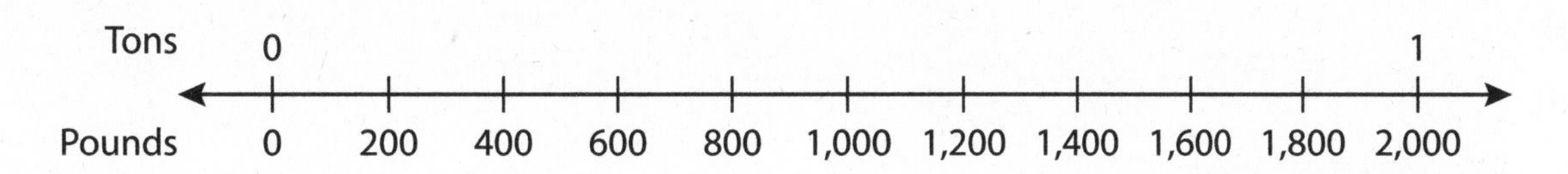

Vehicle: 2 tons

The vehicle weighs ________ pounds.

Whale Shark: 30,000 pounds

The whale shark weighs ________ tons.

Structure How can you convert tons to pounds? How can you convert pounds to tons?

Think and Grow: Convert Customary Weights

Key Idea When finding equivalent customary weights, multiply to convert from a larger unit to a smaller unit. Divide to convert from a smaller unit to a larger unit.

Customary Units of Weight
1 pound (lb) = 16 ounces (oz)
1 ton (T) = 2,000 pounds (lb)

Example Convert $4\frac{1}{4}$ tons to pounds.

4 $\frac{1}{4}$ 2,000 8,000 500

There are ______ pounds in 1 ton.

Because you are converting from a larger unit to a smaller unit, multiply.

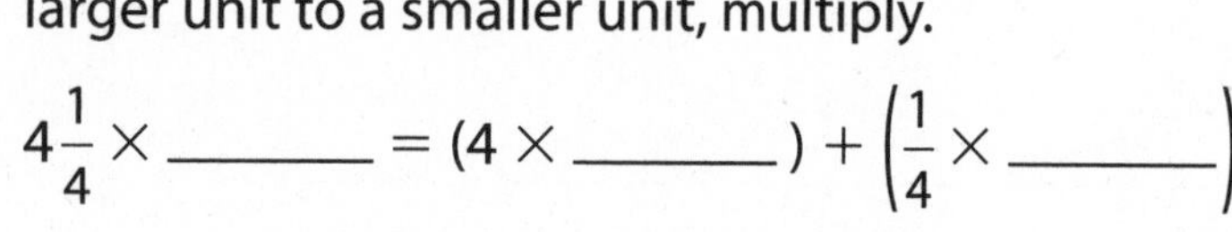

$4\frac{1}{4} \times$ ______ $= (4 \times$ ______ $) + \left(\frac{1}{4} \times\right.$ ______ $\left.\right)$

$=$ ______

So, $4\frac{1}{4}$ tons is ______ pounds.

Example Convert 40 ounces to pounds.

40 oz 16 oz 16 oz 8 oz 1 lb

There are ______ ounces in 1 pound.

Because you are converting from a smaller unit to a larger unit, divide.

$40 \div$ ______ $= \square\frac{\square}{\square}$

So, 40 ounces is ______ pounds.

You could also write the weight as 2 pounds 8 ounces.

Show and Grow *I can do it!*

Convert the weight.

1. 9 T = ______ lb

2. $6\frac{1}{2}$ lb = ______ oz

3. 6,000 lb = ______ T

4. 80 oz = ______ lb

Name ______________________________

Apply and Grow: Practice

Convert the weight.

5. 10,000 lb = ________ T

6. 8 lb = ________ oz

7. 240 oz = ________ lb

8. $7\frac{1}{4}$ T = ________ lb

9. 150 oz = ________ lb ________ oz

10. 32,000 oz = ________ T

Compare.

11. 30 T ◯ 6,000 lb

12. 53 oz ◯ $3\frac{1}{2}$ lb

13. 8 T ◯ 224,000 oz

14. What is the weight of the hippopotamus in tons?

4,000 pounds

15. **MP Reasoning** Compare 10 pounds and 165 ounces using mental math. Explain.

16. **MP Number Sense** Which measurements are equivalent to 60 ounces?

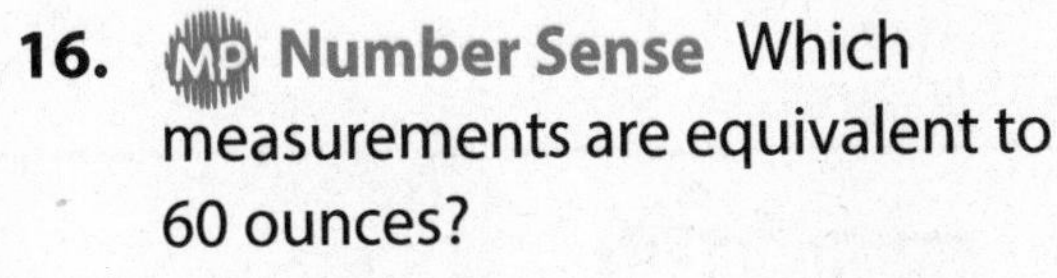

$3\frac{3}{4}$ lb $3\frac{1}{2}$ lb

3 lb 12 oz   3 lb 4 oz

Think and Grow: Modeling Real Life

Example A newborn baby boy weighs 122 ounces. A newborn baby girl weighs 6 pounds 4 ounces. Which baby weighs more? How much more?

Convert the weight of the boy to pounds and ounces.

There are _____ ounces in 1 pound.

$122 \div$ _____ $=$ _____ R _____

$16\overline{)122}$ = 7 R _____
− ☐
☐

The boy weighs _____ pounds _____ ounces.

Subtract the weight of the girl from the weight of the boy.

_____	pounds	_____	ounces
− 6	pounds	4	ounces
☐	pound	☐	ounces

So, the boy weighs _____ pound _____ ounces more than the girl.

Show and Grow *I can think deeper!*

17. Which box of cereal weighs more? How much more?

$1\frac{1}{2}$ pounds

17 ounces

18. A male rhinoceros weighs $2\frac{1}{4}$ tons. Which rhinoceros weighs more? How much more? Write your answer as a fraction in simplest form.

Female rhinoceros: 3,500 lb

19. **DIG DEEPER!** Can all of the passengers listed in the table ride the boat at once? Explain.

Passenger Weights (pounds)			
191	184	150	248
170	215	132	145
265	126	259	175

Name ____________________

Homework & Practice 11.4

Learning Target: Write weights using equivalent customary measures.

Example Convert 12,000 pounds to tons.

There are __2,000__ pounds in 1 ton.

Because you are converting from a smaller unit to a larger unit, divide.

12,000 ÷ __2,000__ = __6__

So, 12,000 pounds is __6__ tons.

Convert the weight.

1. 10 T = ______ lb

2. 32 oz = ______ lb

3. 48,000 lb = ______ T

4. 50 lb = ______ oz

5. $5\frac{3}{4}$ T = ______ lb

6. $8\frac{1}{2}$ lb = ______ oz

7. 168 oz = ______ lb ______ oz

8. 96,000 oz = ______ T

Compare.

9. 16 lb ◯ 32,000 T

10. 128 oz ◯ $8\frac{1}{4}$ lb

11. 11 T ◯ 384,000 oz

12. A newborn puppy weighs 3 pounds 5 ounces. What is the weight of the puppy in ounces?

13. MP **Number Sense** How many tons are equal to 500 pounds? Write your answer as a fraction in simplest form.

Open-Ended Complete the statement.

14. _______ pounds $>$ 72 ounces

15. 13 tons $<$ _______ pounds

16. **Modeling Real Life** An employee at a juice cafe uses 10 ounces of kale and $\frac{3}{4}$ pound of apples to make a drink. Does the employee use more kale or apples? How much more?

17. **Modeling Real Life** You have a 3-pound bag of clay. You use 8 ounces of clay to make an ornament. How many ornaments can you make using all of the clay?

Review & Refresh

Find the product. Check whether your answer is reasonable.

18. $\begin{array}{r} 145 \\ \times \quad 12 \\ \hline \end{array}$

19. $\begin{array}{r} 561 \\ \times \quad 87 \\ \hline \end{array}$

20. $\begin{array}{r} 823 \\ \times \quad 65 \\ \hline \end{array}$

Name ______________________________

Capacity in Customary Units 11.5

Learning Target: Write capacities using equivalent customary measures.

Success Criteria:

- I can compare the sizes of two customary units of capacity.
- I can write a customary capacity using a smaller customary unit.
- I can write a customary capacity using a larger customary unit.

Explore and Grow

Describe the relationship between cups and fluid ounces (fl oz). Then complete the table.

1 cup is __________ times as much as 1 fluid ounce.

Capacity (cups)	Capacity (fluid ounces)
1	
2	
3	
4	

Structure How can you convert cups to fluid ounces? How can you convert fluid ounces to cups?

Think and Grow: Convert Customary Capacities

Key Idea When finding equivalent customary capacities, multiply to convert from a larger unit to a smaller unit. Divide to convert from a smaller unit to a larger unit.

Customary Units of Capacity
1 cup (c) = 8 **fluid ounces** (fl oz)
1 pint (pt) = 2 cups (c)
1 quart (qt) = 2 pints (pt)
1 gallon (gal) = 4 quarts (qt)

Example Convert $6\frac{1}{2}$ cups to fluid ounces.

There are ______ fluid ounces in 1 cup.

Because you are converting from a larger unit to a smaller unit, multiply.

$6\frac{1}{2} \times$ ______ $= (6 \times$ ______$) + \left(\frac{1}{2} \times\right.$ ______$\left.\right)$

$=$ ______

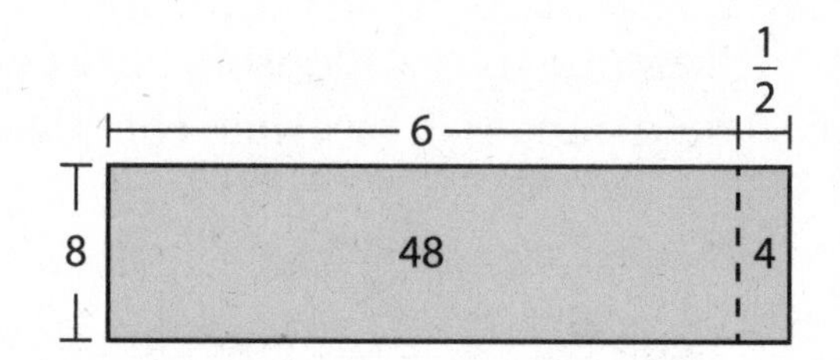

So, $6\frac{1}{2}$ cups is ______ fluid ounces.

Example Convert 14 quarts to gallons.

There are ______ quarts in 1 gallon.

Because you are converting from a smaller unit to a larger unit, divide.

$14 \div$ ______ $= \square\frac{\square}{\square}$

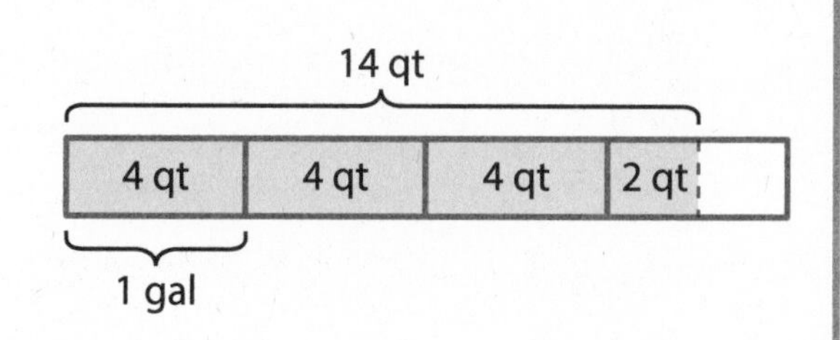

So, 14 quarts is ______ gallons.

Show and Grow *I can do it!*

Convert the capacity.

1. 9 gal = ______ qt

2. 20 pt = ______ c

3. 42 pt = ______ qt

4. 68 qt = ______ gal

Name ________________________________

Apply and Grow: Practice

Convert the capacity.

5. 7 c = _____ fl oz

6. 6 pt = _____ qt

7. 16 qt = _____ gal

8. 15 pt = _____ c

9. $2\frac{1}{4}$ c = _____ fl oz

10. 12 c = _____ qt

Compare.

11. 14 c ◯ 10 pt

12. 38 qt ◯ $8\frac{1}{2}$ gal

13. 4 gal ◯ 32 pt

14. You fill your turtle's aquarium with 40 pints of water. How many gallons of water do you use?

15. MP **Number Sense** Newton's water cooler contains $1\frac{1}{2}$ gallons of water. How many times can he fill his 16-fluid ounce canteen with water from the water cooler? Explain.

16. DIG DEEPER! Order the capacities from least to greatest. Explain how you converted the capacities.

$8\frac{1}{2}$ c	72 fl oz	$7\frac{3}{4}$ c	56 fl oz

Think and Grow: Modeling Real Life

Example A car's engine contains $4\frac{1}{2}$ quarts of oil. Can a mechanic use a 24-cup container to drain all of the oil?

First, convert the quarts of oil to pints.

There are _____ pints in 1 quart.

$4\frac{1}{2} \times$ _____ $= (4 \times$ _____$) + \left(\frac{1}{2} \times \right.$ _____$\left.\right)$

$=$ _____ pints

You can also multiply $4\frac{1}{2}$ by the number of cups in 1 quart to find the answer.

Convert the pints of oil to cups.

There are _____ cups in 1 pint.

$9 \times$ _____ $=$ _____ cups

Compare the cups of oil to the capacity of the container.

_____ ○ 24

So, the mechanic __________ use the container to drain all of the oil.

Show and Grow *I can think deeper!*

17. An adult has 192 fluid ounces of blood in his body. How many pints of blood are in his body?

18. You make $4\frac{1}{2}$ cups of soup. One serving is 12 fluid ounces. How many servings of soup do you make?

19. **DIG DEEPER!** A scientist has two beakers of a solution, one containing 5 cups and the other containing $1\frac{1}{2}$ pints. How many gallons of the solution does the scientist have? Write your answer as a fraction in simplest form.

Name ____________________

Homework & Practice 11.5

Learning Target: Write capacities using equivalent customary measures.

Example Convert 20 pints to quarts.

There are __2__ pints in 1 quart.

Because you are converting from a smaller unit to a larger unit, divide.

$20 \div$ __2__ $=$ __10__

So, 20 pints is __10__ quarts.

Convert the capacity.

1. 9 pt = ____ c

2. 72 fl oz = ____ c

3. 6 c = ____ fl oz

4. $3\frac{3}{4}$ gal = ____ qt

5. $5\frac{1}{2}$ qt = ____ pt

6. 40 pt = ____ gal

7. 64 qt = ____ c

8. 112 fl oz = ____ pt

Compare.

9. 48 qt ◯ 12 gal

10. 24 fl oz ◯ $3\frac{1}{4}$ c

11. 10 qt ◯ 24 c

12. You buy 2 gallons of apple cider. How many cups of apple cider do you buy?

13. MP **Logic** Your friend makes a table of equivalent capacities. Write two pairs of customary units represented by the chart.

_____	_____
1	4
2	8
3	12
4	16

14. DIG DEEPER! Which measurements are greater than 16 pints?

300 fluid ounces   28 cups   10 quarts

1 gallon   275 fluid ounces   35 cups

15. **Modeling Real Life** Your friend buys 8 quarts of frozen yogurt. How many cups of frozen yogurt does she buy?

16. **Modeling Real Life** A recipe calls for $2\frac{1}{4}$ cups of milk. You want to make 2 batches of the recipe. Should you buy a pint, quart, or half gallon of milk?

Review & Refresh

Find the product.

17. $0.5 \times 0.7 =$ _____

18. $46.2 \times 0.68 =$ _____

19. $1.4 \times 0.3 =$ _____

Name ___________________________________

Make and Interpret Line Plots 11.6

Learning Target: Make line plots and use them to solve problems.

Success Criteria:

- I can make a line plot.
- I can interpret a line plot.
- I can use a line plot to solve a real-life problem.

Explore and Grow

Measure and record your height to the nearest quarter of a foot. Collect the heights of all the students in your class and create a line plot of the results.

Construct Arguments Make two conclusions from the line plot.

Think and Grow: Make Line Plots

Example The table shows the amounts of water that 10 students use for a science experiment. Make a line plot to display the data. How many students use more than $\frac{1}{2}$ cup?

Amounts of Water (cup)				
$\frac{7}{8}$	$\frac{5}{8}$	$\frac{3}{4}$	$\frac{1}{2}$	$\frac{5}{8}$
$\frac{1}{4}$	$\frac{1}{2}$	$\frac{5}{8}$	$\frac{3}{4}$	$\frac{3}{8}$

Step 1: Write the data values as fractions with a common denominator.

The denominators of the data values are 2, 4, and 8. Because 2 and 4 are factors of 8, use a denominator of 8.

$\frac{1}{2} = \frac{1 \times 4}{2 \times 4} = \frac{\square}{\square}$ $\frac{1}{4} = \frac{1 \times 2}{4 \times 2} = \frac{\square}{\square}$ $\frac{3}{4} = \frac{3 \times 2}{4 \times 2} = \frac{\square}{\square}$

Step 2: Use a scale on a number line that shows all of the data values.

Step 3: Mark an X for each data value.

Water Amounts for Science Experiments

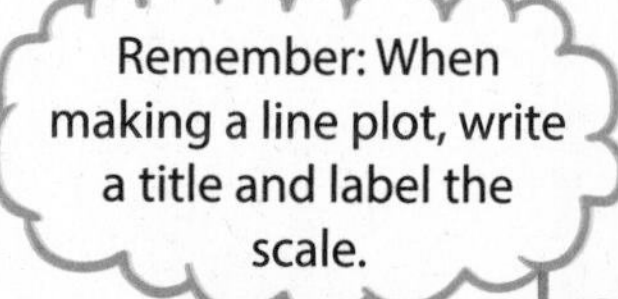

Remember: When making a line plot, write a title and label the scale.

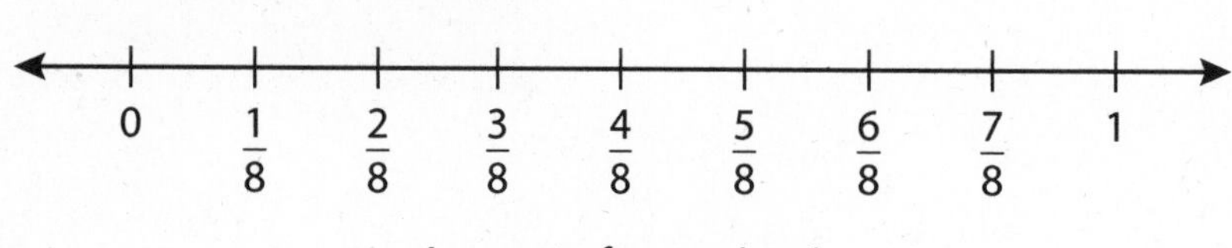

Amount of water (cup)

There are _____ students who use more than $\frac{1}{2}$ cup.

Show and Grow *I can do it!*

1. The table shows the distance your friend swims each day for 10 days. Make a line plot to display the data.

Swimming Distances (mile)				
$\frac{5}{8}$	$\frac{1}{2}$	$\frac{3}{4}$	$\frac{1}{2}$	$\frac{3}{4}$
$\frac{3}{8}$	$\frac{3}{4}$	$\frac{1}{2}$	$\frac{7}{8}$	$\frac{3}{4}$

Swimming Distances

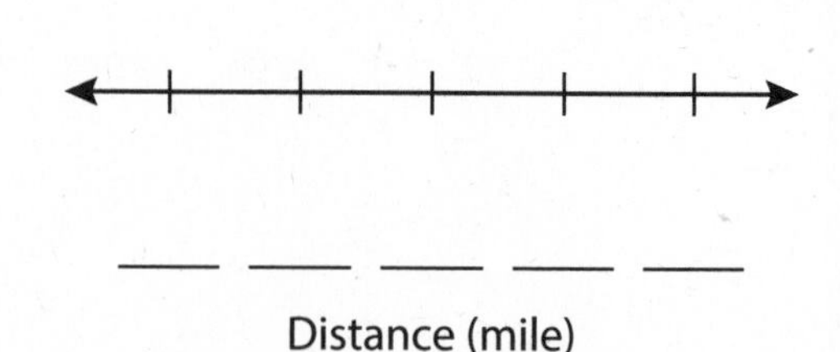

Distance (mile)

How many days does your friend swim $\frac{3}{4}$ mile or more?

Name ______________________________

Apply and Grow: Practice

2. The table shows the amounts of mulch a landscaping company orders on 10 different days. Make a line plot to display the data.

Amounts of Mulch (ton)				
$\frac{1}{4}$	$\frac{7}{8}$	$\frac{1}{2}$	$\frac{3}{4}$	$\frac{7}{8}$
$\frac{7}{8}$	$\frac{3}{4}$	$\frac{3}{4}$	$\frac{7}{8}$	$\frac{1}{2}$

Mulch Amounts for Landscaping

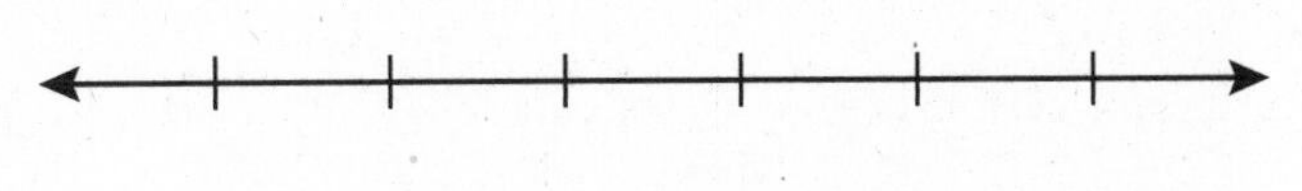

___ ___ ___ ___ ___ ___

Amount of mulch (ton)

On how many days did the landscaping company use more than $\frac{1}{2}$ ton of mulch?

What do you notice about the data?

---

3. DIG DEEPER! Your teacher has the three packages of seeds shown. She divides the first package into bags weighing $\frac{1}{2}$ ounce each. She divides the second package into bags weighing $\frac{1}{4}$ ounce each. She divides the third package into bags weighing $\frac{1}{8}$ ounce each. Find the total number of bags of seeds. Use a line plot to display the results.

2 ounces

2 ounces

2 ounces

Bag Weights

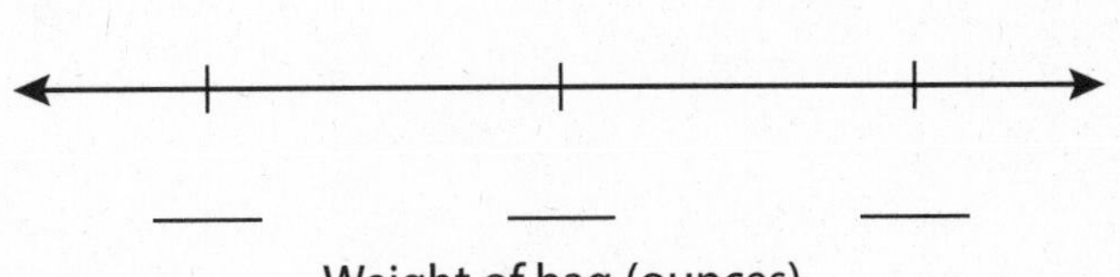

___ ___ ___

Weight of bag (ounces)

Think and Grow: Modeling Real Life

Example You record the amounts of time you skateboard each day for 8 days. Your friend skateboards the same total amount of time, but for an equal number of hours each day. How long does your friend skateboard each day?

Skateboarding Practice

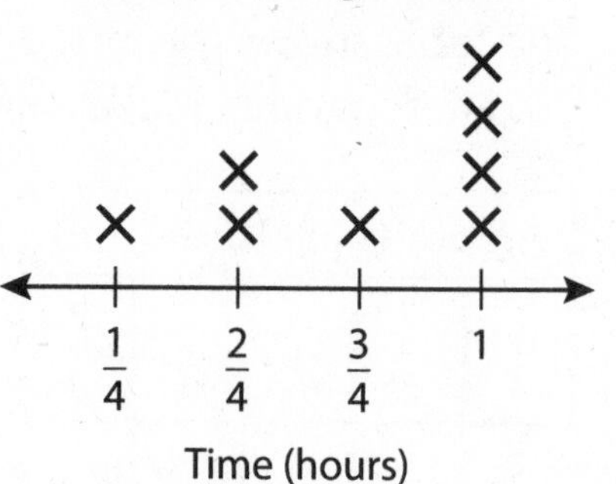

Step 1: You and your friend skateboard the same total amount of time. Use the line graph to find the number of hours you each skateboard.

$\frac{1}{4} + \left(2 \times \frac{2}{4}\right) + \frac{3}{4} + (4 \times 1) = \frac{1}{4} + \_\_\_\_ + \frac{3}{4} + \_\_\_\_$ Multiply.

$= \frac{1}{4} + \frac{3}{4} + \_\_\_\_ + \_\_\_\_$ Commutative Property of Addition

$= \frac{1}{4} + \frac{3}{4} + (\_\_\_\_ + \_\_\_\_)$ Associative Property of Addition

$= \_\_\_\_ + \_\_\_\_$ Add.

$= \_\_\_\_$ Add.

So, you each skateboard for _____ hours.

Step 2: Divide the number of hours by the number of days.

$\_\_\_\_ \div \_\_\_\_ = \frac{\square}{\square}$

So, your friend skateboards for _____ hour each day.

Show and Grow *I can think deeper!*

4. You record the amounts of trail mix you pour into 12 bags. Your friend has the same total amount of trail mix, but equally divides it among 12 bags. How much trail mix does your friend pour into each bag?

Trail Mix Bags

$\frac{1}{2}$ 1 $1\frac{1}{2}$ 2 $2\frac{1}{2}$

Amount of trail mix (cups)

Name ______________________________

Homework & Practice 11.6

Learning Target: Make line plots and use them to solve problems.

Example The table shows the amounts of time that 10 students take to land three balls in a row in a game. Make a line plot to display the data. How many students take less than $\frac{3}{4}$ minute?

Amounts of Time (minute)				
$\frac{1}{2}$	$\frac{3}{4}$	$\frac{1}{2}$	$\frac{1}{4}$	$\frac{3}{4}$
$\frac{1}{4}$	$\frac{1}{2}$	$\frac{3}{4}$	$\frac{1}{2}$	$\frac{1}{2}$

Step 1: Write the data values as fractions with a common denominator.

Step 2: Use a scale on a number line that shows all of the data values.

Step 3: Mark an X for each data value.

Completion Times

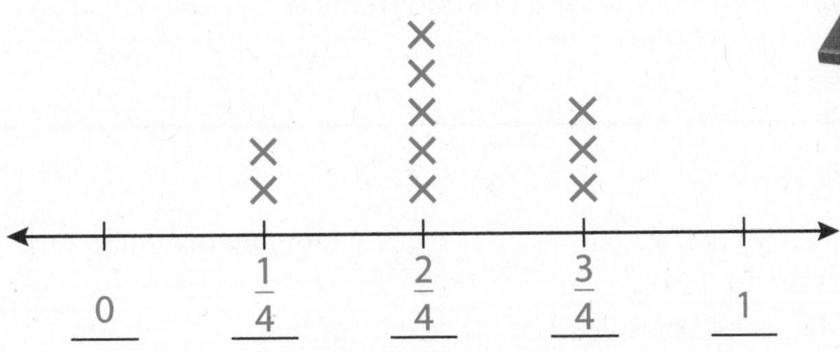

Amount of time (minute)

There are 7 students who take less than $\frac{3}{4}$ minute.

1. The table shows the weights of 10 newborn pygmy marmosets. Make a line plot to display the data. How many pygmy marmosets weigh more than $\frac{1}{2}$ ounce?

Weights (ounce)				
$\frac{1}{2}$	$\frac{5}{8}$	$\frac{7}{8}$	$\frac{3}{4}$	$\frac{7}{8}$
$\frac{7}{8}$	$\frac{3}{4}$	$\frac{5}{8}$	$\frac{7}{8}$	$\frac{3}{4}$

Pygmy Marmoset Weights

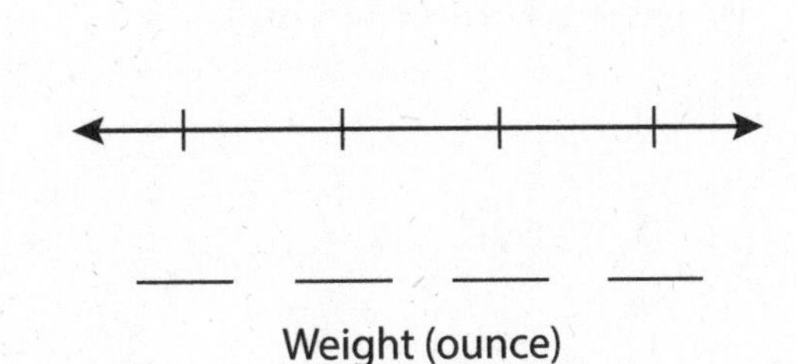

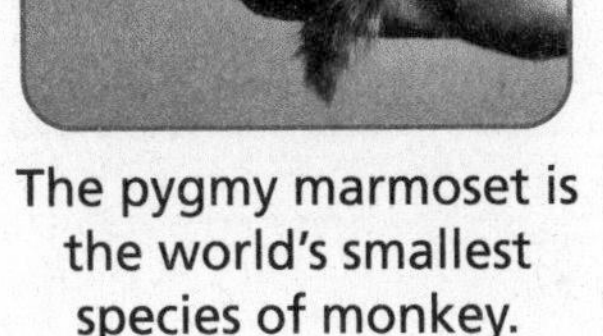

The pygmy marmoset is the world's smallest species of monkey.

Use the table.

2. The table shows the amounts of berries required to make 10 different smoothie recipes. Make a line plot to display the data.

Berries (cup)									
$\frac{3}{4}$	$\frac{1}{2}$	$\frac{1}{8}$	$\frac{1}{2}$	$\frac{3}{4}$	$\frac{1}{2}$	$\frac{1}{4}$	$\frac{3}{4}$	$\frac{3}{4}$	$\frac{1}{4}$

Berry Amounts for Smoothies

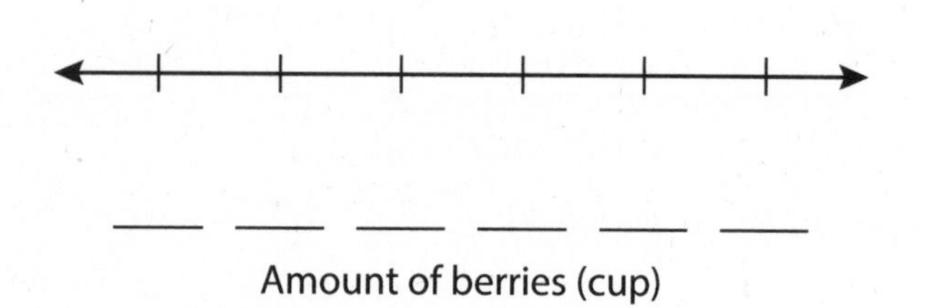

What is the most common amount of berries required?

How many times as many recipes use $\frac{3}{4}$ cup of berries as $\frac{1}{4}$ cup of berries?

3. **DIG DEEPER!** How many total cups of berries are needed to make one of each smoothie?

Use the line plot.

4. **Modeling Real Life** The line plot shows the number of miles you run each day for 10 days. Your friend runs the same total number of miles, but runs an equal number of miles each day. How far does your friend run each day?

Running Distances

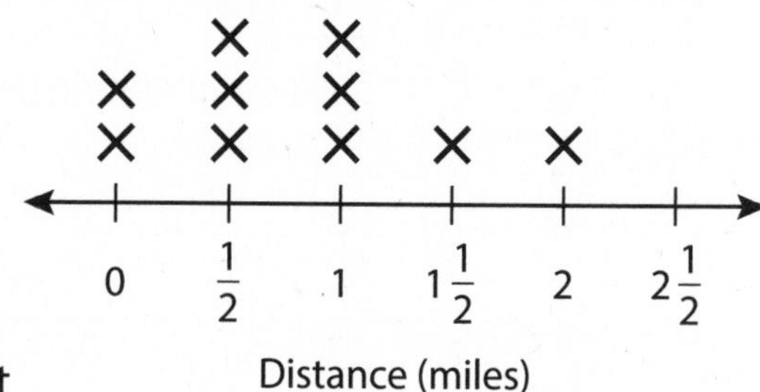

5. **DIG DEEPER!** Your cousin runs a total amount that is 6 times as far as your friend runs in one day. How far does your cousin run?

Review & Refresh

Divide. Then check your answer.

6. $561 \div 7 =$ _____

7. $3{,}029 \div 4 =$ _____

8. $2{,}814 \div 9 =$ _____

Name ______________________________

Problem Solving: Measurement 11.7

Learning Target: Solve multi-step word problems involving units of measure.

Success Criteria:
- I can understand a problem.
- I can make a plan to solve.
- I can solve a problem.

Explore and Grow

Make a plan to solve the problem.

A fruit vendor sells fruit by the pound. You have a tote that can hold up to 4 pounds. A bag of oranges weighs $2\frac{1}{4}$ pounds. A bag of apples weighs 28 ounces. Can your tote hold both bags of fruit? Explain.

MP **Precision** Which bag of fruit is heavier? Explain.

Think and Grow: Problem Solving: Measurement

Example A recipe calls for $2\frac{1}{4}$ cups of milk. You have $\frac{1}{4}$ pint of whole milk and $1\frac{1}{2}$ cups of skim milk. Do you have enough milk for the recipe?

Understand the Problem

What do you know?

- The recipe calls for $2\frac{1}{4}$ cups of milk.
- You have $\frac{1}{4}$ pint of whole milk and $1\frac{1}{2}$ cups of skim milk.

What do you need to find?

- You need to find whether you have enough milk for the recipe.

Make a Plan

How will you solve?

- Convert $\frac{1}{4}$ pint of whole milk to cups.
- Add the amounts of whole milk and skim milk.
- Compare the amount of milk you have to the amount needed.

Solve

Step 1: Convert $\frac{1}{4}$ pint to cups.

There are _____ cups in 1 pint.

$\frac{1}{4} \times$ _____ = _____

You have _____ cup of whole milk.

Step 2: Add the amounts of whole milk and skim milk.

You have _____ cup of whole milk and _____ cups of skim milk.

_____ + _____ = _____

You have _____ cups of milk.

Compare: _____ cups ◯ $2\frac{1}{4}$ cups

So, you __________ have enough milk for the recipe.

Show and Grow *I can do it!*

1. Explain how you can check whether the answer above is reasonable.

Name ____________________

Apply and Grow: Practice

Understand the problem. What do you know? What do you need to find? Explain.

2. Your friend buys 1 pound of walnuts, 6 ounces of peanuts, and $\frac{1}{2}$ pound of cashews. How many ounces do the nuts weigh in all?

3. A bottle of orange juice contains 64 fluid ounces. How many cups of orange juice are in 3 bottles?

Understand the problem. Then make a plan. How will you solve? Explain.

4. Your friend wants to buy curtains that hang from the top of the window to the floor. Curtain lengths are typically measured in inches. What length curtains should he buy?

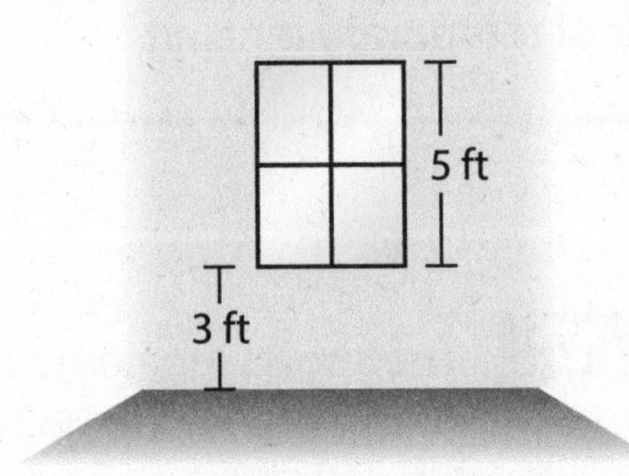

5. Your friend runs a total distance of 1 kilometer at track practice by running 100-meter hurdles. How many times does he run the hurdles?

6. A trailer can carry $13\frac{1}{2}$ tons. It has room to carry 6 cars at once. Can the trailer carry 6 cars that each weigh 3,800 pounds? Explain.

7. **DIG DEEPER!** You walk your dog 4 laps around the block each day. Each block is 400 meters. How many total kilometers do you walk your dog around the block after 35 weeks?

Think and Grow: Modeling Real Life

Example A crew member needs to put a temporary fence around the perimeter of the rectangular football field. How many feet of temporary fencing does the crew member need?

160 ft

120 yd

Think: What do you know? What do you need to find? How will you solve?

Step 1: Convert the length of the field to feet.

There are _____ feet in 1 yard.

$120 \times$ _____ $=$ _____

The length of the field is _____ feet.

Step 2: Use a formula to find the perimeter of the field.

$P = (2 \times \ell) + (2 \times w)$

$= (2 \times$ _____$) + (2 \times$ _____$)$

$=$ _____ $+$ _____

$=$ _____ So, the crew member needs _____ feet of temporary fencing.

Show and Grow *I can think deeper!*

8. An artist puts a wood border around the perimeter of the rectangular mural. How many feet of wood does the artist need?

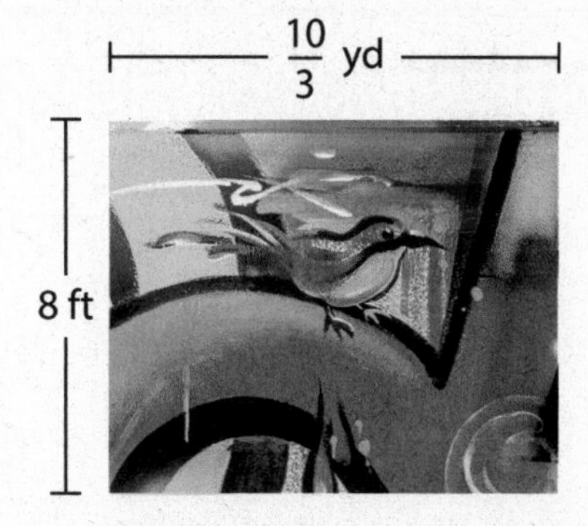

9. **DIG DEEPER!** The sports jug contains 5 gallons of water. The paper cup holds 8 fluid ounces of water. How many paper cups can 3 sports jugs fill? Justify your answer.

Name ___________________________________

Homework & Practice 11.7

Learning Target: Solve multi-step word problems involving units of measure.

Example Descartes eats $2\frac{1}{4}$ pounds of food each week. He buys his food in 6-ounce cans. How many cans should he buy so that he has enough food for the week?

Think: What do you know? What do you need to find? How will you solve?

Step 1: Convert $2\frac{1}{4}$ pounds to ounces to find how many ounces of food Descartes eats each week.

There are __16__ ounces in 1 pound.

$2\frac{1}{4} \times$ __16__ = __36__

Descartes eats __36__ ounces of food each week.

Step 2: Find how many 6-ounce cans he should buy.

__36__ ÷ __6__ = __6__

He should buy __6__ cans.

Descartes should buy __6__ cans so that he has enough food for the week.

Understand the problem. Then make a plan. How will you solve? Explain.

1. A robotic insect has a mass of 80 milligrams. The mass of a quarter is 5.67 grams. How many more grams is the mass of a quarter than the mass of the robotic insect?

2. Newton pours water out of a filled 2-liter beaker. Now it only has 1,025 milliliters of water in it. How many milliliters of water did Newton pour out?

3. You run 5 laps around a track. Each lap is 400 meters. How many total kilometers do you run?

4. Two hotel workers have a total of 30 bags of luggage each weighing 50 pounds. One worker weighs 150 pounds, and the other weighs 210 pounds. Can they transfer themselves and all of the bags in the elevator at once? Explain.

5. DIG DEEPER! You have 84 feet of streamers. You cut 24 pieces that are each $\frac{1}{2}$ yard long. How many feet of streamers do you have left?

6. **Writing** Write and solve a two-step word problem involving units of measure.

7. **Modeling Real Life** You want to hang a wallpaper border around the perimeter of the rectangular bathroom shown. How many yards of wallpaper border do you need?

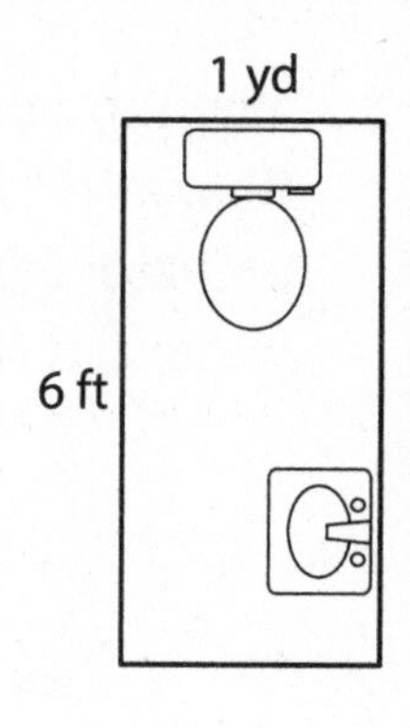

8. DIG DEEPER! You need $\frac{1}{2}$ gallon of fertilizer to cover a lawn. What is the least amount of money that you can pay and have enough fertilizer?

128 fluid ounces
$40

16 fluid ounces
$11

Review & Refresh

Divide. Then check your answer.

9. $25\overline{)5,343}$

10. $24\overline{)2,064}$

Name ______________________________

Performance Task 11

Passenger airliners come in many different sizes. Plane A and Plane B are two different types of wide-body jet airliners.

Plane A

1. The table shows some facts about Plane A.

Plane A	
Length	250 ft 2 in.
Wingspan	224 ft 7 in.
Maximum takeoff weight	493.5 tons

   a. The length of Plane B is 80 yards. Which is longer, Plane A or Plane B? How much longer?

   b. The wingspan of Plane B is $37\frac{1}{12}$ feet longer than the wingspan of the Plane A. What is the wingspan of Plane B?

2. Before an airliner can take off, the pilot has to make sure it weighs less than the maximum takeoff weight.

   a. Plane A weighs 404,600 pounds and can carry at most 422,000 pounds of fuel. How many pounds can the airliner hold in passengers and cargo?

   b. The maximum landing weight of Plane A is 300,000 pounds less than the maximum takeoff weight. Why does an airliner weigh less at the end of a flight than at the beginning?

   c. Plane A uses 20 quarts of fuel for each mile it flies. How many gallons of fuel does the plane use during a 3,200-mile flight?

Plane B
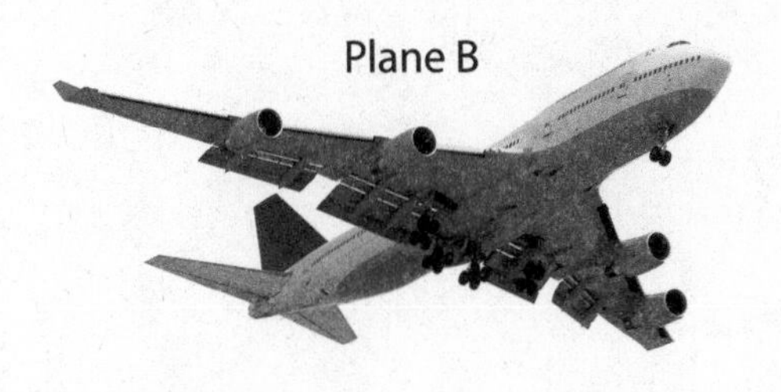

3. Plane B can hold 544 passengers. Plane A can hold $\frac{3}{4}$ of passengers that Plane B can hold.

   a. How many passengers can Plane A hold?

   b. An airline estimates that each passenger weighs about 200 pounds, including carry-on baggage. How much passenger and carry-on weight does the airline estimate for Plane B?

Surround and Capture

Directions:

1. Players take turns.
2. On your turn, place a counter on a yellow hexagon.
3. Solve for the missing measurement and cover the answer with another counter. If you surround a monster, then put a counter on the monster. If you do not surround a monster, then your turn is over.
4. Continue playing until all measurements are covered.
5. The player who captures the most monsters wins!

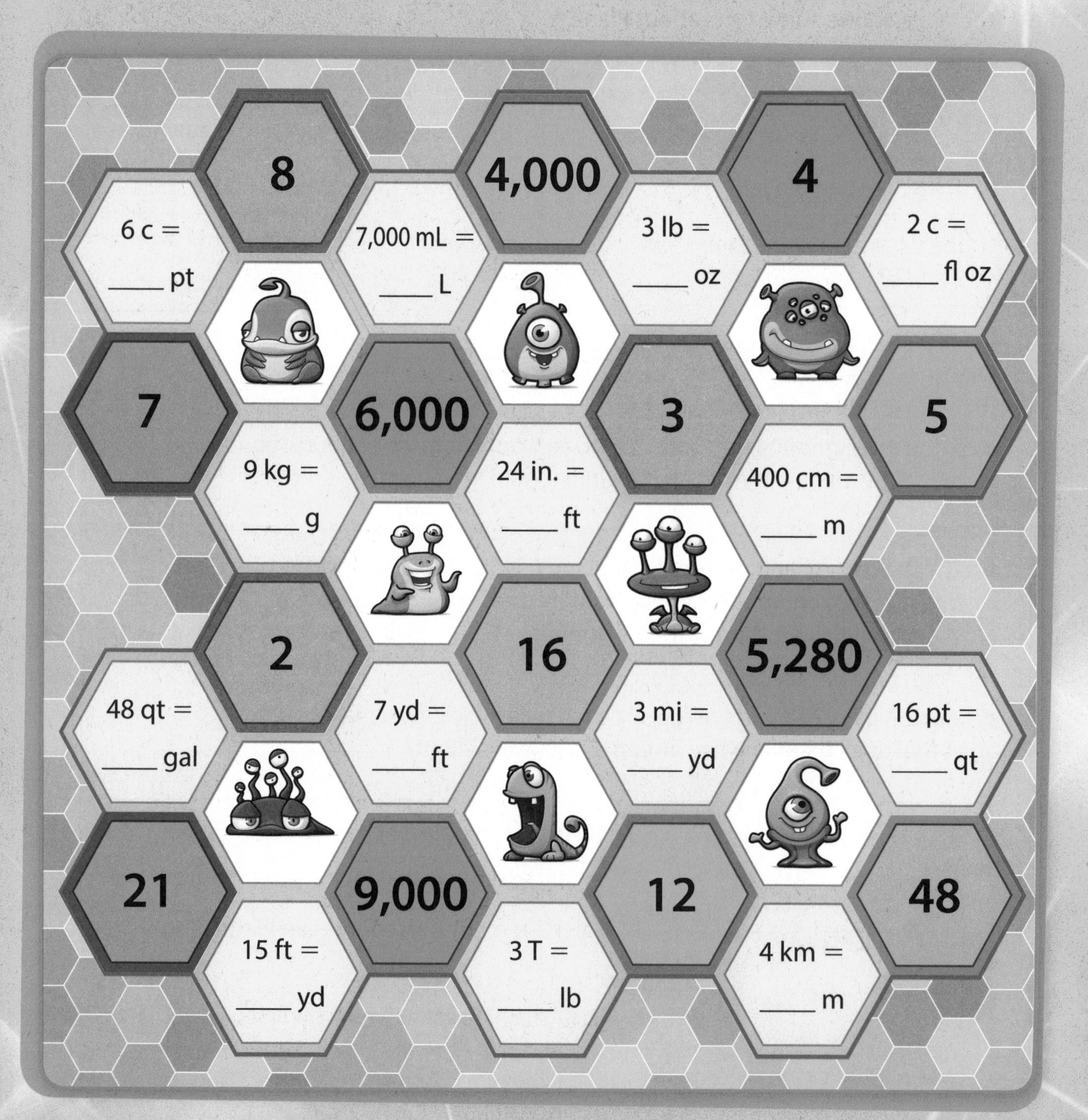

Name ______________________________

Chapter Practice 11

11.1 Length in Metric Units

Convert the length.

1. 4 cm = ______ mm

2. 81 m = ______ cm

3. 0.56 km = ______ cm

4. 9 mm = ______ m

Compare.

5. 73 m ◯ 7.3 km

6. 0.6 cm ◯ 0.06 m

7. 2 mm ◯ 0.2 cm

11.2 Mass and Capacity in Metric Units

Convert the mass.

8. 3 kg = ______ g

9. 0.006 g = ______ mg

10. 70 g = ______ kg

11. 29,000 mg = ______ kg

Convert the capacity.

12. 400 mL = _____ L

13. 10 L = _____ mL

14. 7 mL = _____ L

15. 0.65 L = _____ mL

11.3 Length in Customary Units

Convert the length.

16. 2 mi = _____ yd

17. $14\frac{2}{3}$ yd = _____ ft

18. 103 in. = _____ ft _____ in.

19. 2,340 in. = _____ yd

Compare.

20. $5\frac{2}{3}$ yd ◯ 17 ft

21. 67 in. ◯ 5 ft 10 in.

22. 16 mi ◯ 84,000 ft

11.4 Weight in Customary Units

Convert the weight.

23. $4\frac{1}{2}$ T = _____ lb

24. 100,000 lb = _____ T

Convert the weight.

25. 217 oz = ______ lb ______ oz

26. 956 oz = ______ lb

Compare.

27. $5\frac{1}{4}$ T ◯ 15,000 lb

28. 258 oz ◯ 17 lb 12 oz

29. 192,000 oz ◯ 7 T

30. MP **Number Sense** Which measurements are equivalent to 52 ounces?

3 lb 2 oz $\quad$ $3\frac{1}{4}$ lb $\quad$ 3 lb 4 oz $\quad$ $3\frac{1}{2}$ lb

11.5 Capacity in Customary Units

Convert the capacity.

31. 18 qt = ______ pt

32. $4\frac{3}{4}$ c = ______ fl oz

33. 72 pt = ______ gal

34. 81 qt = ______ gal

Compare.

35. $5\frac{1}{4}$ gal ◯ 21 qt

36. $3\frac{1}{2}$ pt ◯ 9 c

37. 4 qt ◯ 20 c

38. **Modeling Real Life** You have $2\frac{1}{4}$ gallons of apple juice. How many pints of apple juice do you have?

11.6 Make and Interpret Line Plots

39. The table shows the amounts of clay made by 10 students. Make a line plot to display the data.

Amounts of Clay (cup)				
$\frac{1}{2}$	$\frac{5}{8}$	$\frac{1}{2}$	$\frac{3}{4}$	$\frac{1}{2}$
$\frac{5}{8}$	$\frac{3}{4}$	$\frac{3}{4}$	$\frac{7}{8}$	$\frac{3}{4}$

Clay Amounts Made

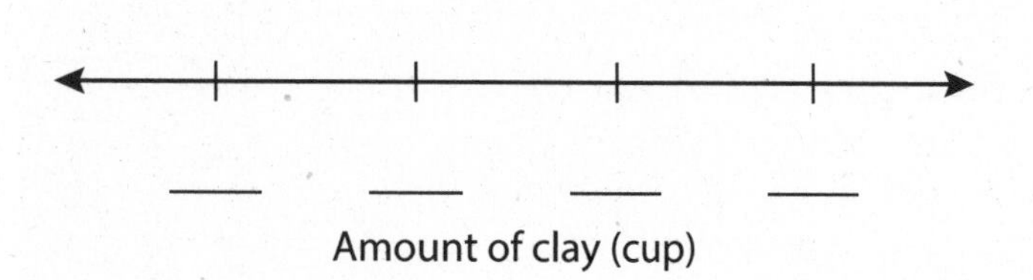

Amount of clay (cup)

How many students made more than $\frac{3}{4}$ cup of clay?

What is the most common amount of clay made?

11.7 Problem Solving: Measurement

40. A recipe calls for $2\frac{3}{4}$ cups of fava beans. You have a $1\frac{1}{4}$-pint can of fava beans and $\frac{1}{2}$ cup of cooked fava beans. Do you have enough fava beans for the recipe?

Ful mudammas, a popular dish in Egypt, contains fava beans.

Name ____________________

Cumulative Practice 1–11

1. Your friend estimates that a bookcase is $2\frac{1}{2}$ feet wide. The actual width is $\frac{2}{3}$ foot longer. What is the width of the bookcase?

Ⓐ $1\frac{1}{6}$ feet

Ⓑ $1\frac{5}{6}$ feet

Ⓒ $2\frac{3}{5}$ feet

Ⓓ $3\frac{1}{6}$ feet

---

2. What is the product of 845 and 237?

Ⓐ 10,140

Ⓑ 32,955

Ⓒ 200,265

Ⓓ 4,655,985

---

3. How many milliliters are equal to 0.6 liter?

Ⓐ 0.006 mL

Ⓑ 0.06 mL

Ⓒ 60 mL

Ⓓ 600 mL

---

4. Which are equivalent to $6 \times \frac{3}{10}$?

▢ $\frac{3}{10} \times \frac{3}{10} \times \frac{3}{10} \times \frac{3}{10} \times \frac{3}{10} \times \frac{3}{10}$

▢ $1\frac{4}{5}$

▢ $1\frac{8}{10}$

▢ $\frac{9}{10}$

▢ $18 \times \frac{1}{10}$

▢ $6 \times \frac{1}{3} \times \frac{1}{10}$

5. To find 34 + (16 + 23), your friend adds 34 and 16. Then she adds 23 to the sum. Which property did she use?

Ⓐ Addition Property of Zero

Ⓑ Associative Property of Addition

Ⓒ Commutative Property of Addition

Ⓓ Distributive Property

---

6. An Eastern Hognose Snake is $2\frac{1}{2}$ feet long. It grows another foot. What is the new length of the snake in inches?

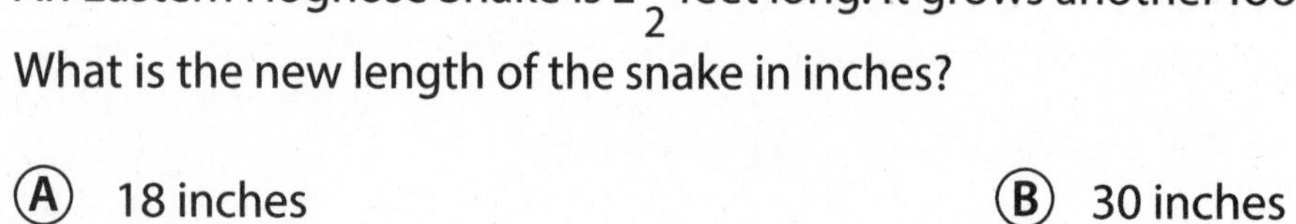

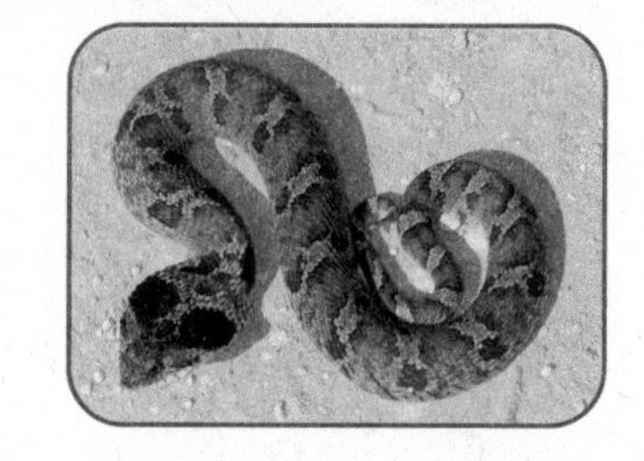

Ⓐ 18 inches

Ⓑ 30 inches

Ⓒ 36 inches

Ⓓ 42 inches

---

7. What common factor should you divide the numerator and denominator of $\frac{16}{24}$ by so that it is in simplest form?

Ⓐ 2

Ⓑ 4

Ⓒ 8

Ⓓ 16

---

8. A salesperson at a fabric store has 30 yards of fabric. He puts the same number of yards of fabric on each of 4 rolls for a display. How many yards of fabric does the salesperson put on each roll?

Ⓐ $\frac{4}{30}$ yard

Ⓑ $\frac{2}{15}$ yard

Ⓒ 7 yards

Ⓓ $7\frac{1}{2}$ yards

---

9. Descartes estimates 96.3 × 42 by rounding each number to the nearest ten. What is Descartes's estimate?

Ⓐ 400

Ⓑ 3,600

Ⓒ 3,840

Ⓓ 4,000

10. The fifth-grade classes are making a mural to hang in the front hallway of the school.

Think Solve Explain

Part A Each class creates a square for the mural that has side lengths of $\frac{1}{2}$ meter. What is the area of each square?

Part B There are 12 classes. What is the area of the entire mural? Explain.

---

11. Which expressions have a quotient of 4.6?

- ◯ 124.2 ÷ 27
- ◯ 55.2 ÷ 12
- ◯ 15.64 ÷ 34
- ◯ 82.8 ÷ 18

---

12. What is the quotient of 5 and $\frac{1}{8}$?

---

13. You need thirty 5-foot pieces of string for a project. A store sells string by the yard. How many yards of string will you need to buy?

Ⓐ 1 yard

Ⓑ 50 yards

Ⓒ 150 yards

Ⓓ 450 yards

14. In which equations does $k = \frac{3}{4}$?

☐ $k \times \frac{1}{5} = \frac{3}{20}$

☐ $\frac{5}{6} \times k = \frac{5}{8}$

☐ $\frac{2}{2} \times \frac{1}{2} = k$

☐ $\frac{3}{6} \times \frac{3}{2} = k$

---

15. Newton brings 3 bags of popcorn that are all the same size to a club. There are 12 people at the club. Each person eats the same amount of popcorn and all of the popcorn is eaten. What fraction of a bag of popcorn does each person eat?

Ⓐ $\frac{1}{4}$

Ⓑ $\frac{12}{3}$

Ⓒ 4

Ⓓ 36

---

16. Evaluate $25.4 - 16.31 + 4.59$.

Ⓐ 4.5

Ⓑ 13.68

Ⓒ 13.70

Ⓓ 15.70

---

17. What is the sum of $\frac{5}{6}$ and $\frac{1}{4}$?

Ⓐ $\frac{1}{4}$

Ⓑ $\frac{7}{12}$

Ⓒ $\frac{3}{5}$

Ⓓ $1\frac{1}{12}$

---

18. What is 40,071 written in word form?

Ⓐ four hundred seventy-one

Ⓑ four thousand, seventy-one

Ⓒ forty thousand, seventy-one

Ⓓ forty thousand, seven hundred ten

Name ________________________________

STEAM Performance Task 1-11

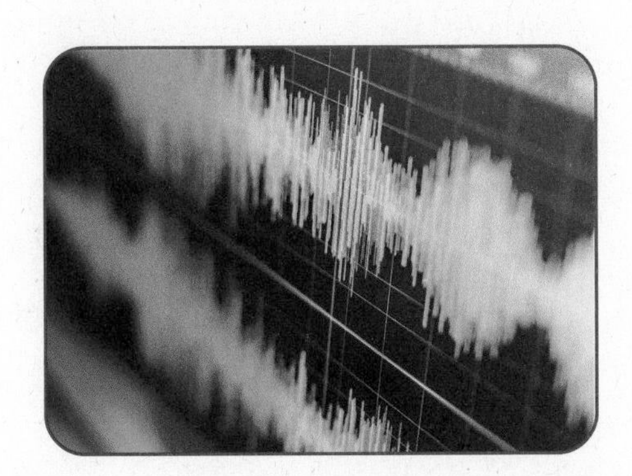

Sound is created from vibrations in the air called sound waves. In music, when you hear different pitches, it is because the sound waves are traveling at different speeds. The frequency of a pitch measures the number of sound waves per second. Higher pitches have higher frequencies, and lower pitches have lower frequencies. Frequencies are measured in Hertz.

1. A_4 is the musical note commonly used to tune instruments. The frequency of A_4 is 440 Hertz, because the sound vibrates 440 times per second.

a. The frequency of A_3 is $\frac{1}{2}$ the frequency of A_4. The frequency of A_2 is $\frac{1}{2}$ the frequency of A_3. What is the frequency of A_2?

b. Is the pitch of A_2 higher or lower than the pitch of A_4? Explain.

c. How can you use the frequency of A_4 to find the frequency of A_5? Explain.

d. The frequency of B_4 is 493.88 Hertz. What is the frequency of B_3?

e. A computer software program can correct the frequency of a sound so it has perfect pitch. A violin plays a note that has a frequency of 255.1 Hertz. Explain how to change the frequency so it has the pitch of B_3.

---

2. Use the Internet or some other resource to learn about how audio processors can help to correct a singer's pitch, or to alter the way a song sounds. Write one interesting thing you learn.

3. You borrow a guitar to learn how to play. Use the table to decide which guitar you should borrow.

	Guitar A	Guitar B
Length	37 inches	40 inches
Weight	$7\frac{5}{16}$ pounds	176 ounces

a. Based on your height, you need a guitar that is close to 1 yard long. Which guitar is closer?

b. You also need a guitar that is close to 8 pounds. Which guitar do you think you should borrow? Explain.

c. Guitar B is called a full-size guitar and Guitar A is called a $\frac{7}{8}$-size guitar. Is the length of Guitar A $\frac{7}{8}$ the length of Guitar B? Explain.

d. The scale length on a guitar affects the pitch. To find the scale length of a guitar, multiply the distance between the nut and the 12th fret by 2. On your guitar, that distance is $12\frac{3}{4}$ inches. What is the scale length of your guitar?

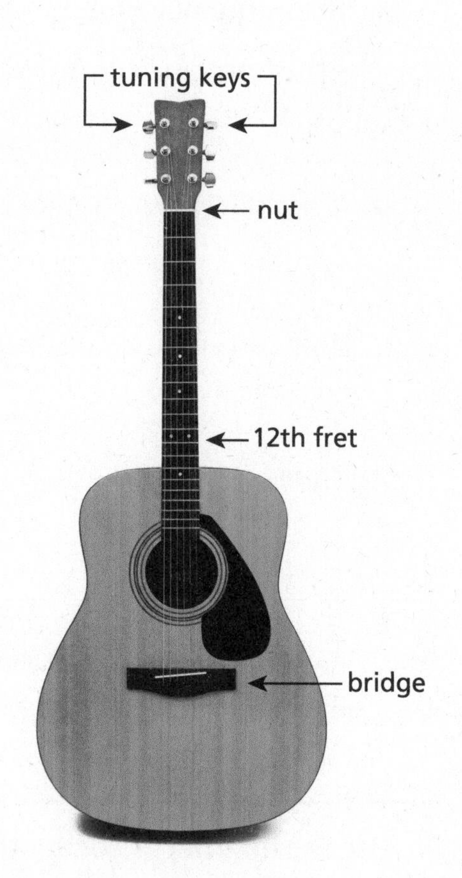

e. The strings on your guitar are $\frac{3}{8}$ inch longer than the scale length to allow you to tune the strings to the correct pitch. What are the string lengths on your guitar?

f. When you tune a string, you adjust it tighter to make the pitch higher, or looser to make the pitch lower. You use a tuning instrument to help you tune your guitar. It says that your A_4 string has a frequency of 436.2 Hertz. How should you adjust the string to get the pitch in tune?

12

Patterns in the Coordinate Plane

- Have you ever seen an animated video?
- You create a computer animation. How can you use a coordinate plane to move an image to different locations on the screen?

Chapter Learning Target:
Understand patterns and the coordinate plane.

Chapter Success Criteria:
- I can identify patterns.
- I can plot points in a coordinate plane.
- I can analyze line graphs.
- I can interpret relationships.

Name ______________________________

12 Vocabulary

Review Words

endpoints
line
line segment
points

Organize It

Use the review words to complete the graphic organizer.

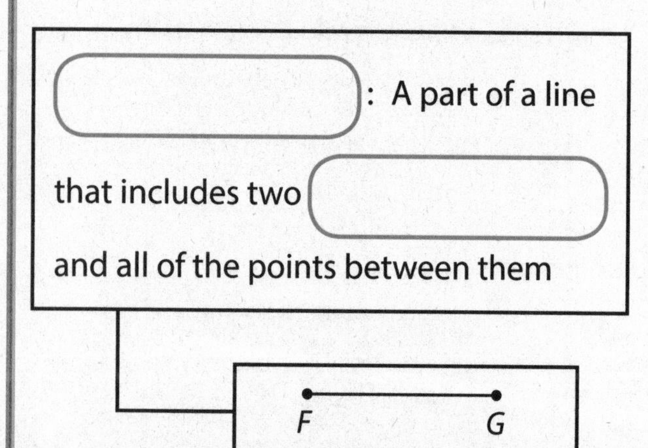

: A part of a line
that includes two
and all of the points between them
F
G

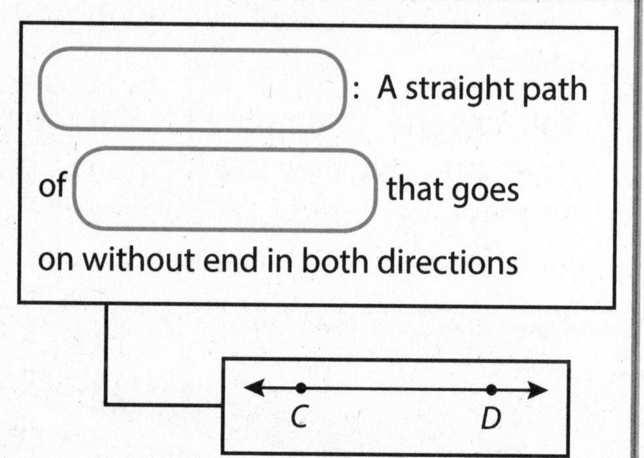

: A straight path
of
that goes
on without end in both directions
C
D

Define It

Use your vocabulary cards to identify the word. Find the word in the word search.

1. The vertical number line in a coordinate plane
2. The point, represented by the ordered pair (0, 0), where the x-axis and the y-axis intersect in a coordinate plane
3. A pair of numbers that is used to locate a point in a coordinate plane
4. The horizontal number line in a coordinate plane

O	I	Q	D	M	Y	J	E	H	R	P
F	N	U	B	S	A	G	W	C	X	S
O	R	D	E	R	E	D	P	A	I	R
I	P	A	X	L	A	S	Q	K	N	E
D	C	H	I	N	V	I	U	G	I	Y
X	M	E	S	R	T	X	F	X	T	O
A	R	G	L	Z	S	A	N	D	J	R
N	O	M	H	K	U	Y	X	T	A	I
S	T	I	P	N	G	S	L	W	P	G
Y	X	J	U	E	T	B	I	O	N	I
R	G	D	X	A	X	I	S	A	S	N

Chapter 12 Vocabulary Cards

coordinate plane

data

line graph

ordered pair

origin

x-axis

x-coordinate

y-axis

Values collected from observations or measurements

Day	1	2	3	4	5
Packages Delivered	128	154	137	168	193

A plane that is formed by the intersection of a horizontal number line and a vertical number line

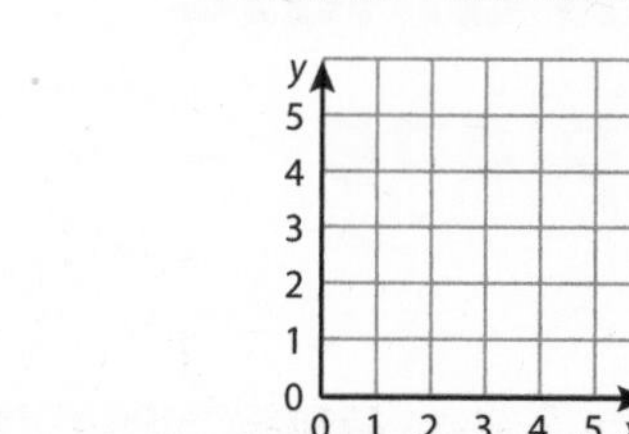

A pair of numbers that is used to locate a point in a coordinate plane

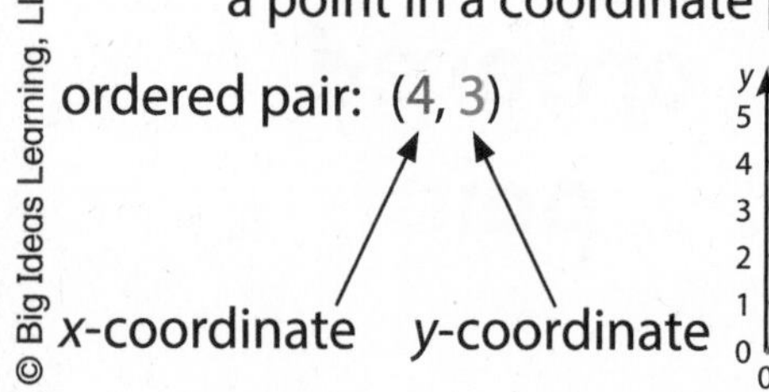

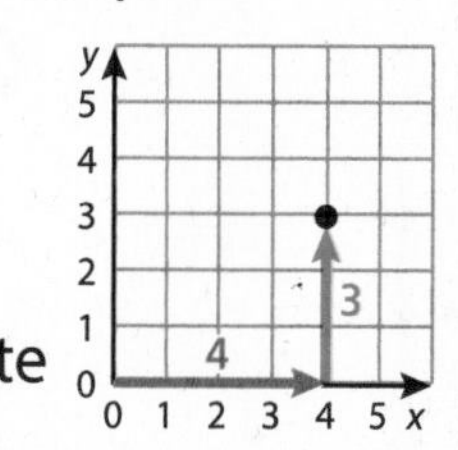

A graph that uses line segments to show how data values change over time

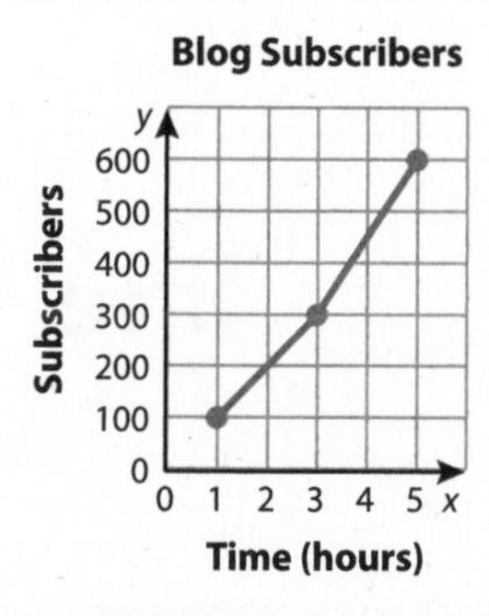

The horizontal number line in a coordinate plane

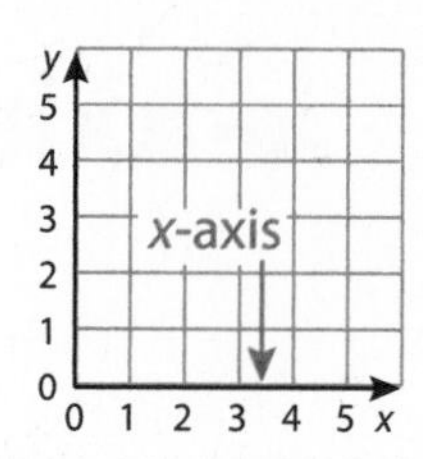

The point, represented by the ordered pair (0, 0), where the *x*-axis and the *y*-axis intersect in a coordinate plane

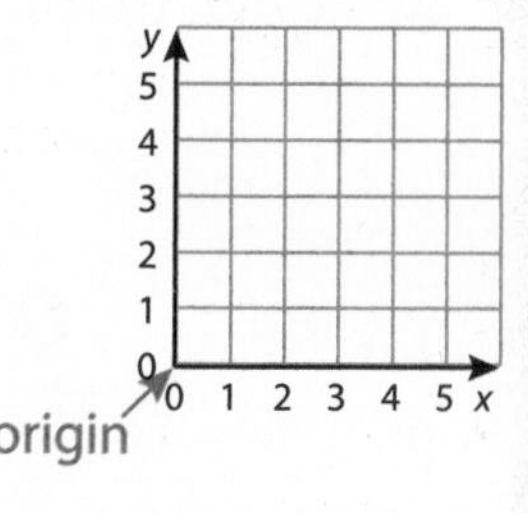

The vertical number line in a coordinate plane

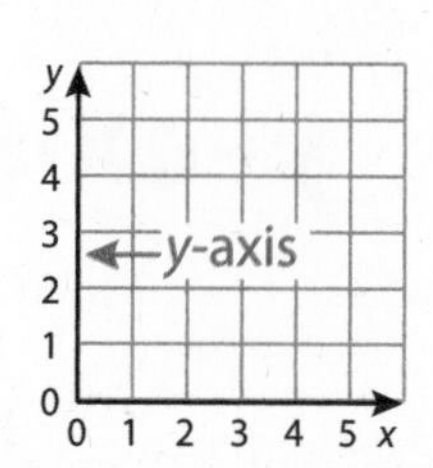

The first number in an ordered pair, which gives the horizontal distance from the origin along the *x*-axis

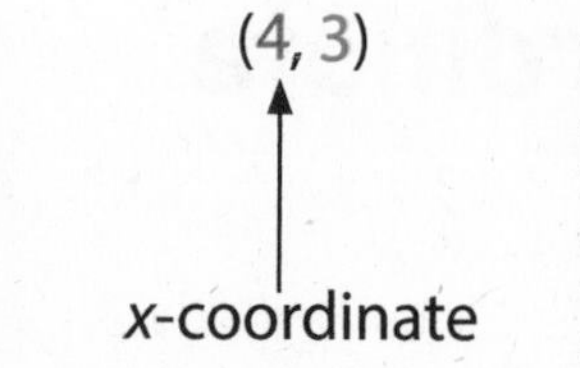

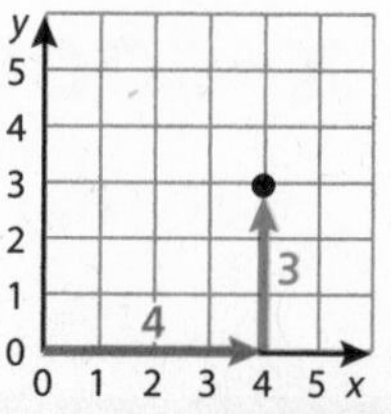

Chapter 12 Vocabulary Cards

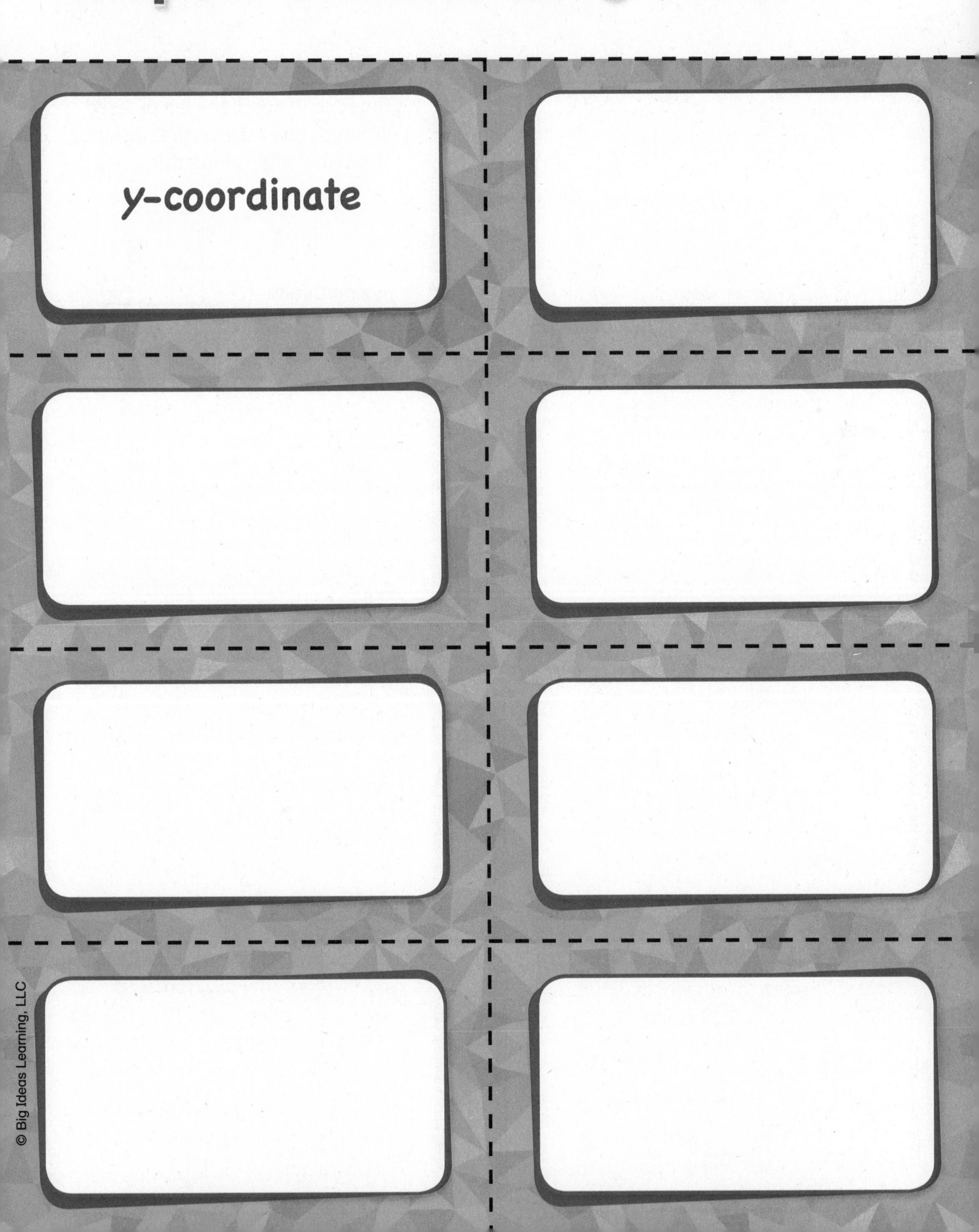

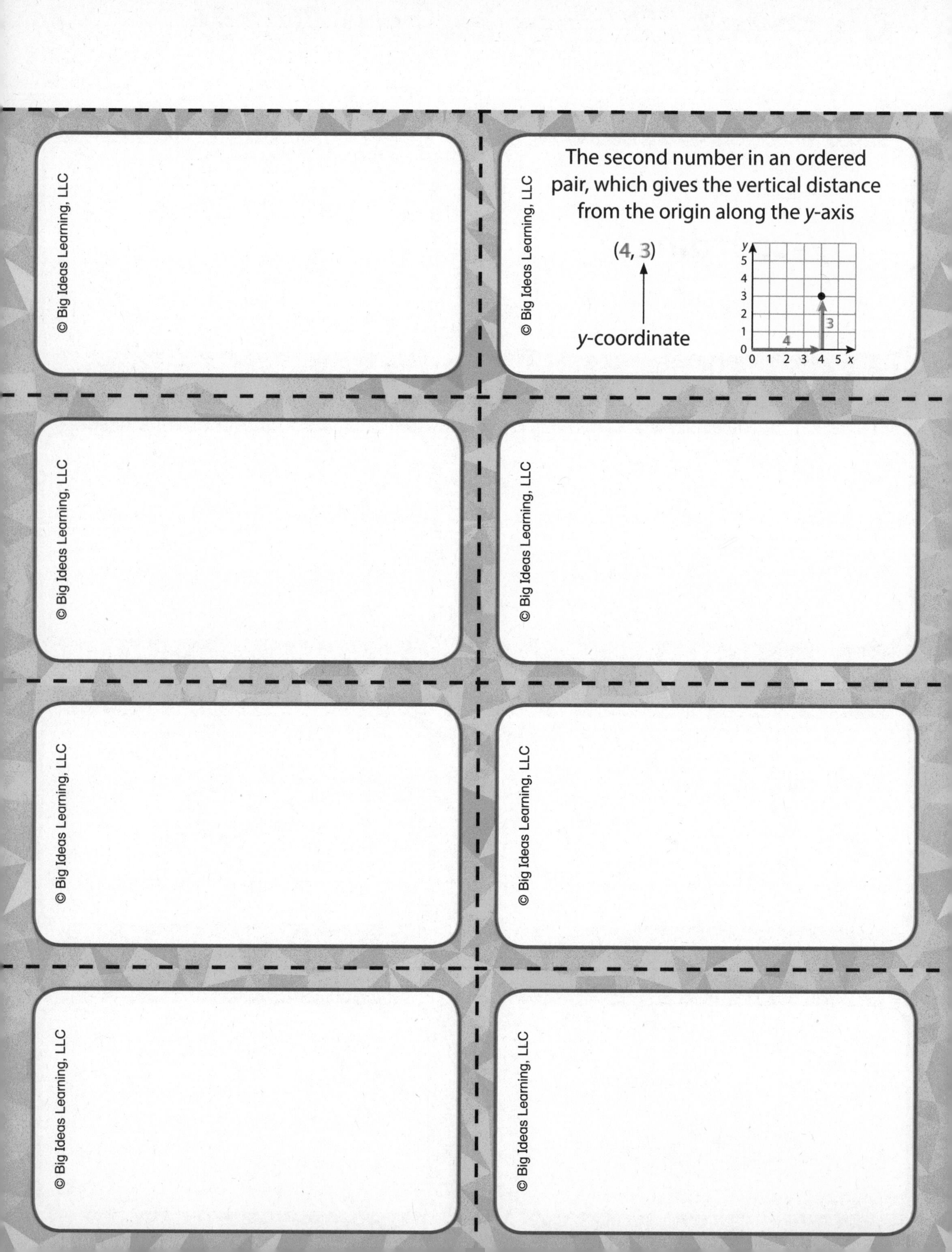

The second number in an ordered pair, which gives the vertical distance from the origin along the y-axis
(4, 3)
y-coordinate

Name ____________________

Plot Points in a Coordinate Plane 12.1

Learning Target: Identify and plot points in a coordinate plane.

Success Criteria:
- I can use an ordered pair to identify the location of a point in a coordinate plane.
- I can plot and label a point in a coordinate plane.

Explore and Grow

Choose a location for your buried treasure on the *My Treasure* grid. Choose a point where two grid lines intersect. An example is shown.

Take turns with a partner guessing the location of each other's buried treasure. Keep track of your guesses on the *My Guesses* grid. After each guess, give a clue to help your partner, such as "my treasure is northwest of your guess." Continue to guess until a treasure is located.

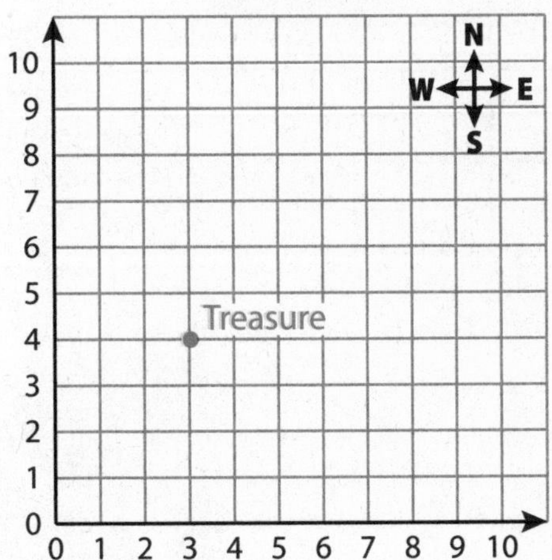

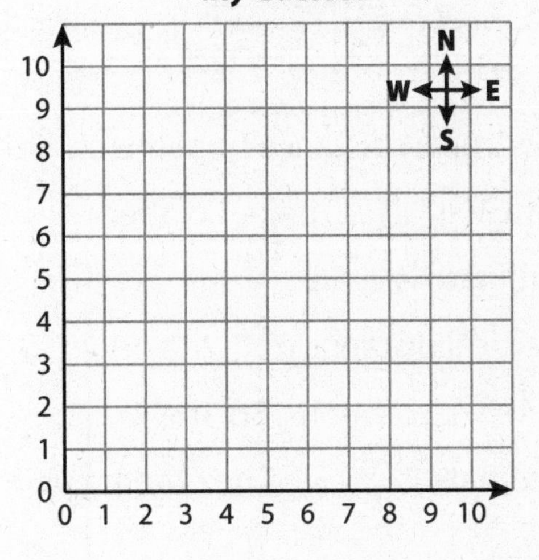

MP **Reasoning** The point where the horizontal number line and the vertical number line intersect is called the *origin*. Why do you think it is called that?

Think and Grow: The Coordinate Plane

Key Idea A **coordinate plane** is formed by the intersection of a horizontal number line and a vertical number line. An **ordered pair** is a pair of numbers that is used to locate a point in a coordinate plane.

ordered pair: (4, 6)

The first number, the **x-coordinate**, gives the horizontal distance from the origin along the *x*-axis.

The second number, the **y-coordinate**, gives the vertical distance from the origin along the *y*-axis.

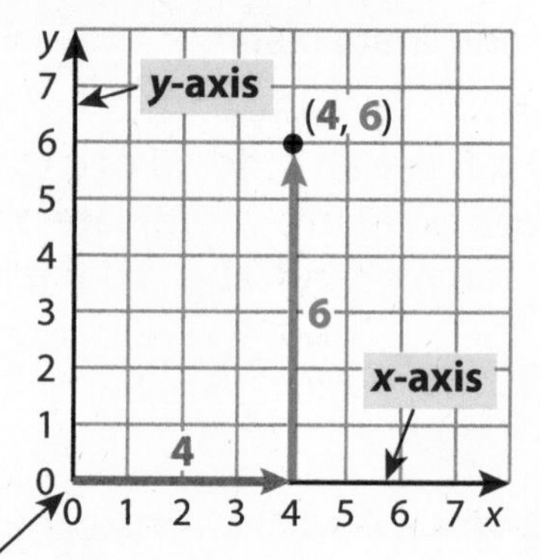

The axes intersect at the **origin**, (0, 0).

Example Write the ordered pair that corresponds to point *M*.

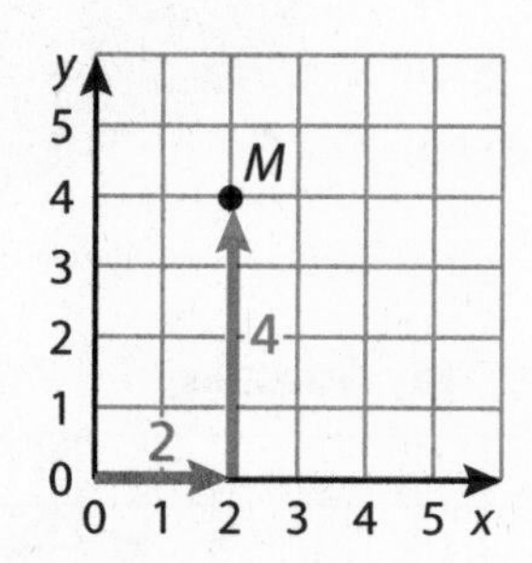

The horizontal distance from the origin to point *M* is _____ units. So, the *x*-coordinate is _____.

The vertical distance from the origin to point *M* is _____ units. So, the *y*-coordinate is _____. The ordered pair is _____.

Example Point *Y* is located at (5, 3). Plot and label the point.

Start at the origin. Move _____ units right and _____ units up. Then plot and label the point.

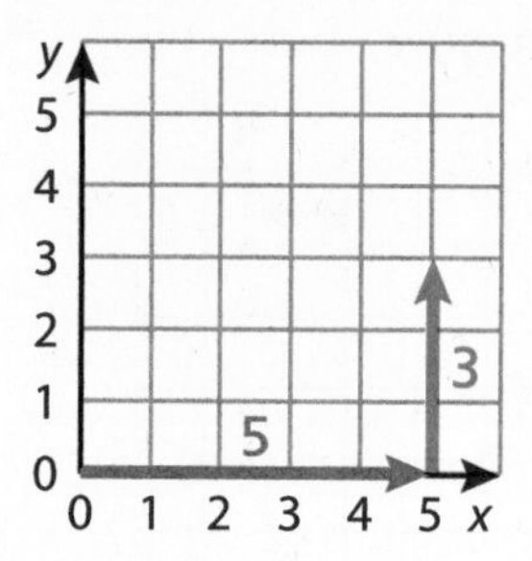

The point can be labeled as *Y*, (5, 3), or *Y*(5, 3).

Show and Grow *I can do it!*

Write the ordered pair corresponding to the point.

1. Point *B*

2. Point *S*

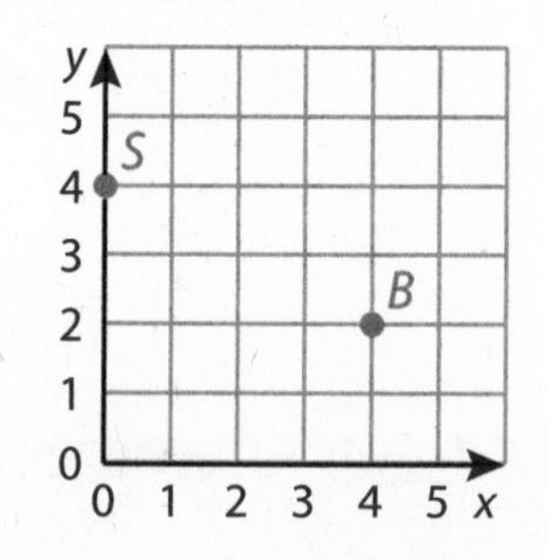

Plot and label the point in the coordinate plane.

3. *F*(5, 4)

4. *P*(3, 0)

Name ______________________________

Apply and Grow: Practice

Use the coordinate plane to write the ordered pair corresponding to the point.

5. Point M

6. Point Q

7. Point N

8. Point R

9. Point P

10. Point T

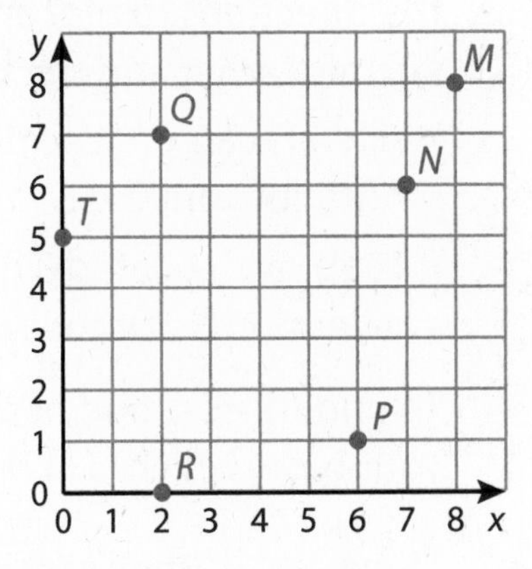

Plot and label the point in the coordinate plane above.

11. $S(0, 3)$

12. $F(2, 5)$

13. $W(0, 0)$

Name the point for the ordered pair.

14. (5, 2)

15. (8, 4)

16. (0, 3)

17. MP **Reasoning** How are the locations of the points $A(0, 4)$ and $B(4, 0)$ different in a coordinate plane?

18. DIG DEEPER! Newton buries a bone in a park at the location shown. How can he use a coordinate plane to describe its location?

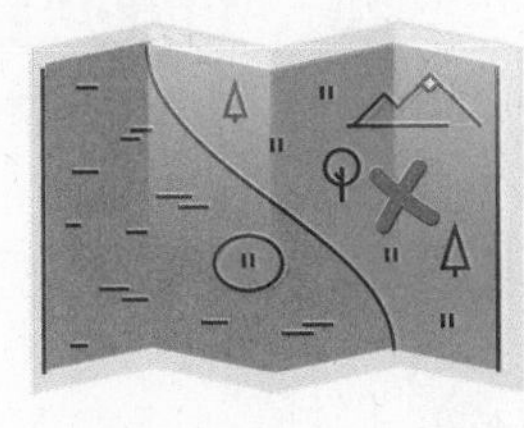

Think and Grow: Modeling Real Life

Example In a video game, you move an aircraft carrier and a tugboat away from your base. Use the directions to plot and label the locations of the aircraft carrier and the tugboat.

- Aircraft carrier: Located 3 miles east and 4 miles north of your base.
- Tugboat: Located 8 miles east and twice as many miles north of your base as the aircraft carrier.

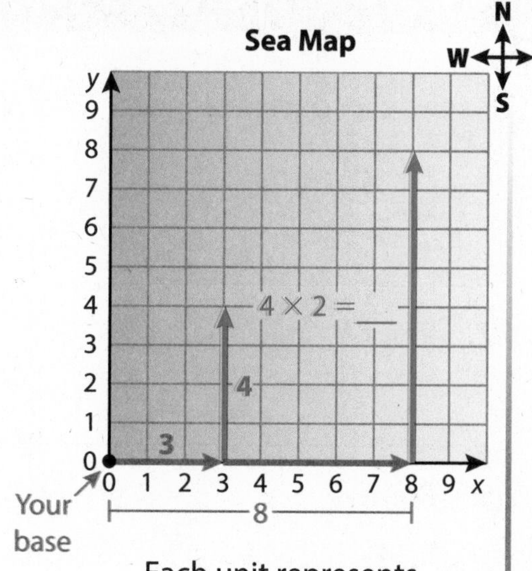

Each unit represents 1 mile.

To find the location of the aircraft carrier, start at your base, which is at the origin.

Move _____ units east, or right, and _____ units north, or up.

Plot and label the point as A(_____, _____).

To find the location of the tugboat, start at your base, which is at the origin.

Move _____ units east, or right, and _____ × _____ = _____ units north, or up.

Plot the label the point as T(_____, _____).

Show and Grow *I can think deeper!*

19. A guidebook describes how to get to various statues in Chicago, Illinois, from Willis Tower. Plot and label the location of each statue on the map.

- Dubuffet's *Monument with Standing Beast*: Walk 2 blocks east and 5 blocks north.
- Miró's *Sun, Moon, and One Star*: Walk twice as many blocks east as you do to get to the *Standing Beast*, and 3 blocks north.

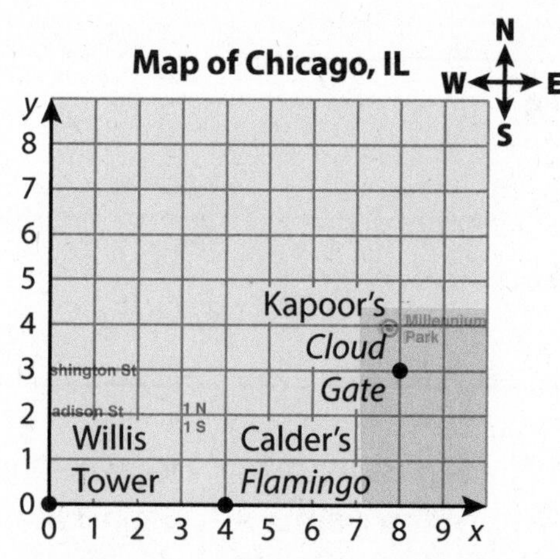

Each unit represents 1 city block.

____

20. **DIG DEEPER!** Which statue is closer to *Sun, Moon, and One Star*, *Cloud Gate* or *Flamingo*? Explain.

Name ______________________

Homework & Practice 12.1

Learning Target: Identify and plot points in a coordinate plane.

Example Write the ordered pair that corresponds to point *A*.

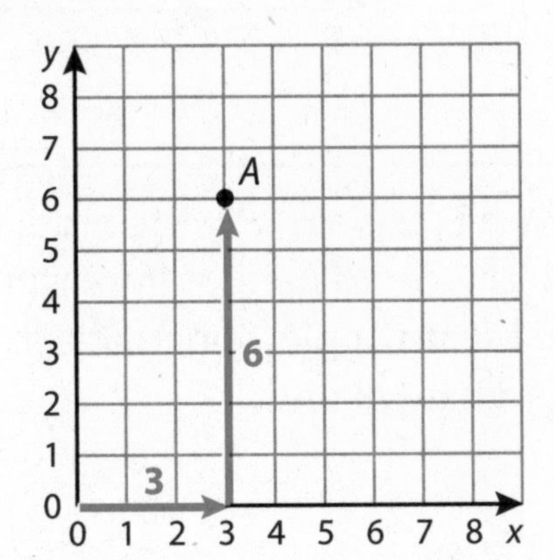

The horizontal distance from the origin to point *A* is 3 units. So, the *x*-coordinate is 3.

The vertical distance from the origin to point *A* is 6 units. So, the *y*-coordinate is 6. The ordered pair is (3, 6).

Example Point *C* is located at (7, 4). Plot and label the point.

Start at the origin. Move 7 units right and 4 units up. Then plot and label the point.

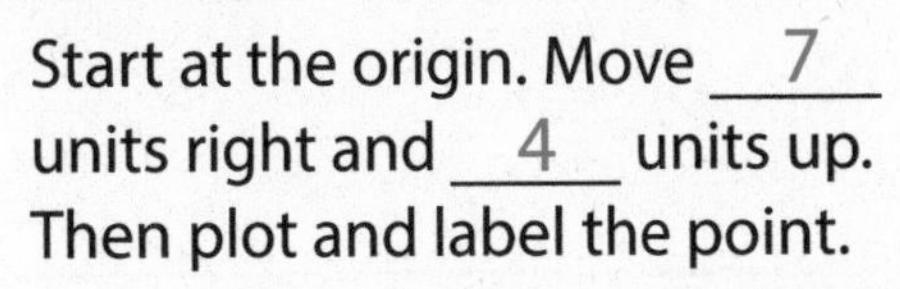

Use the coordinate plane to write the ordered pair corresponding to the point.

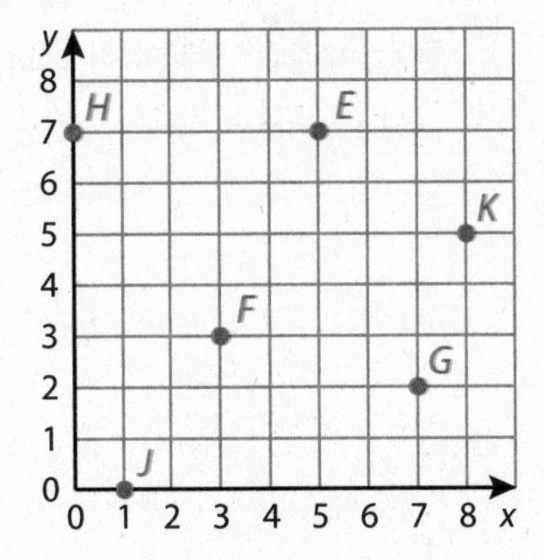

1. Point *E*

2. Point *H*

3. Point *F*

4. Point *J*

5. Point *G*

6. Point *K*

Plot and label the point in the coordinate plane above.

7. *Z*(8, 0)

8. *B*(5, 5)

9. *M*(1, 2)

Name the point for the ordered pair.

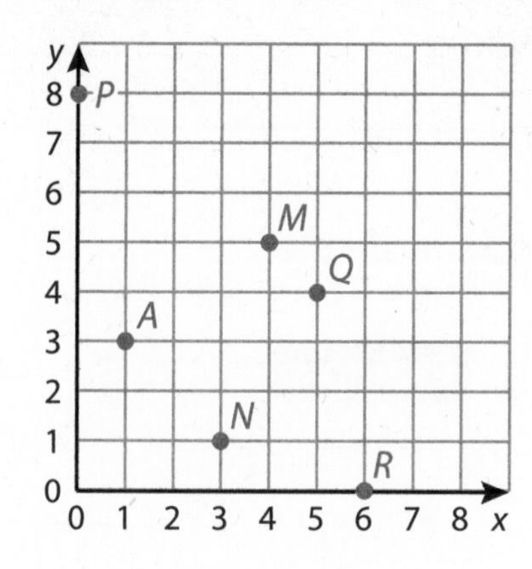

10. (5, 4)

11. (0, 8)

12. (3, 1)

13. **Open-Ended** Use the coordinate plane above. Point T is 3 units from point M. Name two possible ordered pairs for point T.

14. **Writing** Explain why the order of the x- and y-coordinates is important when identifying or plotting points in a coordinate plane.

15. To get from the school to the arcade, you walk 4 blocks east and 3 blocks north. To get from the school to the skate park, your friend walks 2 blocks east and twice as many blocks north as you. Plot and label the locations of the arcade and the skate park.

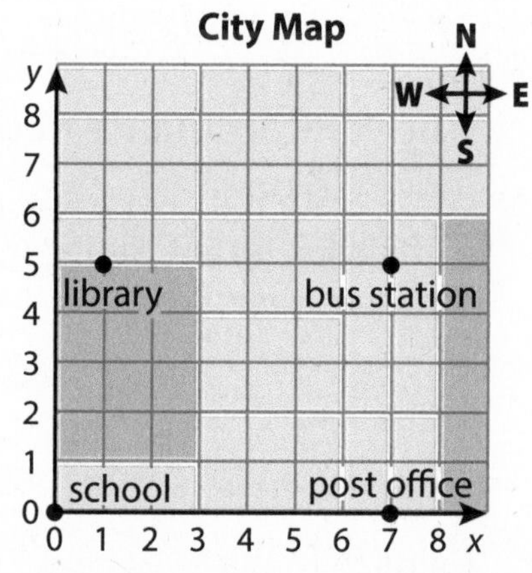

Each unit represents 1 city block.

16. DIG DEEPER! Which building is closer to the bus station, the library or the post office? Explain.

Review & Refresh

Multiply.

17. $1\frac{1}{4} \times 1\frac{1}{3} =$ _____

18. $1\frac{2}{5} \times 2\frac{1}{2} =$ _____

19. $2\frac{5}{6} \times 3\frac{7}{8} =$ _____

Name ____________________

Relate Points in a Coordinate Plane

12.2

Learning Target: Relate points and find distances in a coordinate plane.

Success Criteria:

- I can explain the relationship between two points that have the same *x*-coordinates or *y*-coordinates.
- I can count grid lines to find the distance between two points.
- I can use subtraction to find the distance between two points.

Explore and Grow

Plot and label the points in the coordinate plane.

A(4, 2) *B*(1, 1)

C(10, 2) *D*(9, 6)

E(0, 6) *F*(5 , 7)

G(1, 9) *H*(5, 10)

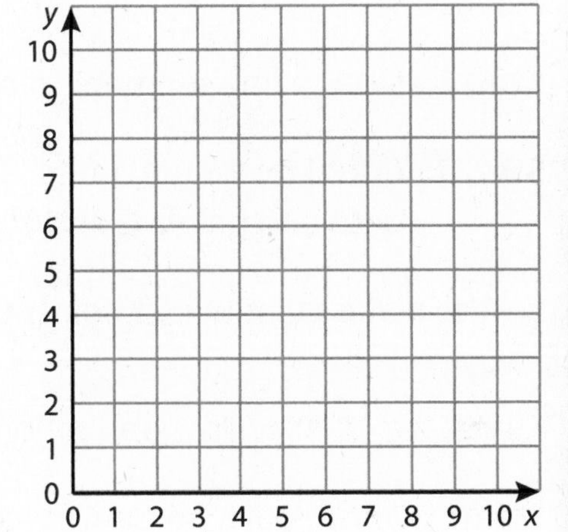

Draw a line segment to connect each pair of points.

A and *C* *B* and *G*

D and *E* *F* and *H*

Plot and label more points that lie on the line segments you drew. What do you notice about the coordinates?

MP **Construct Arguments** How can you find the distance between each pair of points? Explain your reasoning.

Think and Grow: Relate Points in a Coordinate Plane

Key Idea Points on a horizontal line have the same y-coordinates. Points on a vertical line have the same x-coordinates.

You can count units or use subtraction to find the distance between two points when they lie on the same horizontal line or vertical line.

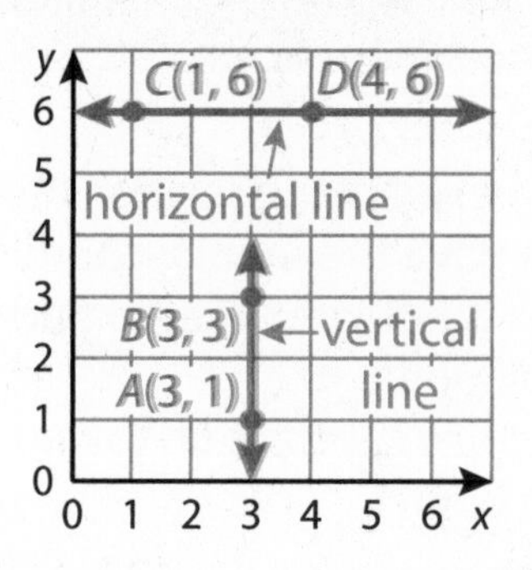

Example Find the distance between points G and H.

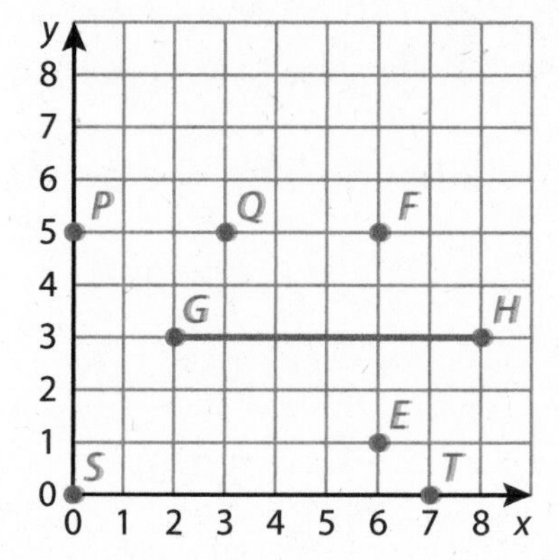

One Way: Count units.

Step 1: Identify the locations of the points: Point G is located at (2, 3). Point H is located at (8, 3).

Step 2: Draw a line segment to connect the points.

Step 3: Count horizontal units: There are ______ units between points G and H.

So, the distance between points G and H is ______.

Another Way: Use subtraction.

Points G and H have the same y-coordinates. They lie on a horizontal line. Subtract the x-coordinates to find the distance.

$8 - 2 =$ ______

So, the distance between points G and H is ______.

Show and Grow I can do it!

Find the distance between the points in the coordinate plane above.

1. E and F

2. P and Q

3. S and T

Name ______________________________

Apply and Grow: Practice

Find the distance between the points in the coordinate plane.

4. E and F

5. 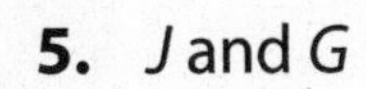 J and G

6. F and K

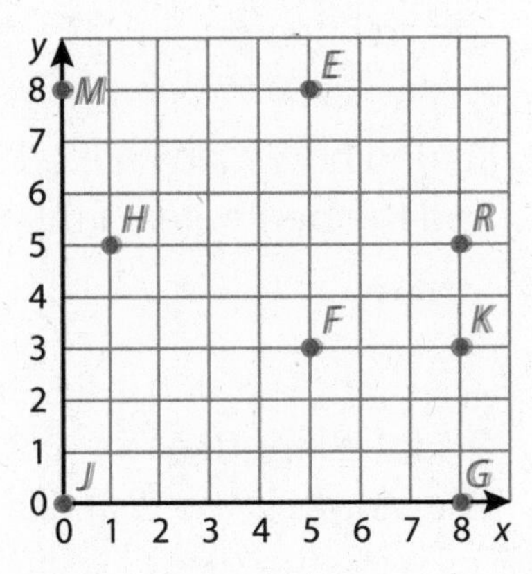

7. Which is longer, $\overline{JM}$ or $\overline{HR}$?

Find the distance between the points.

8. (1, 7) and (7, 7)

9. (0, 1) and (3, 1)

10. (0, 0) and (6, 0)

A line passes through the given points. Name two other points that lie on the line.

11. (0, 6) and (5, 6)

12. (4, 2) and (4, 8)

13. (3, 3) and (3, 6)

14. **YOU BE THE TEACHER** Newton plots the points A(2, 7) and B(6, 7) and connects them with a line segment. Descartes says that (10, 7) also lies on the line segment. Is he correct? Explain.

15. **DIG DEEPER!** Which pair of points does *not* lie on a line that is parallel to the x-axis? Explain.

(3, 1) and (7, 1) (8, 4) and (2, 4)

(1, 2) and (1, 6) (0, 5) and (6, 5)

Think and Grow: Modeling Real Life

Example An archaeologist uses rope to section off a rectangular dig site. How many meters of rope does the archaeologist use?

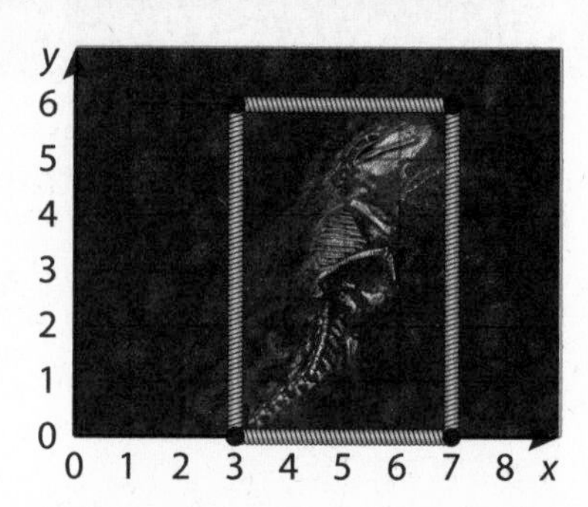

Each unit represents 1 meter.

To find how many meters of rope the archaeologist uses, find the perimeter of the rectangular dig site.

Find the length of the site.
The length is the distance between the points (_____, _____) and (_____, _____).

ℓ = _____ − _____ = _____ meters

Find the width of the site.
The width is the distance between the points (_____, _____) and (_____, _____).

w = _____ − _____ = _____ meters

Use a formula to find the perimeter of the site.

$P = (2 \times \ell) + (2 \times w)$ = (2 × _____) + (2 × _____)

= _____ + _____

= _____

So, the archaeologist uses _____ meters of rope.

Show and Grow *I can think deeper!*

16. The owner of an animal shelter uses fencing to create a rectangular dog pen. How many yards of fencing does the owner use?

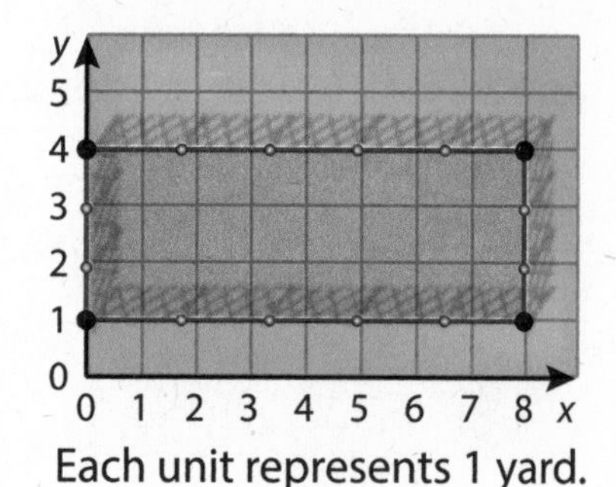

Each unit represents 1 yard.

17. **DIG DEEPER!** You run 5 laps around the edges of the volleyball court. How far do you run in feet? in yards?

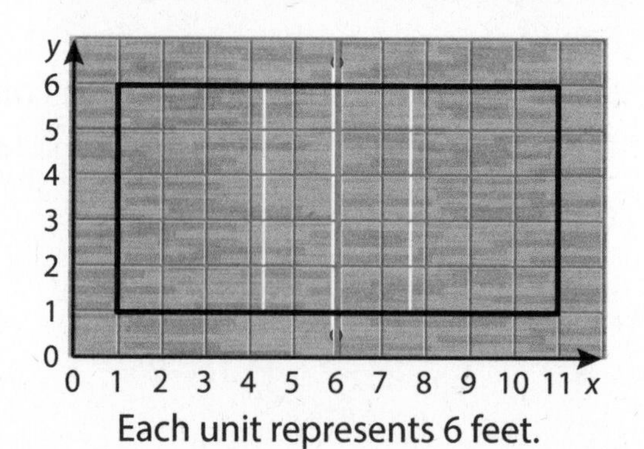

Each unit represents 6 feet.

Name ______________________________

Homework & Practice 12.2

Learning Target: Relate points and find distances in a coordinate plane.

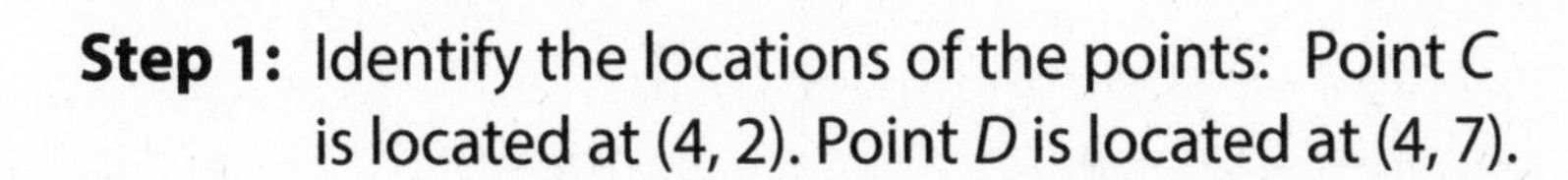

Example Find the distance between points C and D.

One Way: Count units.

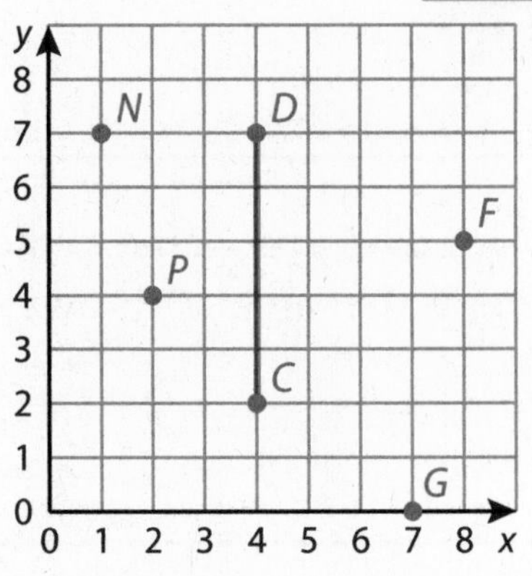

Step 1: Identify the locations of the points: Point C is located at (4, 2). Point D is located at (4, 7).

Step 2: Draw a line segment to connect the points.

Step 3: Count vertical units: There are 5 units between points C and D.

So, the distance between points C and D is 5 units.

Another Way: Use subtraction.

Points C and D have the same x-coordinates. They lie on a vertical line. Subtract the y-coordinates to find the distance.

$7 - 2 = \underline{5}$

So, the distance between points C and D is 5 units.

Find the distance between the points in the coordinate plane.

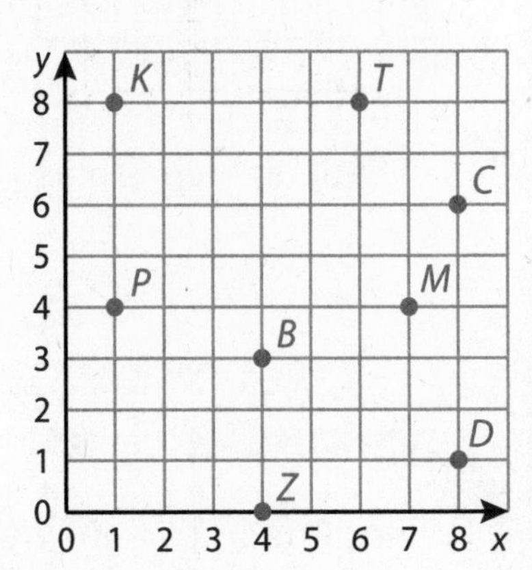

1. P and M

2. B and Z

3. K and T

4. Which is longer, $\overline{CD}$ or $\overline{KP}$?

Find the distance between the points.

5. (1, 5) and (6, 5)

6. (3, 4) and (3, 6)

7. (0, 2) and (0, 9)

A line passes through the given points. Name two other points that lie on the line.

8. (6, 0) and (6, 7)

9. (5, 3) and (1, 3)

10. (2, 2) and (2, 9)

11. MP **Structure** Name four different points that are 3 units away from (5, 4).

12. MP **Number Sense** Which point is farther from (3, 4)? Explain.

J(3, 9) *K*(9, 4)

13. Modeling Real Life A farmer builds a coop for his chickens. He uses poultry netting to enclose the coop. How many feet of netting does he use?

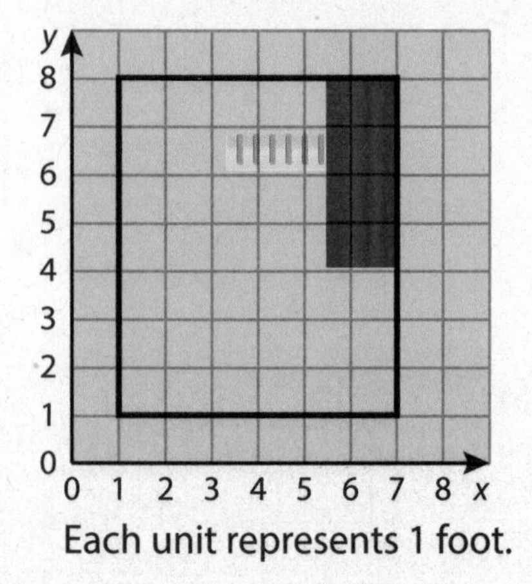

Each unit represents 1 foot.

14. Modeling Real Life A giant chessboard is painted on the ground in a park. How many square yards of space does the chessboard occupy?

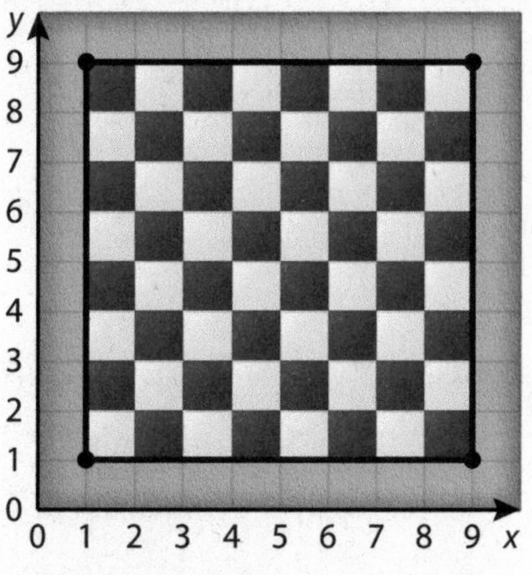

Each unit represents 1 yard.

Review & Refresh

Find the quotient. Then check your answer.

15. 23.6 ÷ 4 = _____

16. 36.9 ÷ 3 = _____

17. 114.87 ÷ 7 = _____

Name ______________________________

Draw Polygons in a Coordinate Plane 12.3

Learning Target: Draw and identify polygons in a coordinate plane.

Success Criteria:

- I can draw polygons in a coordinate plane.
- I can identify polygons in a coordinate plane.
- I can draw a symmetric shape in a coordinate plane given one half of the shape and a line of symmetry.

Explore and Grow

Plot and label three points in which two of the ordered pairs have the same *x*-coordinates and two of the ordered pairs have the same *y*-coordinates.

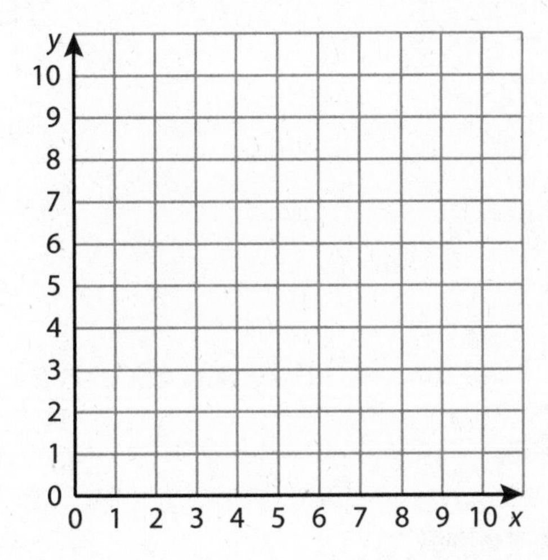

The points represent the vertices of a polygon. Describe the polygon.

Structure Explain how you can plot another point above to form a rectangle.

Think and Grow: Draw Polygons in a Coordinate Plane

Key Idea You can use ordered pairs to represent vertices of polygons. To draw a polygon in a coordinate plane, plot and connect the vertices.

Example The vertices of a polygon are *A*(2, 2), *B*(3, 5), *C*(6, 6), and *D*(6, 2). Draw the polygon in a coordinate plane. Then identify it.

Step 1: Plot and label the vertices.

Step 2: Draw line segments to connect the points.

Be sure to connect the points in order to draw the polygon.

When naming a polygon, be sure to write the vertices in order.

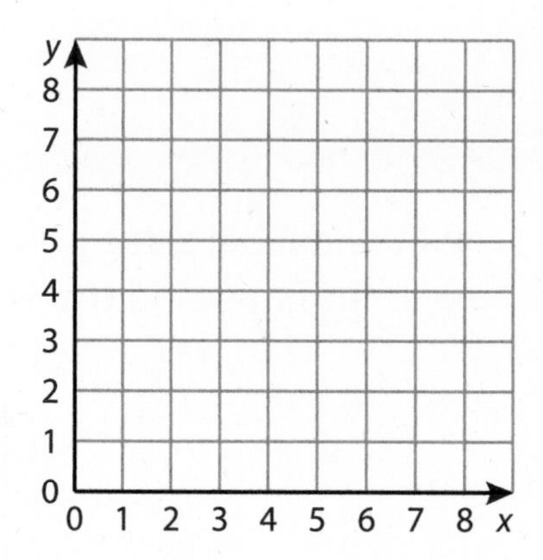

Polygon *ABCD* is a ______________.

Show and Grow I can do it!

Draw the polygon with the given vertices in a coordinate plane. Then identify it.

1. *J*(0, 8), *K*(4, 7), *L*(5, 0)

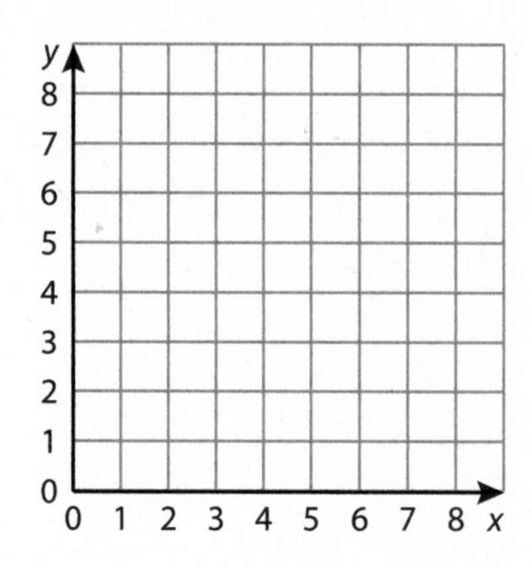

2. *P*(1, 4), *Q*(2, 7), *R*(6, 7), *S*(7, 4), *T*(4, 1)

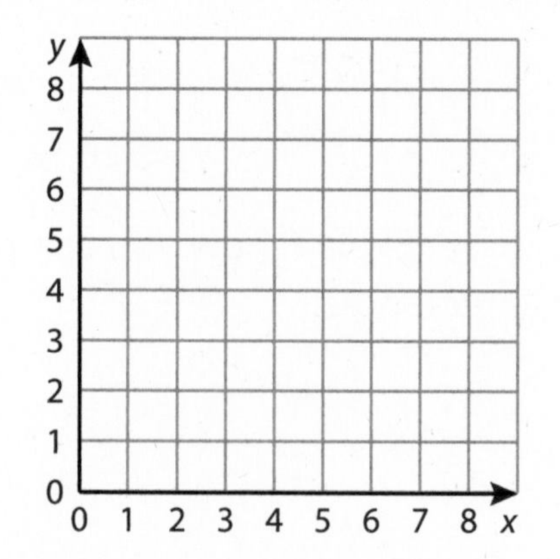

Name ___________________________________

Apply and Grow: Practice

Draw the polygon with the given vertices in a coordinate plane. Then identify it.

3. $C(1, 6), D(4, 6), E(4, 1), F(1, 1)$

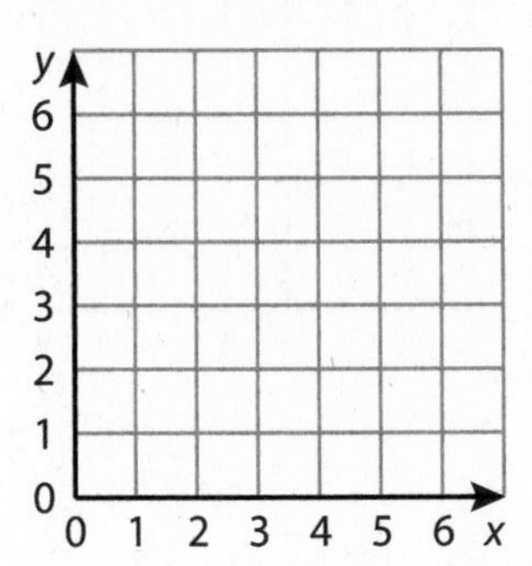

4. $J(2, 2), K(2, 4), L(4, 6), M(6, 4), N(6, 2), P(4, 1)$

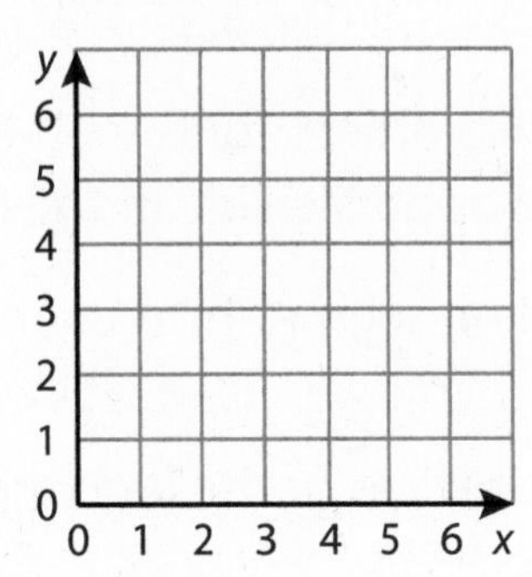

Identify the polygon with the given vertices.

5. $A(2, 6), B(6, 2), C(3, 2)$

6. $G(0, 3), H(6, 3), J(4, 1), K(2, 1)$

7. $P(1, 1), Q(1, 6), R(6, 6), S(6, 1)$

8. $X(0, 0), Y(0, 7), Z(2, 0)$

Plot (6, 3), (6, 8), and (9, 3) in a coordinate plane. Plot another point to form the given quadrilateral. Name the point.

9. rectangle

10. trapezoid

11. Open-Ended Write four ordered pairs that represent the vertices of a square.

12. YOU BE THE TEACHER Your friend draws the polygon shown. She names the polygon *EFDC*. Is your friend correct? Explain.

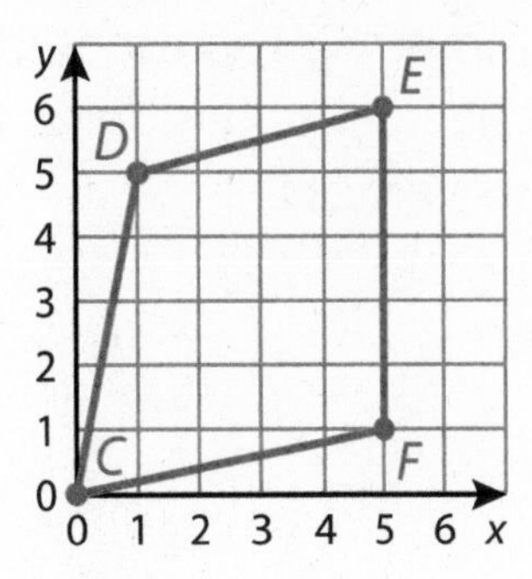

Think and Grow: Modeling Real Life

Example You and a friend use computer software to create a symmetric company logo using a coordinate plane. Your friend completes one half of the logo as shown. Draw the other half. Then list the vertices of the logo.

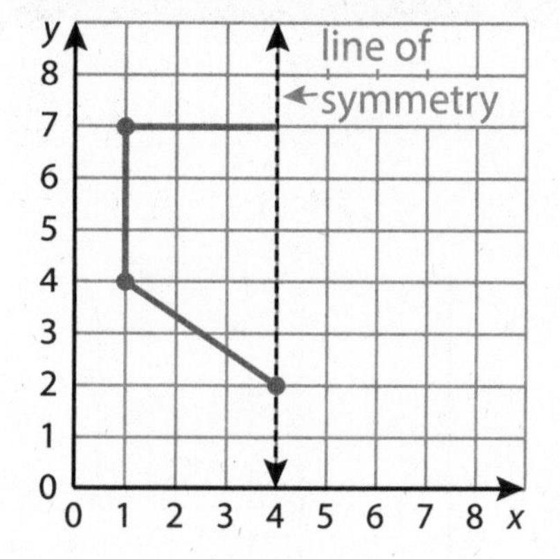

Step 1: Plot the vertices for the other half of the logo on the opposite side of the line of symmetry.

Step 2: Draw line segments to connect the points.

The vertices of the logo are _____, _____, _____, _____, and _____.

Show and Grow *I can think deeper!*

Draw the other half of the symmetric logo. Then list its vertices.

13.

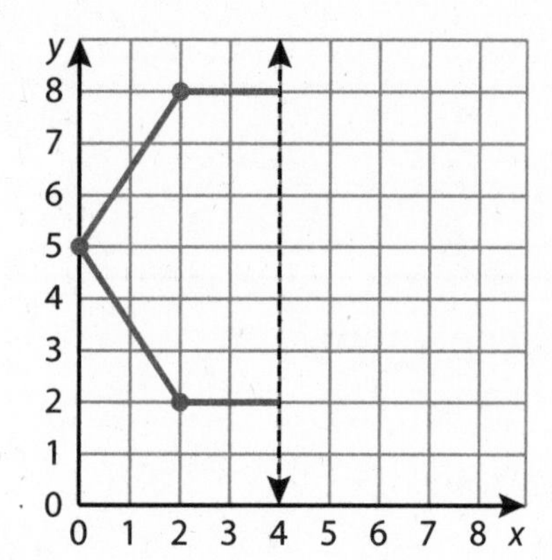

14.

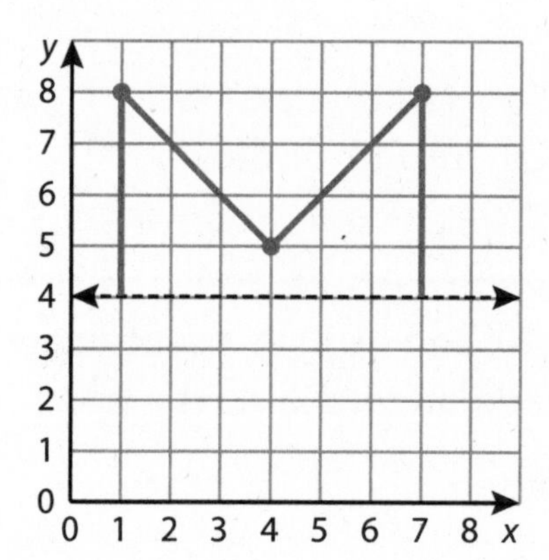

15. **DIG DEEPER!** One half of the design for a symmetric flower garden is shown in the coordinate plane. The line of symmetry is represented by the walkway. Draw the other half of the design for the flower garden. Then list its vertices.

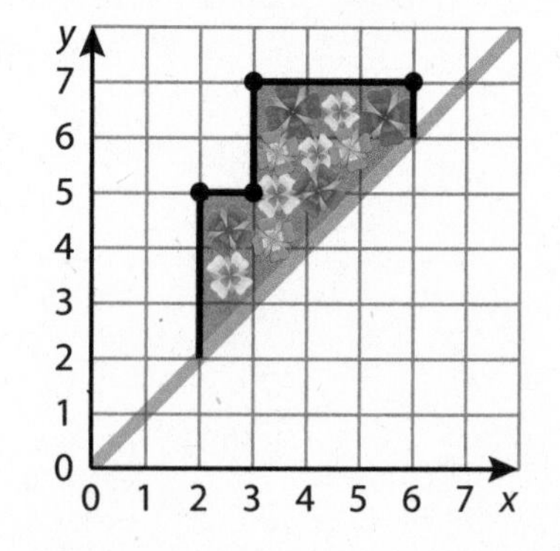

Name ____________________

Homework & Practice 12.3

Learning Target: Draw and identify polygons in a coordinate plane.

Example The vertices of a polygon are $A(2, 3)$, $B(3, 5)$, $C(5, 5)$, $D(6, 3)$, $E(5, 1)$ and $F(3, 1)$. Draw the polygon in a coordinate plane. Then identify it.

Step 1: Plot and label the vertices.

Step 2: Draw line segments to connect the points.

Be sure to connect the points in order to draw the polygon.

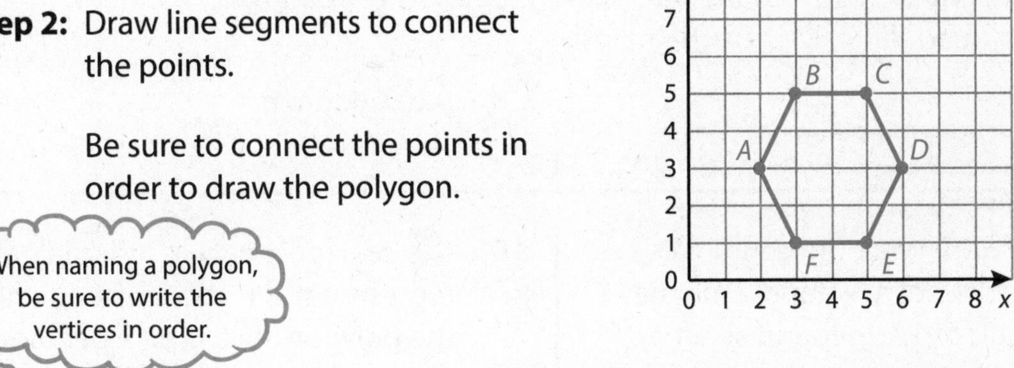

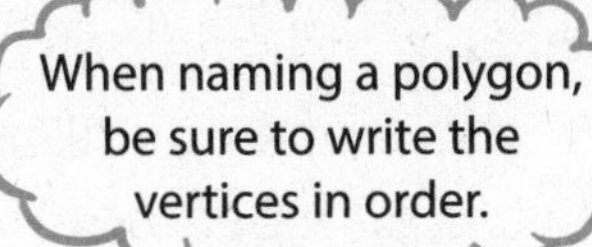

When naming a polygon, be sure to write the vertices in order.

Polygon *ABCDEF* is a hexagon.

Draw the polygon with the given vertices in a coordinate plane. Then identify it.

1. $A(2, 3)$, $B(2, 6)$, $C(5, 6)$, $D(5, 3)$

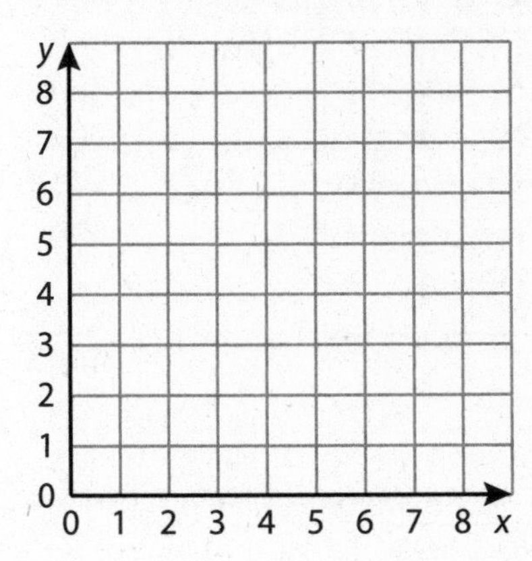

2. $J(3, 2)$, $K(3, 5)$, $L(6, 5)$

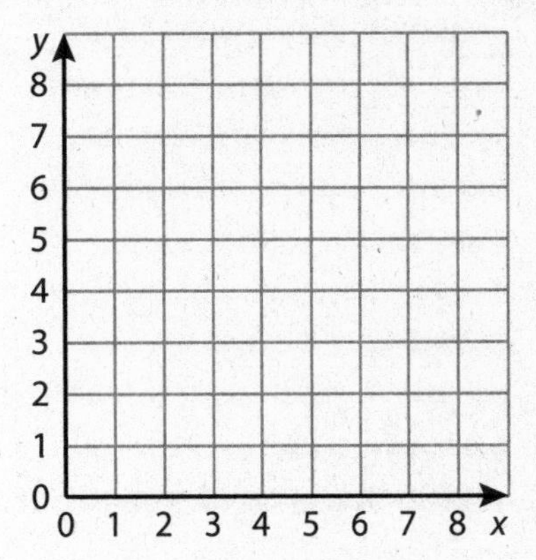

Identify the polygon with the given vertices.

3. $M(2, 6), N(4, 4), P(4, 0), Q(2, 2)$

4. $A(1, 2), B(1, 6), C(4, 6), D(6, 4), E(4, 2)$

5. $P(4, 1), Q(0, 1), R(1, 4), S(5, 5)$

6. $E(1, 2), F(1, 3), G(6, 3), H(6, 2)$

Plot (1, 2), (4, 2), and (3, 4) in a coordinate plane. Plot another point to form the given quadrilateral. Name the point.

7. trapezoid

8. parallelogram

9. **Open-Ended** Write the coordinates of the vertices of a rectangle that has a perimeter of 12 units and an area of 5 square units.

10. MP **Reasoning** Five ordered pairs represent the vertices of a polygon. Will the polygon always be a pentagon?

11. **Modeling Real Life** Draw the other half of the symmetric logo. Then list its vertices.

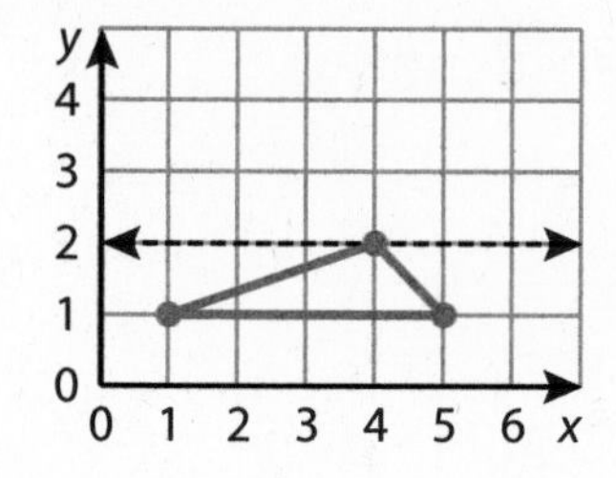

12. **DIG DEEPER!** You complete one fourth of an image with graphic design software. The computer generates the rest of the image with the two lines of symmetry. Draw the rest of the image.

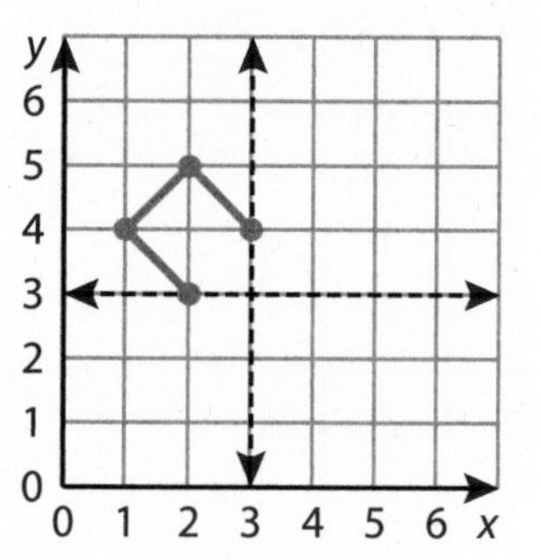

Review & Refresh

Estimate the sum or difference.

13. $\frac{13}{12} - \frac{5}{6}$

14. $\frac{9}{16} + \frac{4}{10}$

15. $\frac{19}{20} - \frac{7}{8}$

Name ______________________

Graph Data 12.4

Learning Target: Graph and interpret data in a coordinate plane.

Success Criteria:

- I can use ordered pairs to represent data.
- I can graph data in a coordinate plane.
- I can interpret data shown in a coordinate plane.

Explore and Grow

The table shows the amount of snow that falls each day for 7 days. Show how you can use ordered pairs in the coordinate plane to represent this information. Explain.

Day	1	2	3	4	5	6	7
Amount (inches)	2	4	6	6	8	10	14

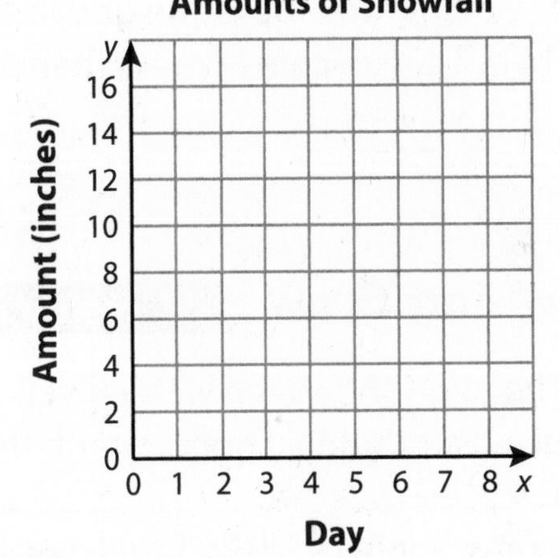

What conclusions can you make from your data display?

Reasoning On Day 8, 1 inch of snow falls. How can you represent this information in the coordinate plane?

Think and Grow: Graph Data

Key Idea **Data** are values collected from observations or measurements. You can use a coordinate plane to graph and interpret two categories of related data.

Example The table shows how many gold bars you collect at each level of a video game. Graph the data in a coordinate plane. In how many levels do you collect more than 30 gold bars?

Level	1	2	3	4	5	6
Gold Bars Collected	25	40	45	0	40	15

Step 1: Write the ordered pairs from the table.

(1, 25), (2, 40), (3, 45), (4, 0), (5, 40), (6, 15)

Step 2: For each axis, choose appropriate numbers to represent the data in the table.

Step 3: Write a title for the graph and label each axis.

Step 4: Plot a point for each ordered pair.

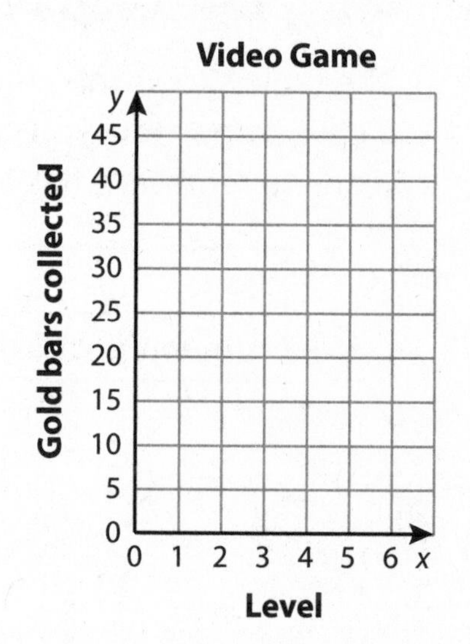

Three points are above the grid line that represents 30 bars. So, you collect more than 30 gold bars in

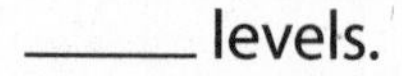 levels.

Show and Grow *I can do it!*

1. The table shows the water levels of a portion of a river during a flood. Graph the data.

Time (hours)	Water Level (feet)
1	5
2	8
3	10
4	11
5	7
6	5

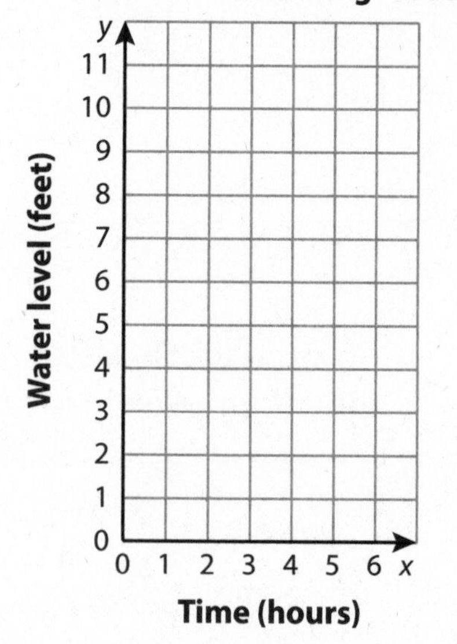

What does the point (5, 7) represent?

Name ______________________________

Apply and Grow: Practice

2. The table shows how many cars a salesman sells in each of 6 months. Graph the data.

Month	1	2	3	4	5	6
Cars Sold	7	10	6	11	10	10

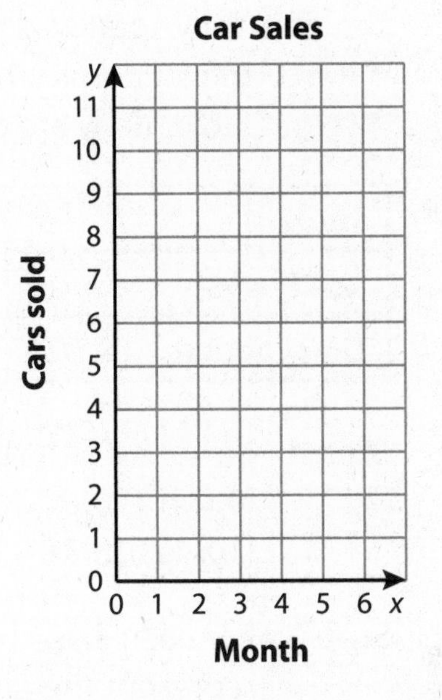

What does the point (1, 7) represent?

What is the difference of the greatest number of cars sold and the least number of cars sold? Explain.

Use the graph.

3. The graph shows how many receiving yards a football player has in each of seven games. How many receiving yards does he have in Game 3?

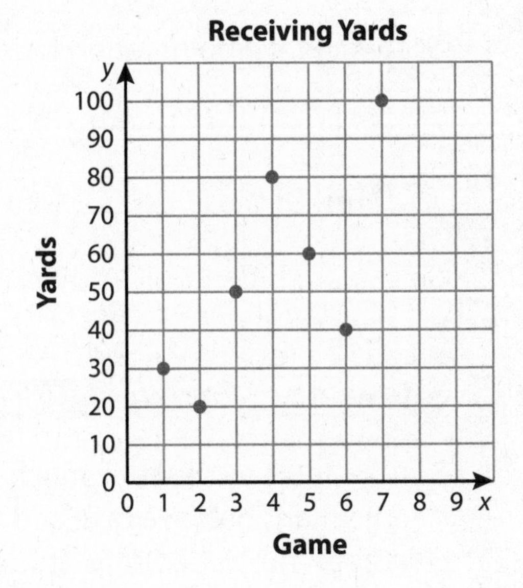

How many times as many receiving yards does he have in Game 4 as in Game 2?

In how many games does he have more than 40 receiving yards? fewer than 40 receiving yards?

---

4. **DIG DEEPER!** The player has 75 receiving yards in Game 8. The player has $\frac{1}{5}$ of this number of receiving yards in Game 9. Graph the data in the coordinate plane above.

Think and Grow: Modeling Real Life

Example The table shows the ages of eight students and the time they spend on the Internet for 1 week. Graph the data. Of the students who spend more than 15 hours on the Internet, how many are older than 10?

Student	A	B	C	D	E	F	G	H
Age (years)	9	10	11	10	10	8	11	12
Time (hours)	15	21	18	16	12	18	9	22

Step 1: Write the ordered pairs from the table.
(9, 15), (10, 21), (11, 18), (10, 16),
(10, 12), (8, 18), (11, 9), (12, 22)

Step 2: For each axis, choose appropriate numbers to represent the data in the table.

Step 3: Write a title for the graph and label each axis.

Step 4: Plot a point for each ordered pair.

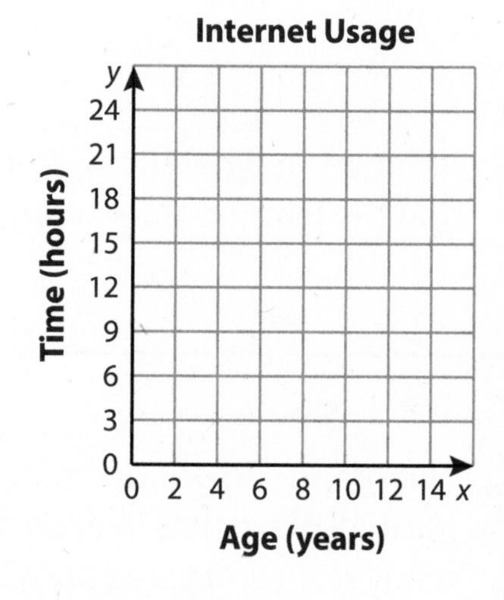

Five points are above the grid line that represents

15 hours. Of those, ______ points represent students older than 10.

So, ______ students older than 10 spend more than 15 hours on the Internet.

Show and Grow *I can think deeper!*

5. The table shows how much five students sleep the night before a quiz and their quiz scores. Graph the data. Of the students who sleep more than 7 hours, how many score higher than 8 points?

Student	A	B	C	D	E
Time (hours)	7	8	9	8	7
Score (points)	8	8	9	10	9

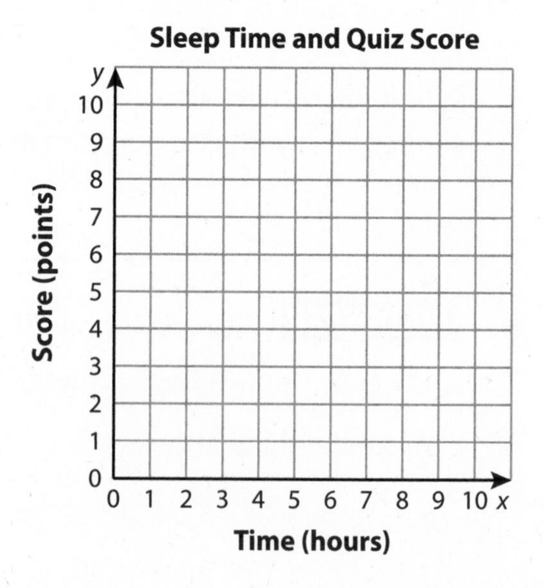

Name ________________________________

Homework & Practice 12.4

Learning Target: Graph and interpret data in a coordinate plane.

Example The table shows how many jumping jacks you do on each of 6 days. Graph the data.

Day	1	2	3	4	5	6
Jumping Jacks	25	35	50	20	30	45

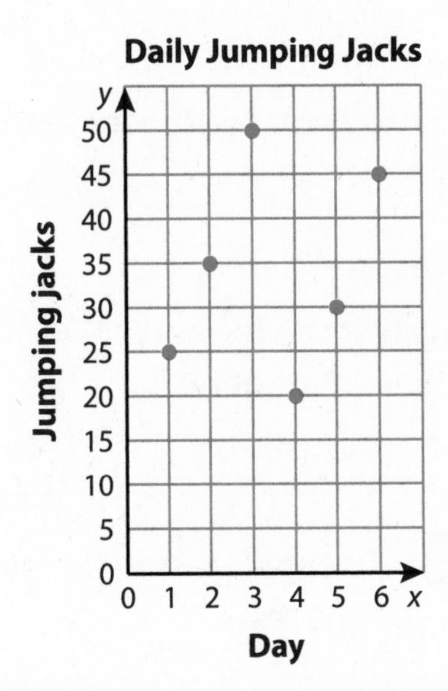

What does the point (3, 50) represent?

On Day 3, you do 50 jumping jacks.

1. The table shows how many students are in a choir club in each of 6 years. Graph the data.

Year	1	2	3	4	5	6
Students	15	20	25	30	30	35

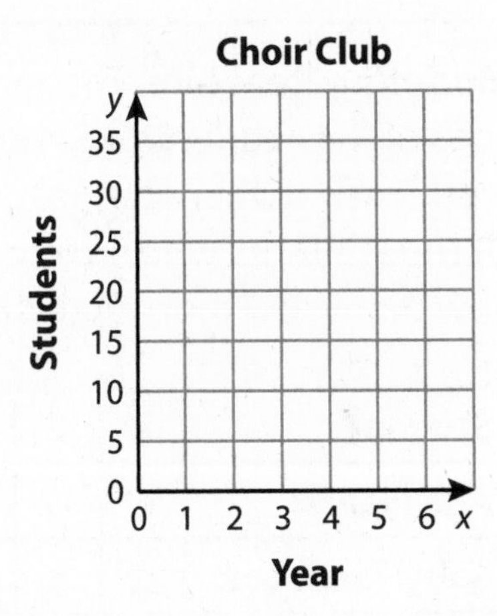

What does the point (2, 20) represent?

What is the difference of the greatest number of students and least number of students? Explain.

Use the graph.

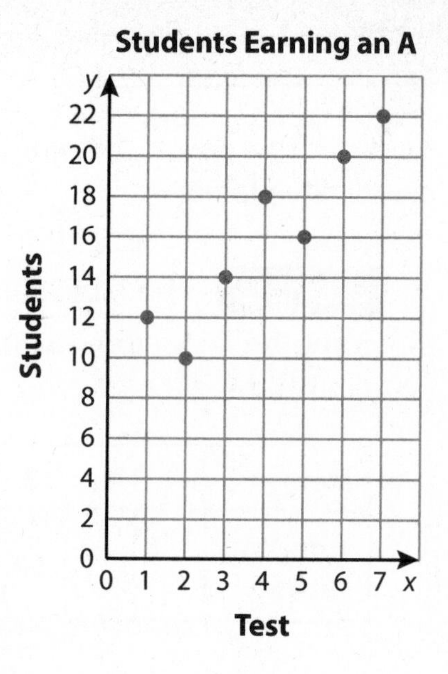

2. The graph shows how many students earn an A on each of seven tests. How many students earn an A on Test 4?

How many times as many students earn an A on Test 6 as on Test 2?

On how many tests do fewer than 20 students earn an A? more than 20 students?

---

3. **DIG DEEPER!** Twenty-five students take Test 1. How many students do *not* earn an A on the test?

---

4. **Modeling Real Life** The table shows the ages of five students and how many baby teeth each of them has lost. Graph the data. Of the students who are older than 10 years, how many lost more than 18 baby teeth?

Student	A	B	C	D	E
Age (years)	10	12	11	10	11
Baby Teeth Lost	18	20	18	14	20

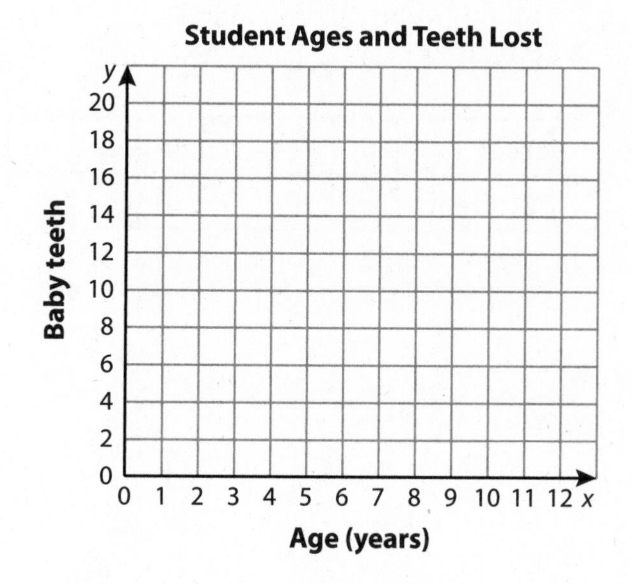

Review & Refresh

Find the quotient. Then check your answer.

5. $5 \div 0.8 =$ ______

6. $91.2 \div 15 =$ ______

7. $14.4 \div 3.2 =$ ______

Name ________________________________

Make and Interpret Line Graphs 12.5

Learning Target: Make and interpret line graphs.

Success Criteria:

- I can make a line graph.
- I can interpret a line graph.

Explore and Grow

The table shows the heights of a bamboo plant over several days. Show how you can use a coordinate plane to represent this information. Explain.

Day	1	3	5	7	9
Height (inches)	4	8	16	22	36

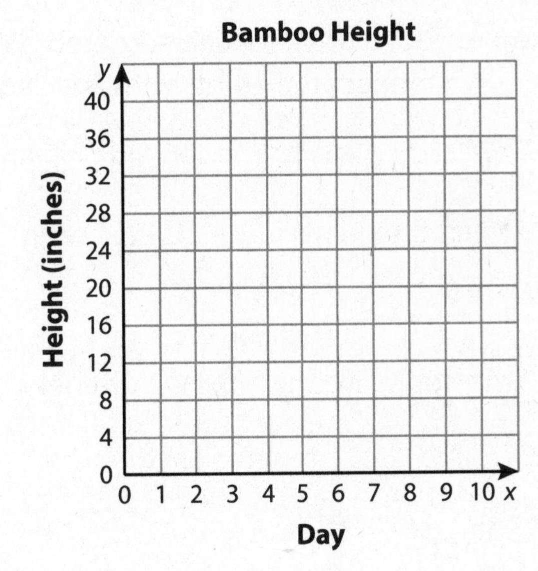

How can you use your graph to estimate the height of the plant on Day 4? Explain.

MP **Reasoning** What could the height of the bamboo plant be on Day 10? Explain your reasoning.

Think and Grow: Make and Interpret Line Graphs

Key Idea A **line graph** is a graph that uses line segments to show how data values change over time.

Example The table shows the weights of a dog over 6 months. Make a line graph of the data. Between which two months of age does the dog gain the most weight?

Age (months)	1	2	3	4	5	6
Weight (pounds)	10	20	30	50	55	58

Step 1: Write the ordered pairs from the table.

(1, 10), (2, 20), (3, 30), (4, 50), (5, 55), (6, 58)

Step 2: For each axis, choose appropriate numbers to represent the data in the table.

Step 3: Write a title for the graph and label each axis.

Step 4: Plot a point for each ordered pair. Then connect the points with line segments.

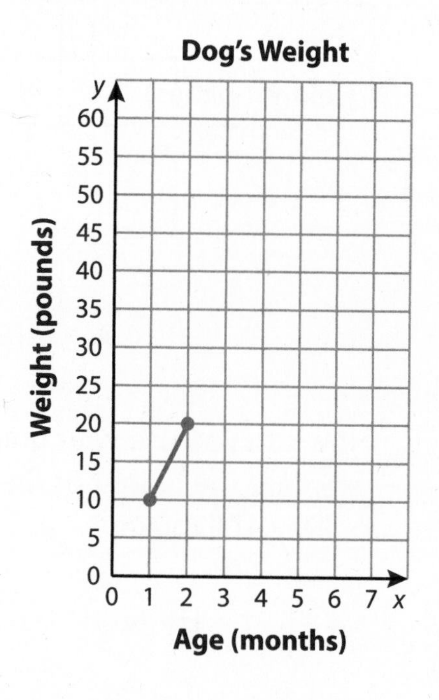

The greatest difference in weights occurs between the points (______, ______) and (______, ______).

So, the dog gains the most weight between ______ and ______ months of age.

Show and Grow I can do it!

Use the graph above.

1. Between which two months of age does the dog gain the least amount of weight? Explain.

2. How much do you think the dog weighs when it is 7 months of age? Explain your reasoning.

Name ____________________

Apply and Grow: Practice

Use the graph.

3. The table shows the height of a seedling over 7 days. Make a line graph of the data.

Day	0	1	2	3	4	5	6	7
Height (centimeters)	0	2	4	8	18	22	24	26

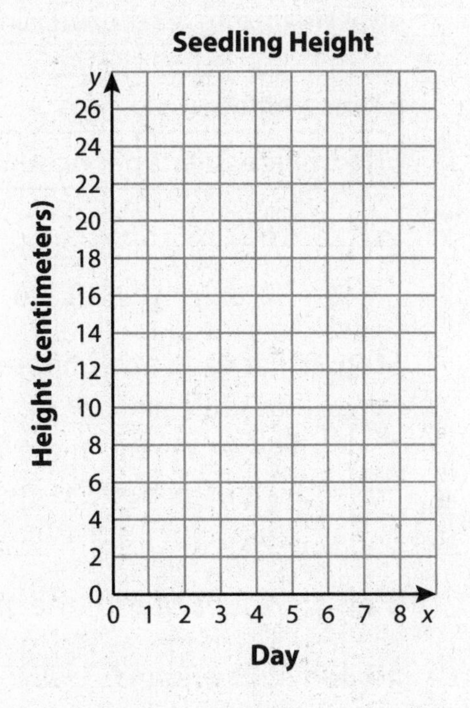

Between which two days did the seedling grow the most? Explain.

How tall do you think the seedling will be after 8 days? Explain.

____

4. MP **Reasoning** Interpret the point (0, 0) in the context of the situation.

____

Use the graph.

5. The graph shows the total numbers of likes a social media page has over 8 days. How many likes does the page have after 4 days?

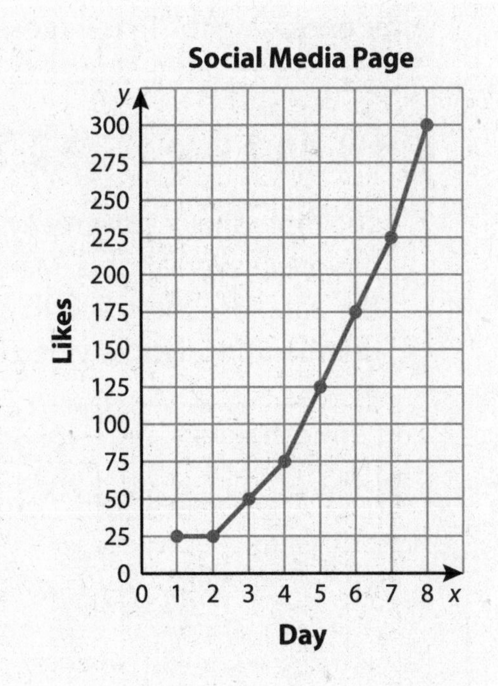

What is the difference of likes on Day 7 and Day 3?

____

6. **DIG DEEPER!** You track the likes between Days 7 and 8 by each hour. Does the total number of likes at every hour fall between 225 and 300? Explain.

Think and Grow: Modeling Real Life

Example The table shows your heart rate during an exercise routine. Make a line graph of the data. Use the graph to estimate your heart rate after exercising for 15 minutes.

Time (minutes)	0	10	20	30	35
Heart Rate (beats per minute)	80	110	140	148	135

Step 1: Write the ordered pairs from the table.
(0, 80), (10, 110), (20, 140), (30, 148), (35, 135)

Step 2: For each axis, choose appropriate numbers to represent the data in the table. You can show a break in the vertical axis between 0 and 80 because there are no data values between 0 and 80.

Step 3: Write a title for the graph and label each axis.

Step 4: Plot a point for each ordered pair. Then connect the points with line segments.

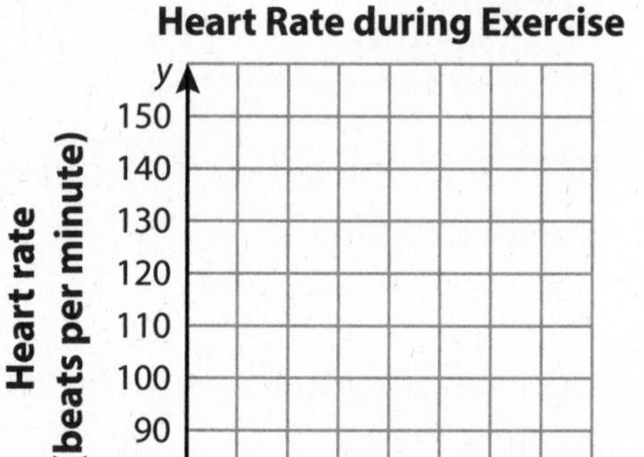

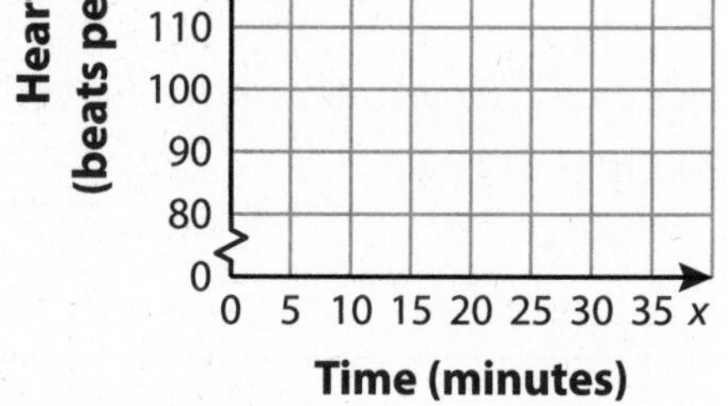

Use the line segment that connects (10, 110) and (20, 140) to estimate your heart rate after exercising for 15 minutes.

After exercising for 15 minutes, your heart rate is about _____ beats per minute.

Show and Grow *I can think deeper!*

7. The table shows how many views a video has over several hours. Make a line graph of the data. Use the graph to estimate the total number of views the video has after 2 hours.

Time (hours)	1	3	5	6	7
Views	500	600	810	1,000	1,290

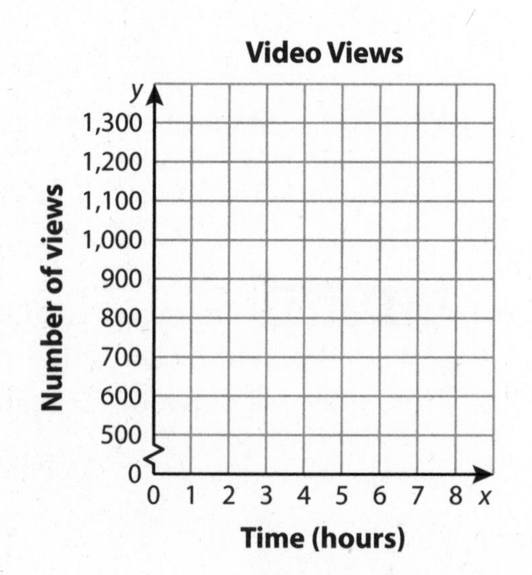

Name ______________________________

Homework & Practice 12.5

Learning Target: Make and interpret line graphs.

Example The table shows the temperatures of a city over several hours during a snowstorm. Make a line graph of the data.

Time (hours)	1	2	3	4	5	6	7
Temperature (degrees Fahrenheit)	30	28	24	24	22	21	20

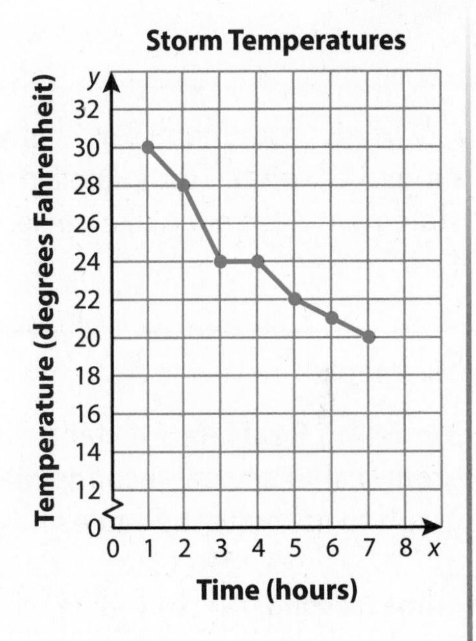

Between which two hours does the temperature decrease the most? Explain.

The greatest difference in temperatures occurs between the points (2, 28) and (3, 24). So, the temperature decreases the most between Hours 2 and 3.

Estimate the temperature at 4 hours and 30 minutes.

23 degrees Fahrenheit

1. The table shows the total number of pieces of beach glass you find during an hour at the beach. Make a line graph of the data.

Time (minutes)	10	20	30	40	50	60
Number of pieces	4	10	20	26	29	35

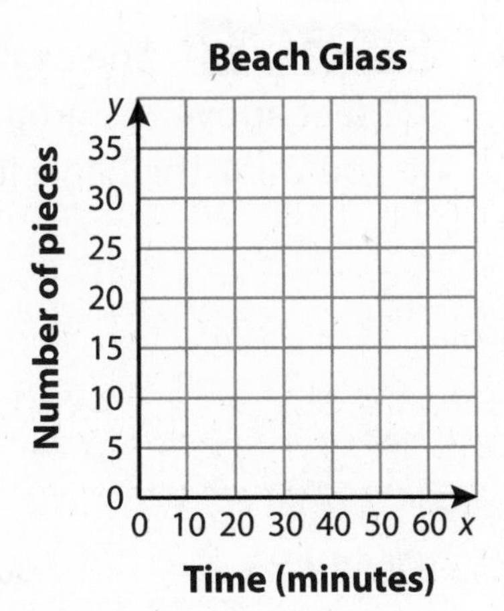

Between which two times did you find the most pieces of beach glass? Explain.

Estimate how many pieces you had after 25 minutes.

Use the graph.

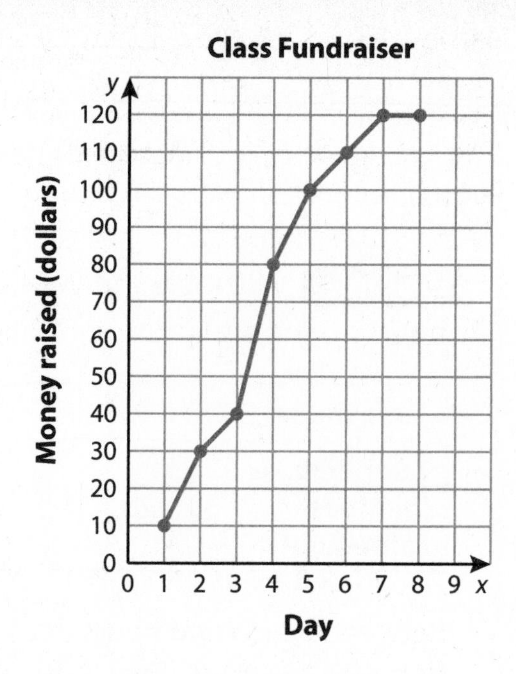

2. The graph shows the total amounts of money your class raises over 8 days. How much money does your class raise after 6 days?

How much money does your class raise between Days 2 and 7?

3. MP **Logic** Your friend says that your class raises $115 after 9 days. Explain why your friend's statement does *not* make sense.

Use the graph.

4. **Modeling Real Life** The table shows a bald eagle's heights above the ground after several seconds. Make a line graph of the data. Use the graph to estimate the eagle's height above the ground after 6 seconds.

Time (seconds)	1	2	5	7	8
Height Above Ground (feet)	40	80	200	240	260

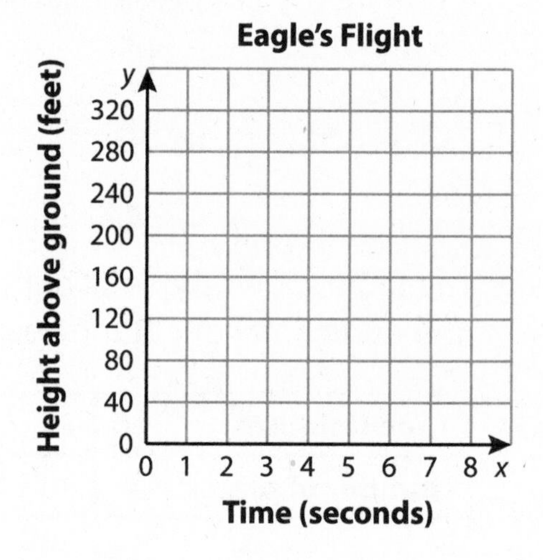

5. **DIG DEEPER!** The eagle flies past her nest, which is 120 feet above the ground. After how many seconds do you think the eagle flies past her nest? Explain.

Review & Refresh

Divide.

6. $1 \div \frac{1}{8} =$ 8

7. $3 \div \frac{1}{4} =$ 12

8. $5 \div \frac{1}{10} =$ 50

Name ________________________________

Numerical Patterns 12.6

Learning Target: Create and describe numerical patterns.

Success Criteria:

- I can create a numerical pattern.
- I can describe features of a numerical pattern.
- I can describe the relationship between two numerical patterns.

Explore and Grow

Newton saves $10 each month. Descartes saves $30 each month. Complete each table. What patterns do you notice?

Newton:

Month	0	1	2	3	4	5
Amount Saved	$0	$10				

Descartes:

Month	0	1	2	3	4	5
Amount Saved	$0	$30				

Repeated Reasoning How much will Newton have saved when Descartes has saved $300? Explain your reasoning.

Think and Grow: Numerical Patterns

Example You use 2 pounds of beef to make a batch of empanadas. Each batch makes eight servings. Complete the rule that relates the number of servings to the number of pounds of beef.

Batches	0	1	2	3	4	5
Beef (pounds)	0	2	4			
Number of Servings	0	8	16			

Step 1: Create each pattern and complete the table.

Use the rule "Add ____" to find the number of pounds of beef.

0, 2, 4, ____, ____, ____

Use the rule "Add ____" to find the number of servings.

0, 8, 16, ____, ____, ____

Step 2: Write ordered pairs that relate the number of servings to the number of pounds of beef.

(0, 0), (8, 2), (16, 4), ____, ____, ____

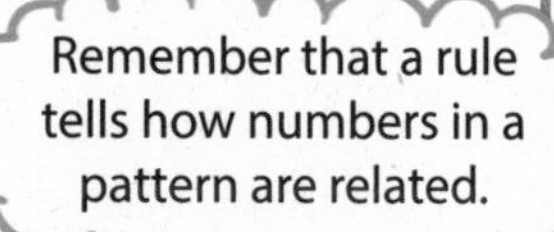

Remember that a rule tells how numbers in a pattern are related.

Step 3: Write a rule. As you make each batch, the number of pounds of beef is always ____ as much as the number of servings.

So, divide the number of servings by ____ to find the number of pounds of beef.

Show and Grow *I can do it!*

1. Use the given rules to complete the table. Then complete the rule that relates the number of hours worked to the amount earned.

	Days	1	2	3	4	5
Add 5.	Hours Worked	5	10			
Add 40.	Amount Earned	$40	$80			

Multiply the number of hours worked by ____ to find the amount earned.

Name ______________________________

Apply and Grow: Practice

Use the given rules to complete the table. Then complete the rule.

2.

	Gallons of Lemonade	1	2	3	4	5
Add 14.	**Water (cups)**	14	28			
Add 2.	**Lemon Juice (cups)**	2	4			

Divide the number of cups of water by _____ to find the number of cups of lemon juice.

3.

	Days	1	2	3	4	5
Add 15.	**Push-Ups**	15	30			
Add 30.	**Sit-Ups**	30	60			

Multiply the number of push-ups by _____ to find the number of sit-ups.

4. Complete the rule. Then use the rule to complete the table.

Multiply the amount of money that Newton saves by _____ to find the amount of money that Descartes saves.

Weeks	0	1	2	3	. . .	9
Newton's Savings	$0	$4	$8	$12	. . .	$36
Descartes's Savings	$0	$12	$24	$36	. . .	

5. MP **Structure** The ordered pairs (3, 2), (6, 4), and (9, 6) relate the number of avocados to the number of plum tomatoes in a guacamole recipe. Use the relationship to complete the table.

Batches	1	2	3	. . .	6
Avocados	3			. . .	
Tomatoes	2			. . .	

Think and Grow: Modeling Real Life

Example For each $1 bill you pay, you get 4 tokens and can play 2 games. You have 60 tokens. How many games can you play?

Think: What do you know? What do you need to find? How will you solve?

Use a rule to create each pattern. Use a table to organize the information.

Add ______.

Add ______.

Amount Exchanged	$1	$2	$3	$4
Tokens	4			
Games	2			

Write ordered pairs that relate the number of tokens to the number of games you can play.

(4, 2), ______, ______, ______

Write a rule. The number of games you can play is always ______ as much as the number of tokens.

So, divide the number of tokens by ______ to find the number of games you can play.

60 ÷ ______ = ______

So, you can play ______ games.

Show and Grow *I can think deeper!*

6. Each day, you read 33 pages and your friend reads 11 pages. How many pages does your friend read when you read 396 pages?

7. **DIG DEEPER!** Each pack of trading cards has 1 hero card, 5 combination cards, and 30 action cards. You buy packs of trading cards and get 35 combination cards. How many hero cards and action cards do you get?

Name ___________________________________

Homework & Practice 12.6

Learning Target: Create and describe numerical patterns.

Example A soccer team has two games and four practices each week. Complete the table. How does the number of games relate to the number of practices?

	Weeks	0	1	2	3	4	5
Add 2.	**Games**	0	2	4	6	8	10
Add 4.	**Practices**	0	4	8	12	16	20

Write ordered pairs that relate the number of games to the number of practices.

(0, 0), (2, 4), (4, 8), (6, 12), (8, 16), (10, 20)

Write a rule. As each week is completed, the number of games is always $\frac{1}{2}$ as much as the number of practices.

So, multiply the number of games by 2 to find the number of practices.

Use the given rules to complete the table. Then complete the rule.

1.

	Students	1	2	3	4	5
Add 10.	**Candles Sold**	10	20			
Add 80.	**Money Raised**	$80	$160			

Multiply the number of candles sold by ______ to find the amount of money raised.

2.

	Containers	1	2	3
Add 30.	**Servings**	30		
Add 600.	**Pretzels**	600		

Multiply the number of servings by ______ to find the number of pretzels.

3. Complete the rule. Then use the rule to complete the table.

Divide the number of contestants by ______ to find the number of winners.

Games	1	2	3	. . .	15
Contestants	8	16	24	. . .	120
Winners	2	4	6	. . .	

4. **DIG DEEPER!** Draw Figure 4. How many red squares are in Figure 8? How many yellow squares are in Figure 8? Explain your reasoning.

Figure 1

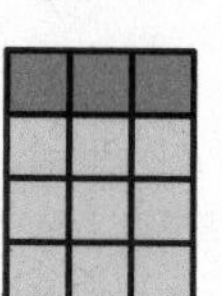

Figure 2

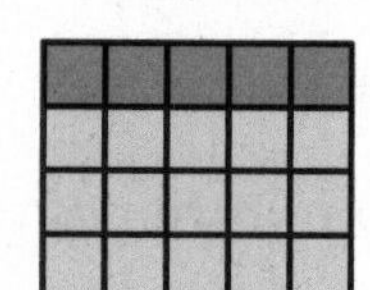

Figure 3

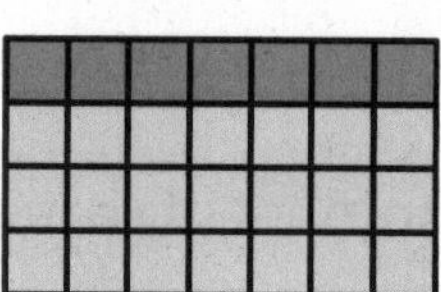

Figure 4

5. **Modeling Real Life** Each person at a baseball game receives 3 raffle tickets and a $2 certificate for the team store. A group of people receives 39 raffle tickets. How much money in certificates does the group receive?

6. **DIG DEEPER!** Write a rule that relates the number of months to the cost of a gym membership. What is the cost of a 1-year membership?

One-time fee is $25.
Then pay just $15
each month!
MEMBERSHIP
SPECIAL

Review & Refresh

Convert the mass.

7. 7 g = ______ mg

8. 92 g = ______ kg

Convert the capacity.

9. 800 mL = ______ L

10. 3 L = ______ mL

Name ____________________

Graph and Analyze Relationships 12.7

Learning Target: Use a graph to describe the relationship between two numerical patterns.

Success Criteria:

- I can generate two numerical patterns.
- I can use two numerical patterns to write and plot ordered pairs in a coordinate plane.
- I can use a graph to describe the relationship between two numerical patterns.

Explore and Grow

Complete each table and graph the data in the coordinate plane. What do you notice about the points?

Yards	1	2	3	4	5
Feet	3				

Gallons	1	2	3	4	5
Pints	8				

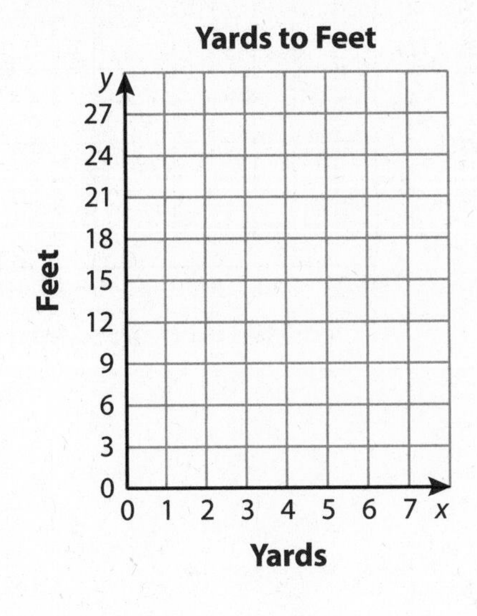

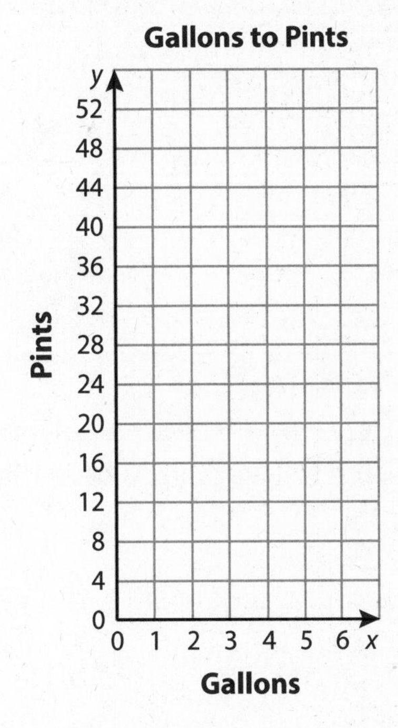

MP **Structure** How can you use the graphs to find the number of feet in 7 yards and the number of pints in 6 gallons? Explain your reasoning.

Think and Grow: Graph and Analyze Relationships

Example For each glass of iced tea Newton makes, he uses 2 spoonfuls of iced-tea mix and 10 fluid ounces of water. Newton uses 16 spoonfuls of iced-tea mix. How many fluid ounces of water does he use?

Step 1: Find the first several numbers in the numerical patterns for the amounts of iced-tea mix and water.

Iced-Tea Mix (spoonfuls)	2	4				
Water (fluid ounces)	10	20				

Step 2: Write the ordered pairs from the table.

(2, 10), (4, 20), ______,

______, ______, ______

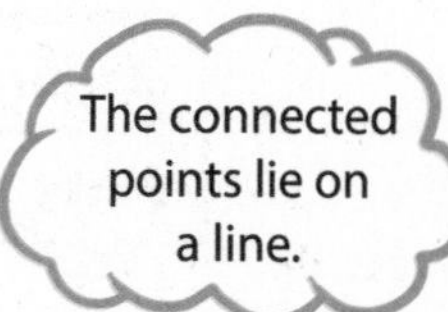
The connected
points lie on
a line.

Step 3: Plot the ordered pairs. Connect the points with line segments.

Newton's Iced Tea

Water (fluid ounces): y-axis 0, 10, 20, 30, 40, 50, 60, 70, 80, 90

Iced-tea mix (spoonfuls): x-axis 0, 2, 4, 6, 8, 10, 12, 14, 16, 18

Because the ordered pairs follow a pattern, you can extend the line to the point where the x-coordinate is 16.

When the x-coordinate is 16, the y-coordinate is ______.

So, Newton uses ______ fluid ounces of water.

Show and Grow *I can do it!*

1. Use the graph above. Newton uses 18 spoonfuls of iced-tea mix. How many fluid ounces of water does he use? Explain your reasoning.

Name ___

Apply and Grow: Practice

Use the given information to complete the table. Describe the relationship between the two numerical patterns and plot the points.

2. A slime recipes calls for 120 milliliters of vegetable oil and 30 grams of cornstarch. You measure 600 milliliters of vegetable oil. How many grams of cornstarch do you need?

Vegetable Oil (milliliters)	120	240		
Cornstarch (grams)	30	60		

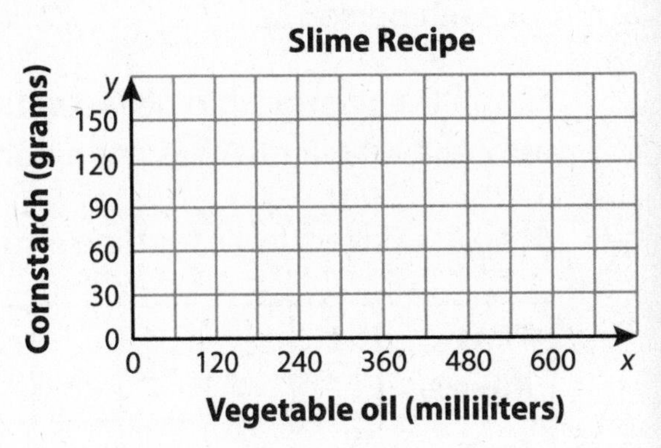

---

3. A sponsor donates $5 for every 4 laps walked around a track. How much money does the sponsor donate for 28 laps walked?

Laps	4	8	12			
Money (dollars)	5	10	15			

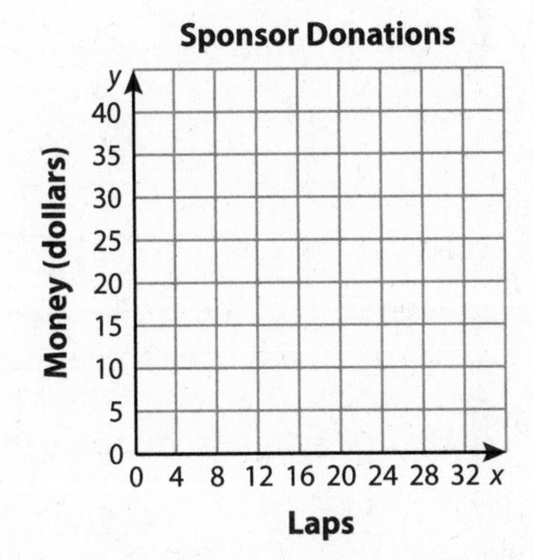

---

4. **Writing** How can you use the graph to determine the number of cups in 4 gallons?

5. **MP Number Sense** What does the ordered pair (0, 0) represent in the graph?

6. **DIG DEEPER!** Use the graph to determine the number of cups in $2\frac{1}{2}$ gallons.

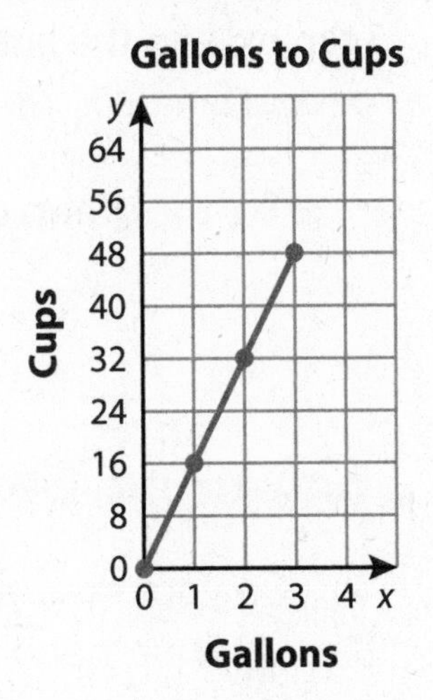

Think and Grow: Modeling Real Life

TRAMPOLINE Park A
$15 per hour

TRAMPOLINE Park B
$10 per half hour

*Price is per person for up to 4 hours.

Example Some friends plan to go to a trampoline park for 2 hours. They want to go to the park that costs less money. Which park should they choose? What is the cost for each person?

Graph the relationship between time and cost at both parks. Park A has been done for you.

Step 1: Make a table for time and cost at Park B.

Time (hours)	$\frac{1}{2}$	1	$1\frac{1}{2}$	2	$2\frac{1}{2}$	3	$3\frac{1}{2}$	4
Cost (dollars)	10	20						

Step 2: Write the ordered pairs from the table.

$\left(\frac{1}{2}, 10\right)$, (1, 20), _____, _____,

_____, _____, _____, _____

Step 3: Plot the ordered pairs. Connect the points with line segments.

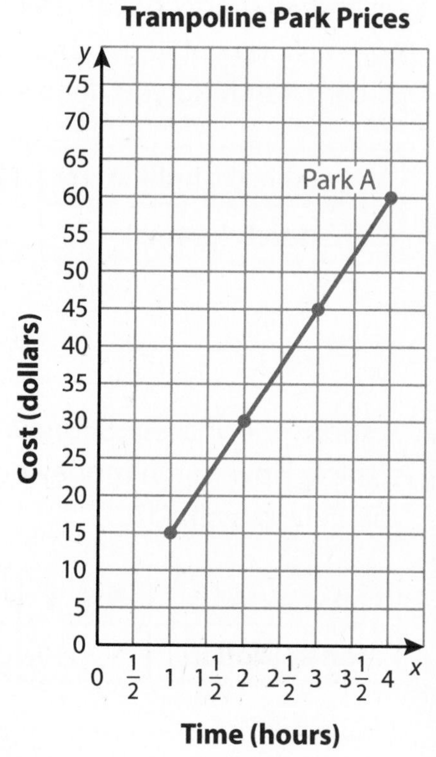

Use the graph to compare the costs for 2 hours at the parks.

Park	x-coordinate	y-coordinate
A	2	_____
B	2	_____

_____ < _____

So, the group of friends should choose Trampoline Park _____.

The cost for each person is $_____.

Show and Grow *I can think deeper!*

7. On your map, every centimeter represents 20 kilometers. On your friend's map, every 2 centimeters represents 50 kilometers. On whose map does 6 centimeters represent a greater distance? How much greater? Explain.

Name ______________________________

Homework & Practice 12.7

Learning Target: Use a graph to describe the relationship between two numerical patterns.

Example A parking meter requires $0.25 for every 20 minutes of parking. You put $1.50 into the meter. How long can you park your car before the meter expires?

Step 1: Find the first several numbers in the numerical patterns for the amount of money and the parking time.

Money (dollars)	0.25	0.50	0.75	1.00	1.25
Parking Time (minutes)	20	40	60	80	100

Step 2: Write the ordered pairs from the table.

(0.25, 20), (0.50, 40), (0.75, 60), (1.00, 80), (1.25, 100)

Step 3: Plot the ordered pairs. Connect the points with line segments.

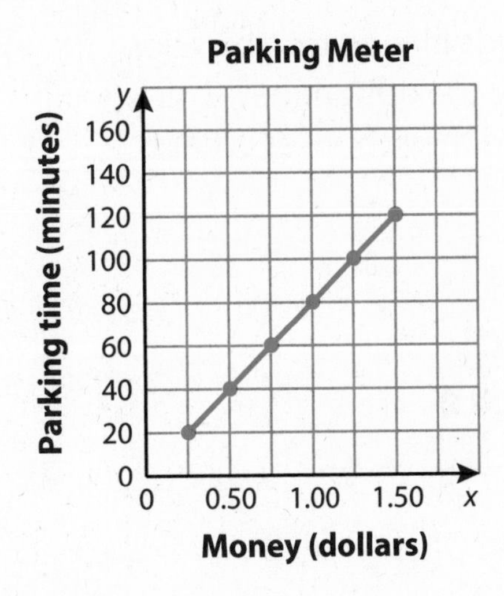

Because the ordered pairs follow a pattern, you can extend the line to the point where the *x*-coordinate is 1.50.

When the *x*-coordinate is 1.50, the *y*-coordinate is 120.

So, you can park your car 120 minutes before the meter expires.

1. Use the graph above. You plan to park your car for 140 minutes. How much money do you put into the meter?

2. A boxer exercises by jumping rope. He completes 150 repetitions every minute. He completes 750 repetitions. For how many minutes does he jump rope?

Time (minutes)	1	2		
Repetitions	150	300		

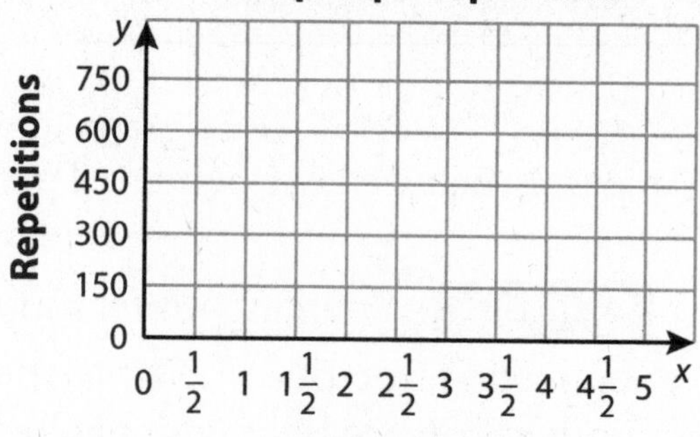

3. YOU BE THE TEACHER Your friend says a baker makes 60 plain bagels in 5 hours. Is your friend correct? Explain.

Time (hours)	1	2	3
Bagels Made	12	24	36

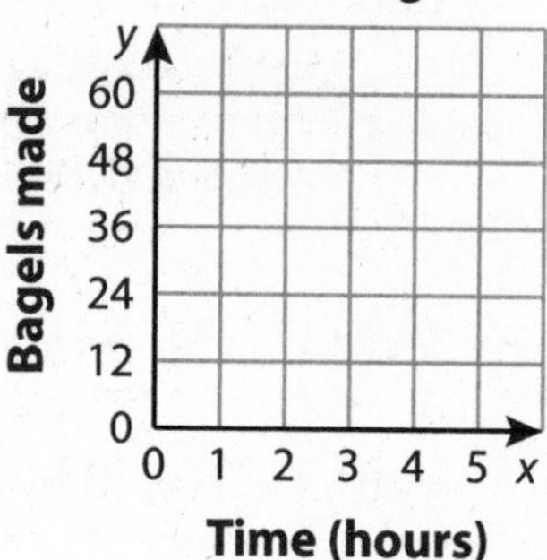

4. Modeling Real Life Some friends plan to rent bicycles for 6 hours. They want to choose the option that costs less money. Which option should they choose? What is the cost for each person?

BIKE RENTALS

OPTION A
$6 per hour
for each person

OPTION B
$15 per 3 hours
for each person

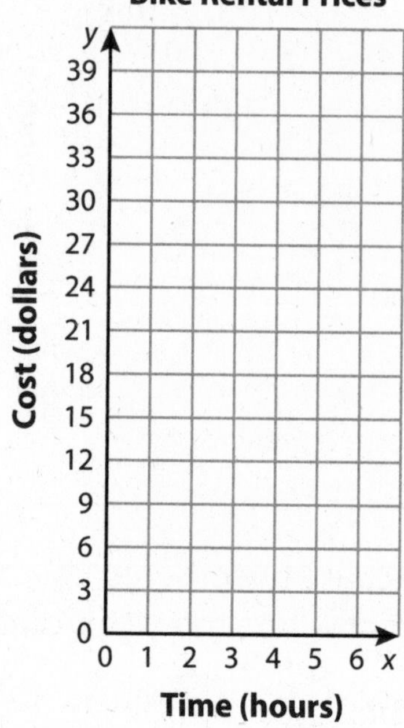

Review & Refresh

Convert the length.

5. $3\frac{1}{3}$ ft = _____ in.

6. 6 mi = _____ yd

Name ____________________

Performance Task 12

You use a series of commands on an app to create an animation of Descartes dancing and jumping.

1. **a.** Complete the animation commands to move Descartes. Plot the points to show his movement.

Animation Commands
Start at (4, 2).
Glide to (☐, ☐).
Glide to (☐, ☐).
Glide to (☐, ☐).
Glide to (☐, ☐).
Move to (4, 2).

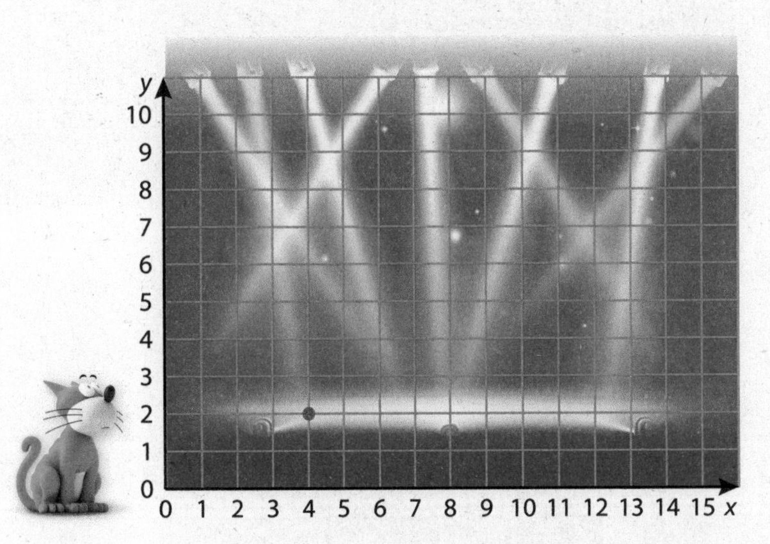

b. Connect the points. Describe the animation in your own words.

---

2. You play the animation commands to make Descartes dance.

a. It takes 4 seconds for Descartes to move through the animation commands 1 time. Complete the table and graph the data in the coordinate plane.

Number of Times Played	1	2	3	4	5
Time (seconds)	4				

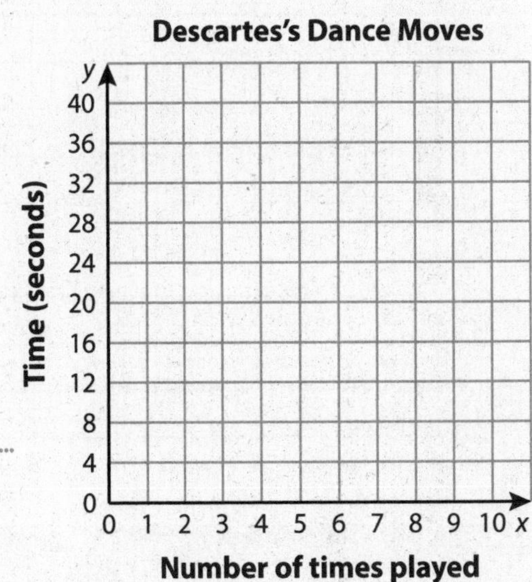

b. You want Descartes to dance for an exact number of seconds. How can you find the number of times to play the animation commands? Use an example to justify your reasoning.

Treasure Hunt

Directions:

1. Each player arranges four Treasure Hunt Gold Bars on the *My Treasure* coordinate plane, horizontally or vertically.
2. On your turn, name an ordered pair in the *Partner's Treasure* coordinate plane. If your partner says you found part of a gold bar, then plot the ordered pair in red. Otherwise, plot the ordered pair in black. Your turn is over.
3. On your partner's turn, if your partner finds part of a gold bar, then plot a red *X* on the ordered pair in the *My Treasure* coordinate plane. Tell your partner when all parts of a gold bar have been found.
4. The first player to find all parts of the partner's gold bars wins!

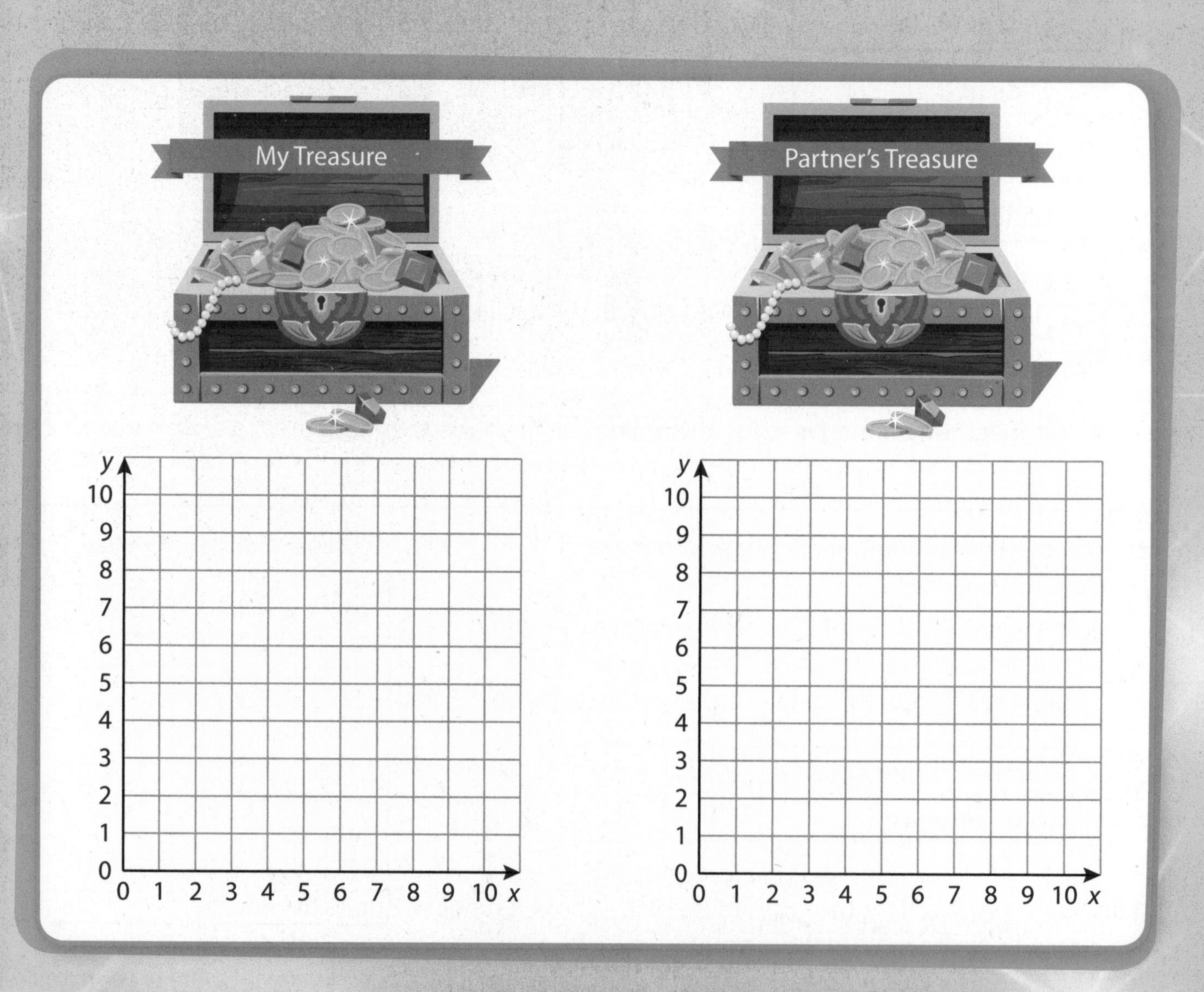

Name ______________________________

Chapter Practice 12

12.1 Plot Points in a Coordinate Plane

Use the coordinate plane to write the ordered pair corresponding to the point.

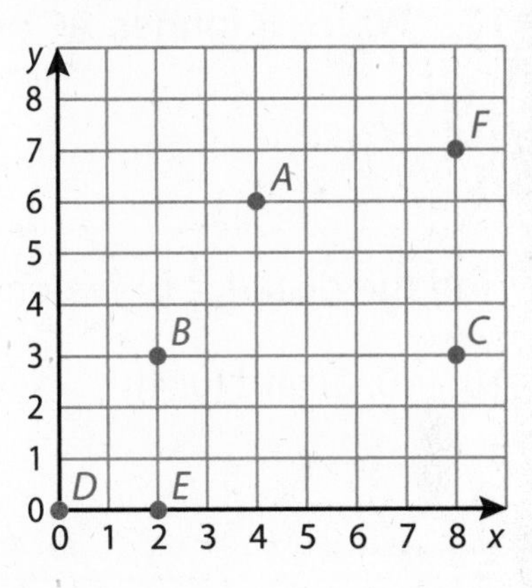

1. Point *A*

2. Point *D*

3. Point *B*

4. Point *E*

5. Point *C*

6. Point *F*

Plot and label the point in the coordinate plane above.

7. $N(3, 0)$

8. $P(1, 5)$

9. $R(2, 1)$

Name the point for the ordered pair.

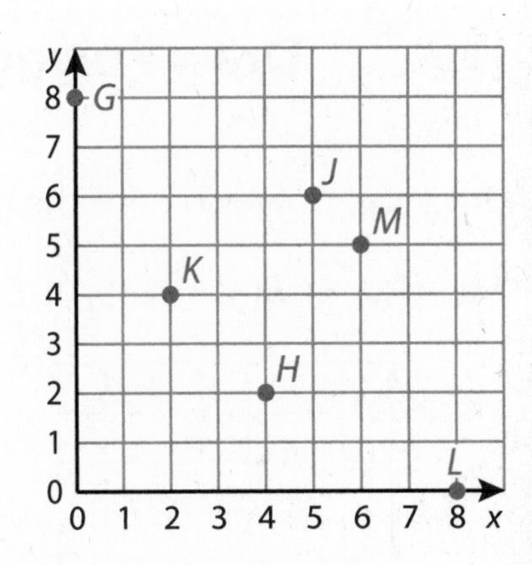

10. (0, 8)

11. (5, 6)

12. (4, 2)

13. **Open-Ended** Use the coordinate plane above. Point *S* is 2 units from point *J*. Name two possible ordered pairs for Point *S*.

12.2 Relate Points in a Coordinate Plane

Find the distance between the points in the coordinate grid.

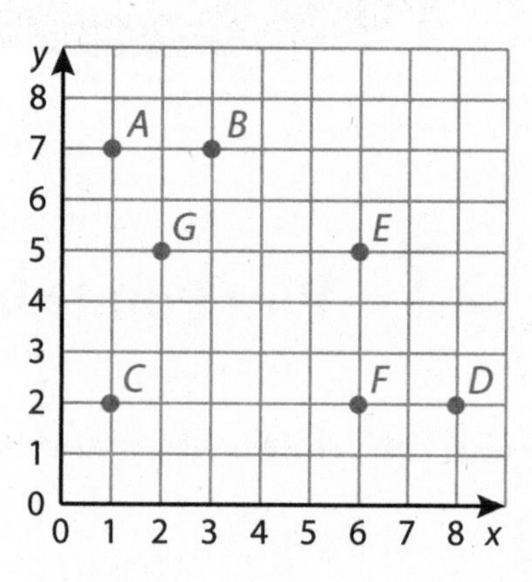

14. *A* and *B*

15. *E* and *F*

16. *C* and *D*

17. Which is longer, $\overline{AC}$ or $\overline{GE}$?

Find the distance between the points.

18. (0, 0) and (0, 4)

19. (3, 2) and (3, 9)

20. (0, 5) and (7, 5)

A line passes through the given points. Name two other points that lie on the line.

21. (0, 1) and (0, 7)

22. (5, 2) and (5, 8)

23. (6, 3) and (0, 3)

12.3 Draw Polygons in a Coordinate Plane

Draw the polygon with the given vertices in a coordinate plane. Then identify it.

24. *A*(2, 5), *B*(5, 5), *C*(5, 0), *D*(2, 0)

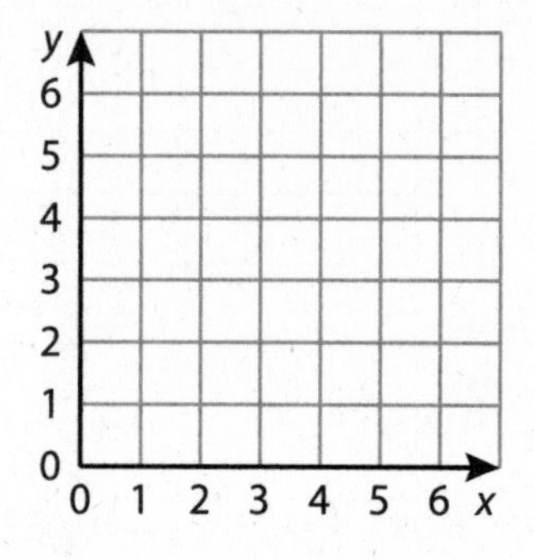

25. *D*(1, 3), *E*(1, 5), *F*(3, 6), *G*(5, 5), *H*(5, 3), *J*(3, 2)

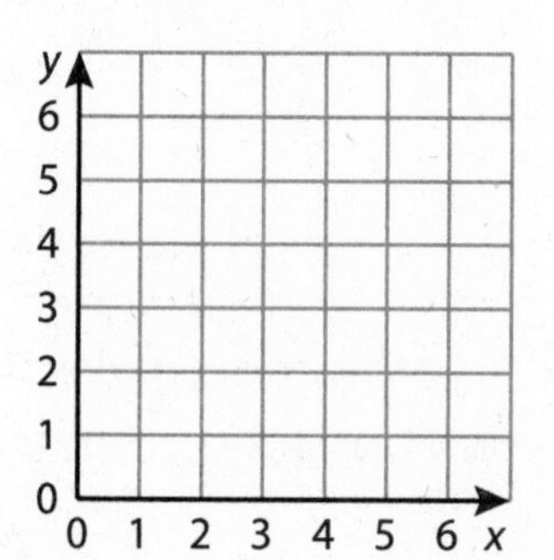

12.4 Graph Data

26. The table shows how many home runs your team scores in each of six kickball games. Graph the data.

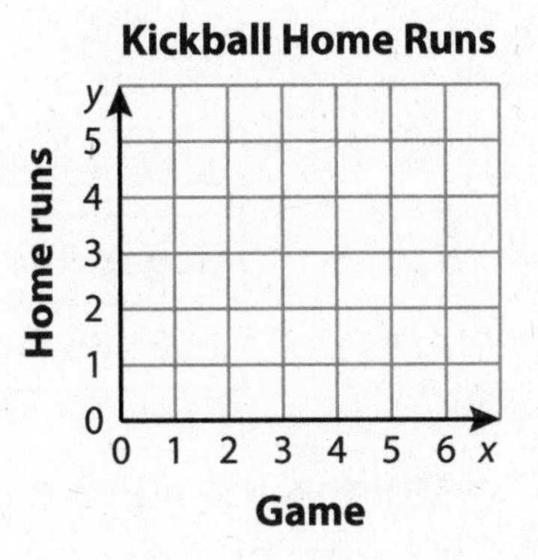

Game	1	2	3	4	5	6
Home Runs	2	1	5	4	2	3

What does the point (6, 3) represent?

What is the difference of the greatest number of home runs and the least number of home runs? Explain.

12.5 Make and Interpret Line Graphs

27. The table shows the total numbers of coupon books you sell over 7 days. Make a line graph of the data.

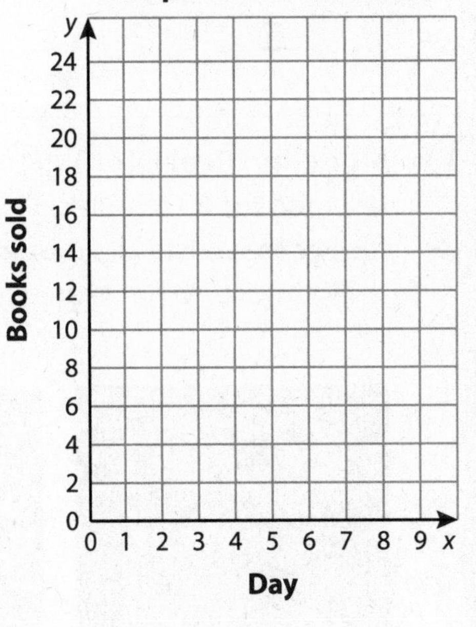

Day	1	2	3	4	5	6	7
Books Sold	0	2	5	8	12	18	23

On which day do you sell the most books? Explain.

How many books do you think you sell after 9 days? Explain.

12.6 Numerical Patterns

28. Use the given rules to complete the table. Then complete the rule.

	Servings of Dip	1	2	3	4	5
Add 5.	Sugar-free Pudding Mix (ounces)	5	10			
Add 15.	Pumpkin (ounces)	15	30			

Multiply the number of ounces of pudding mix by ______ to find the number of ounces of pumpkin.

Graph and Analyze Relationships

29. An employee earns $80 every 8 hours. How much money does she earn after 40 hours?

Time (hours)	8	16		
Amount Earned (dollars)	80	160		

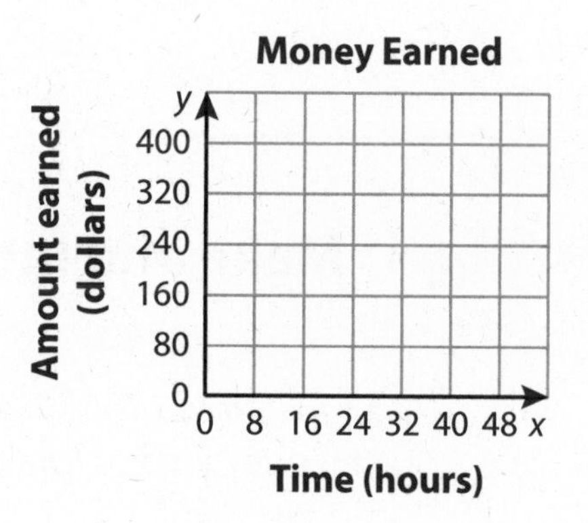

30. Modeling Real Life A group of friends wants to play laser tag for 60 minutes. They want to go to the facility that costs less money. Which facility should they choose? What is the cost for each person?

Facility A

Facility B

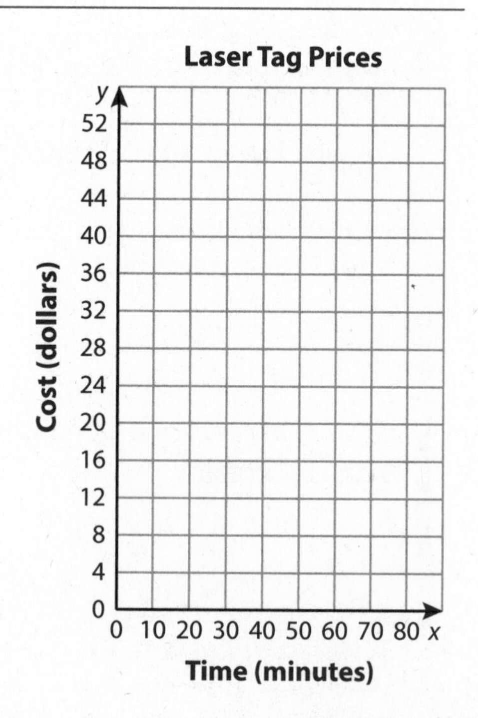

13 Understand Volume

- Why would you need to know the volume of an elevator?
- You ride in an elevator in the shape of a rectangular prism. How can you use the dimensions of the elevator to find its volume?

Chapter Learning Target:
Understand volume.

Chapter Success Criteria:
- I can define volume.
- I can describe volume.
- I can compare volumes.
- I can apply the volume formula.

Name ___________

13 Vocabulary

Review Words
area
perimeter

Organize It

Use the review words to complete the graphic organizer.

______ of a Rectangle	______ of a Rectangle
• The distance around a rectangle • $P = (2 \times \ell) + (2 \times w)$	• The amount of surface a rectangle covers • $A = \ell \times w$

Define It

Use your vocabulary cards to complete the puzzle.

Across

1. A cube that measures one unit on each side

Down

2. A measure of the amount of space that a solid figure occupies

3. A unit used to measure volume

4. The bottom face of a right rectangular prism

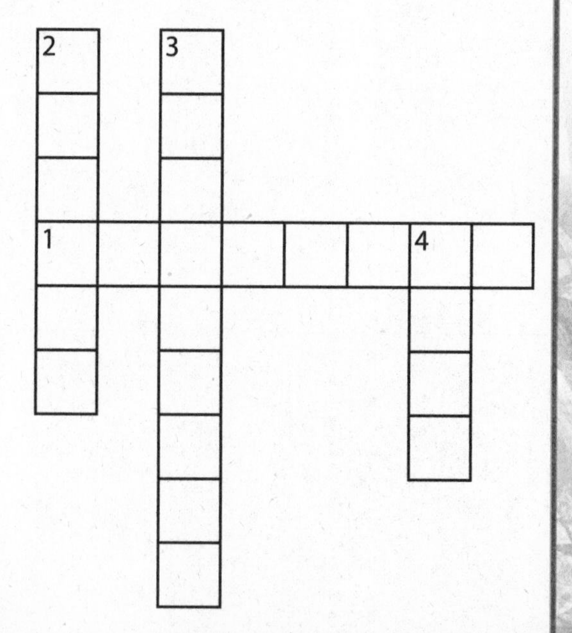

Chapter 13 Vocabulary Cards

base

composite figure

cubic unit

right
rectangular prism

unit cube

volume

A figure that is made of two or more solid figures

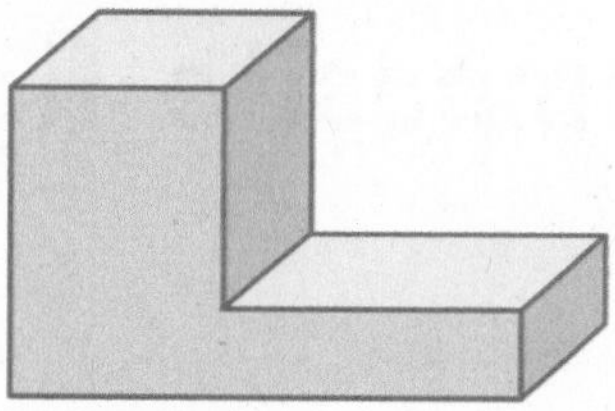

The bottom face of a right rectangular prism

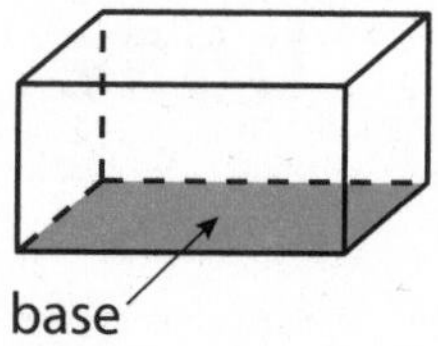

base

A solid figure with six rectangular faces

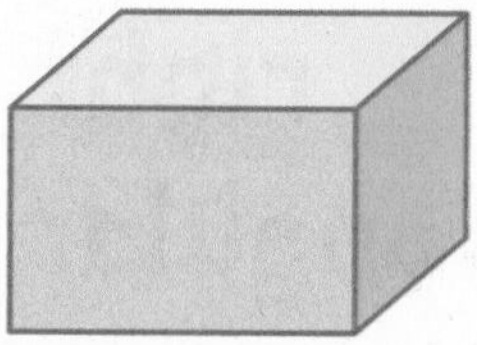

A unit used to measure volume

cubic centimeter
cubic inch
cubic foot

A measure of the amount of space that a solid figure occupies

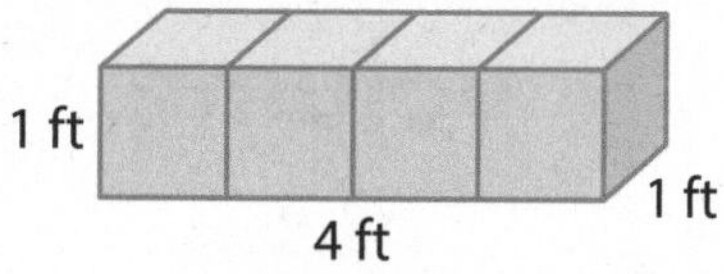

1 ft
4 ft
1 ft

The volume of the figure is 4 cubic feet.

A cube that measures one unit on each side

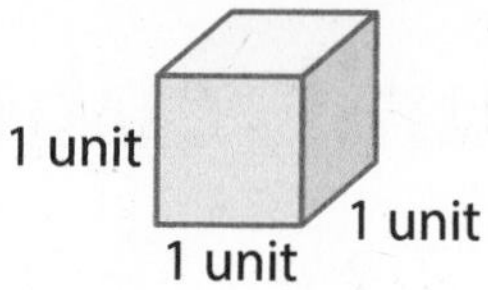

1 unit
1 unit
1 unit

Name ______________________________

Understand the Concept of Volume 13.1

Learning Target: Count to find volumes of solid figures.

Success Criteria:

- I can count the number of unit cubes in a figure.
- I can tell the volume of a solid figure in cubic units.
- I can identify units as cubic inches, cubic feet, or cubic centimeters.

Explore and Grow

Use centimeter cubes to make each figure.

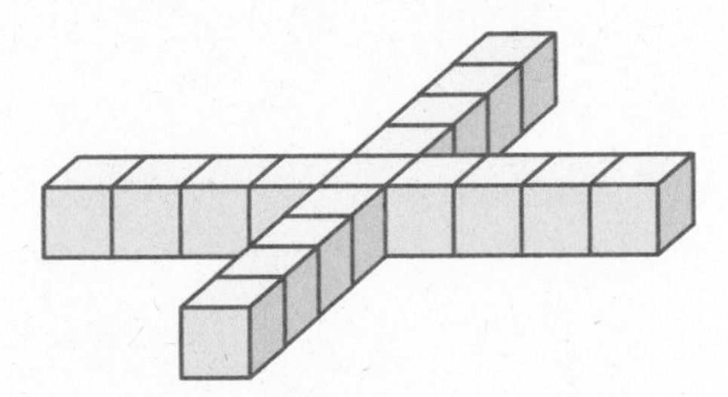

Figure 1

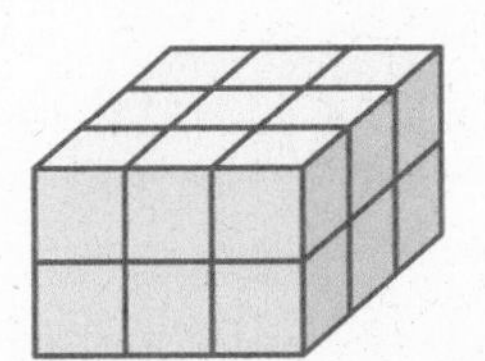

Figure 2

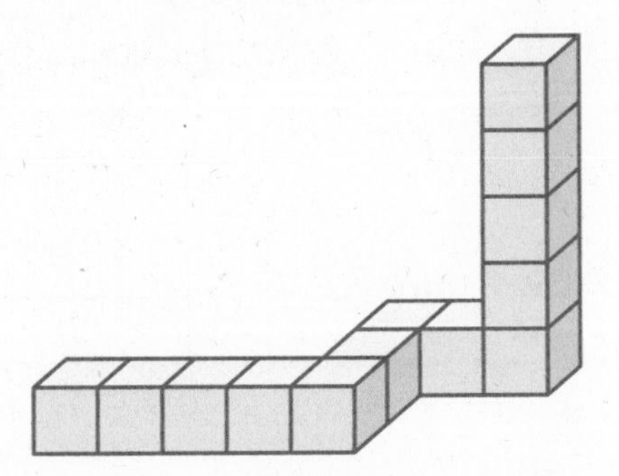

Figure 3

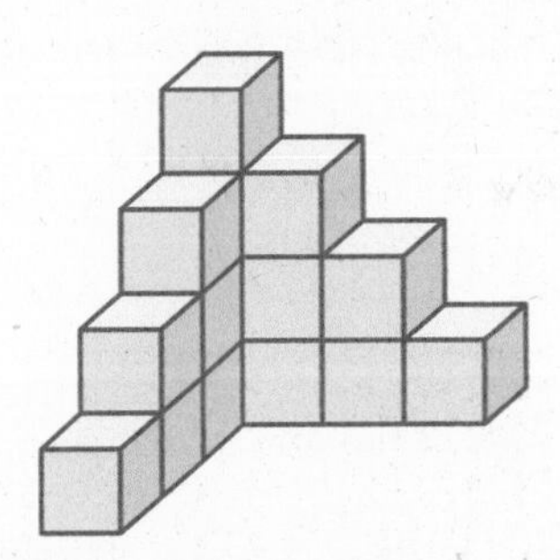

Figure 4

Which figure takes up the most space? Which figure takes up the least space? How do you know?

Structure Can two different figures take up the same amount of space? Explain.

Think and Grow: Use Unit Cubes to Find Volume

Key Idea **Volume** is a measure of the amount of space that a solid figure occupies. The volume of a unit cube is 1 **cubic unit**. You can count unit cubes to find the volume of a solid figure.

unit cube

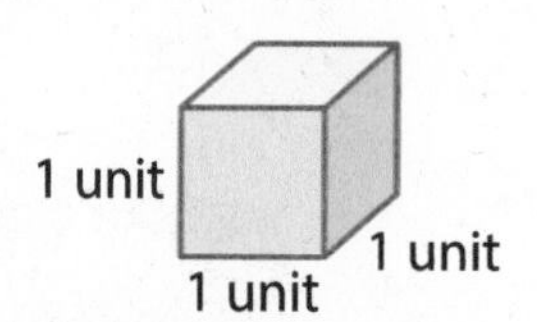

Example Find the volume of the figure.

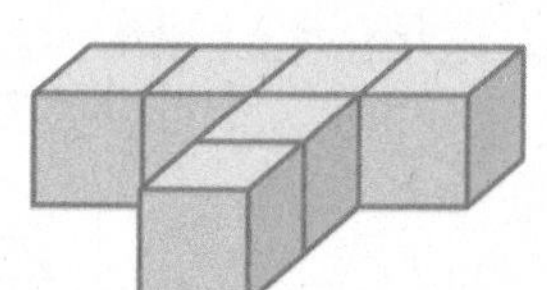

The figure is made of _____ unit cubes.

So, the volume of the figure is _____ cubic units.

Example Find the volume of the figure.

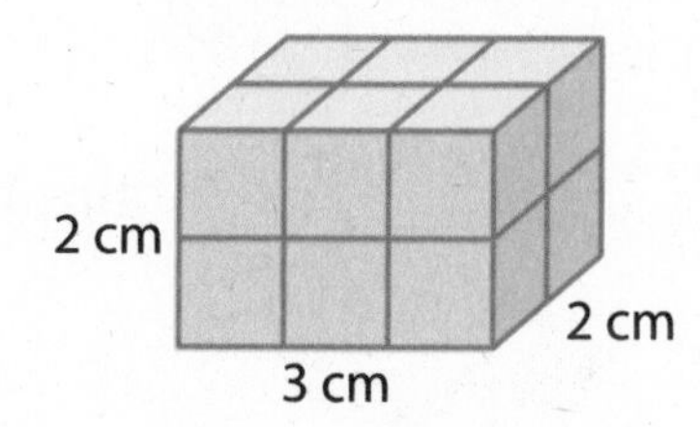

Each unit cube has an edge length of ______________.

So, each unit cube has a volume of ______________.

The figure is made of _____ unit cubes.

So, the volume of the figure is ______________.

Show and Grow *I can do it!*

Find the volume of the figure.

1. Volume = _____ cubic units

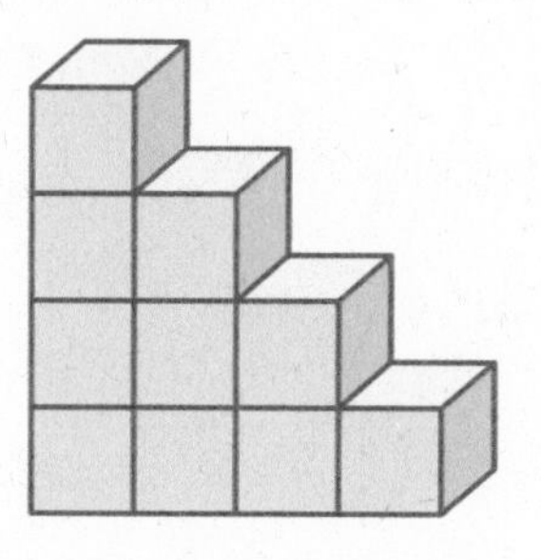

2. Volume = ______________

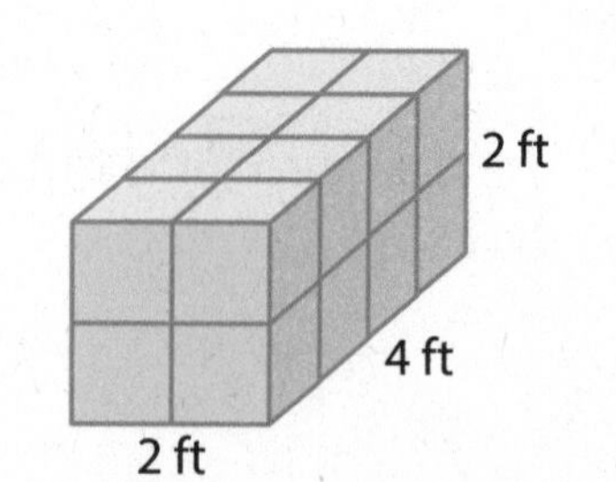

Name ______________________________

Apply and Grow: Practice

Find the volume of the figure.

3. Volume = ______________

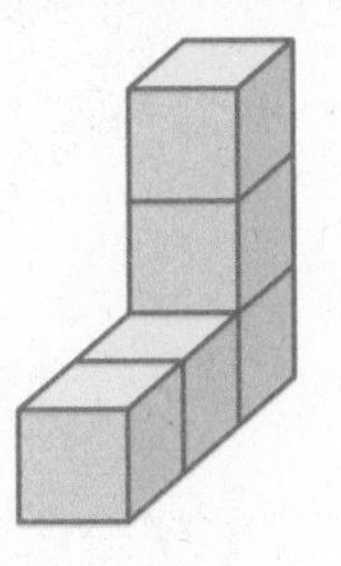

4. Volume = ______________

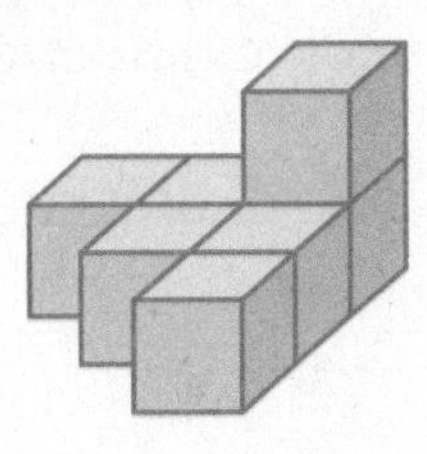

5. Volume = ______________

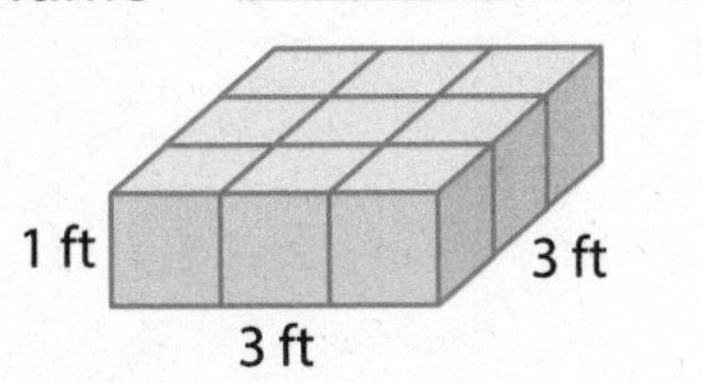

6. Volume = ______________

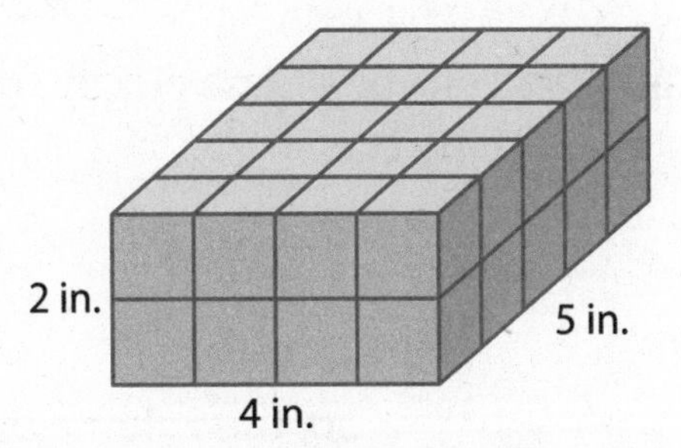

7. Cube-shaped shipping boxes, like the one shown, are packed into a van. They form a solid without any gaps or overlaps that is 3 feet long, 2 feet wide, and 3 feet tall. What is the volume of the solid formed?

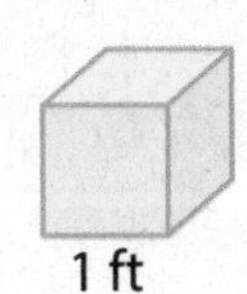

8. **YOU BE THE TEACHER** Your friend says the volume of the figure is 12 cubic centimeters. Is your friend correct? Explain.

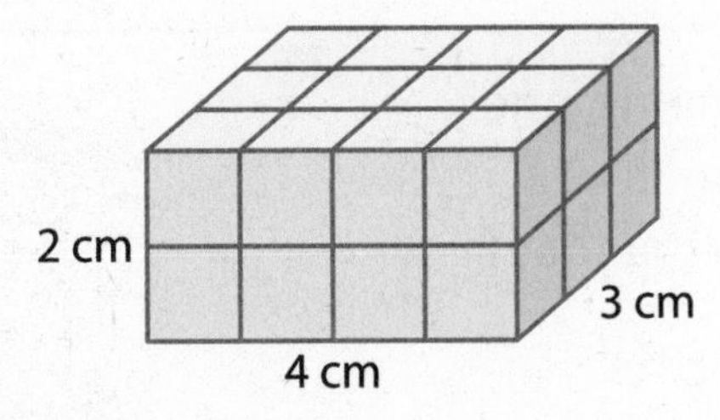

9. **DIG DEEPER!** What is the volume of the figure formed by the missing cubes?

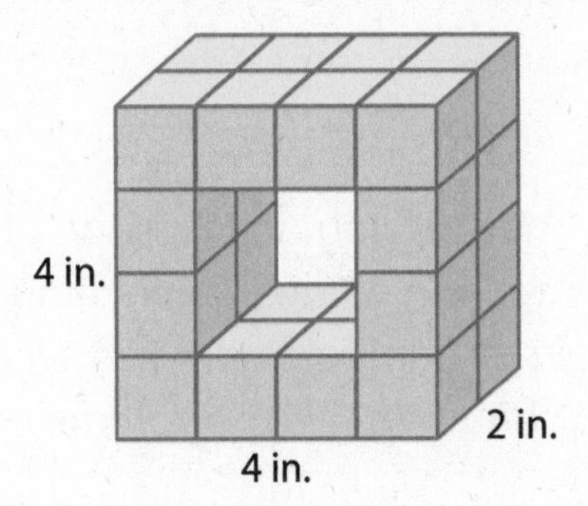

Think and Grow: Modeling Real Life

Example The puzzle is made of unit cubes. A box can hold 20 of the puzzles without any gaps or overlaps. What is the volume of the box?

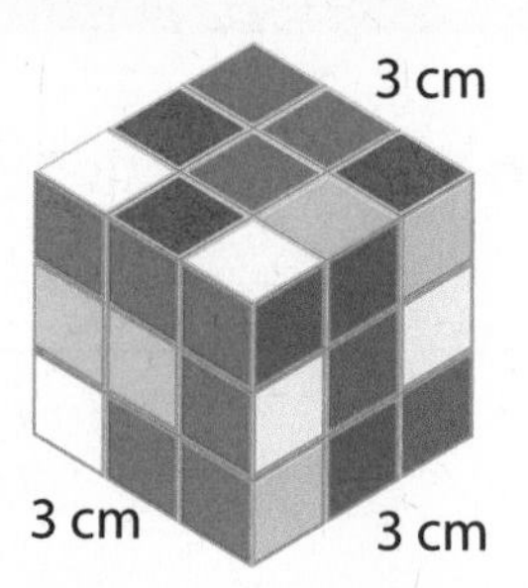

Find the volume of one puzzle.

The puzzle is made of unit cubes that each have an edge length

of ______________.

So, each unit cube has a volume of ______________.

One puzzle is made of _____ unit cubes.

So, the volume of one puzzle is ______________.

Multiply the volume of one puzzle by the number of puzzles that can fit in the box.

_____ × _____ = _____

So, the volume of the box is ______________.

Area is measured in square units, or units².
Volume is measured in cubic units, or units³.

Show and Grow *I can think deeper!*

10. A container holds 8 of the snowglobe boxes shown, without any gaps or overlaps. What is the volume of the container?

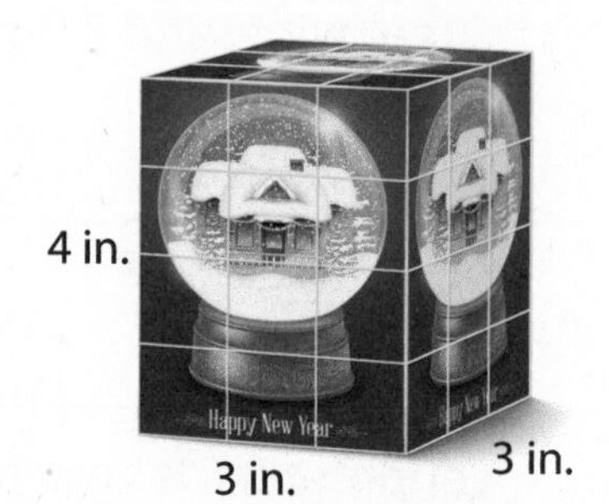

11. The shelving unit is made of six small compartments, one medium compartment, and one large compartment. Each small compartment has a length, a width, and a height of 1 foot. What is the volume of the shelving unit?

Name ____________________

Homework & Practice 13.1

Learning Target: Count to find volumes of solid figures.

Example Find the volume of the figure.

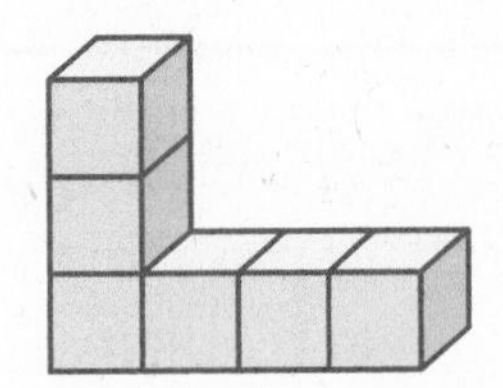

The figure is made of __6__ unit cubes.

So, the volume of the figure is __6__ cubic units.

Example Find the volume of the figure.

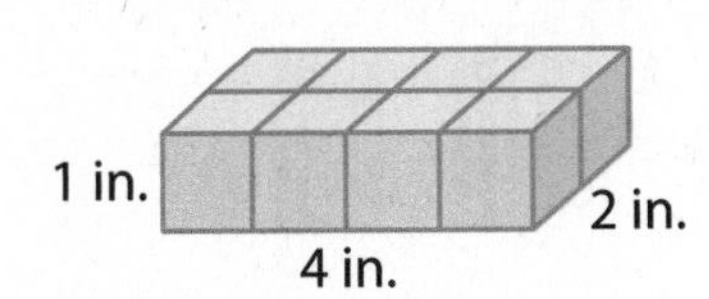

Each unit cube has an edge length of __1 inch__.

So, each unit cube has a volume of __1 cubic inch__.

The figure is made of __8__ unit cubes.

So, the volume of the figure is __8 cubic inches__.

Find the volume of the figure.

1. Volume = ____________

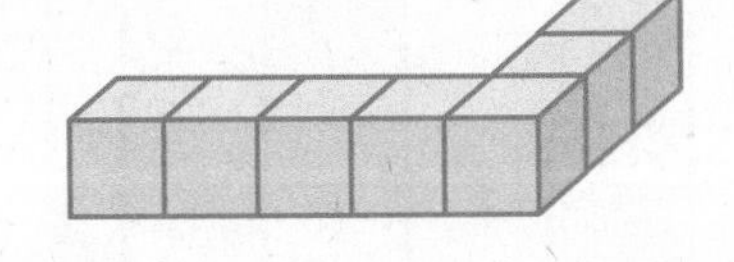

2. Volume = ____________

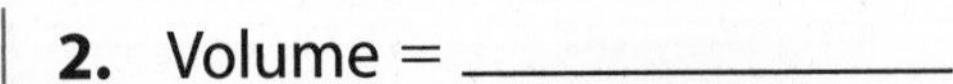

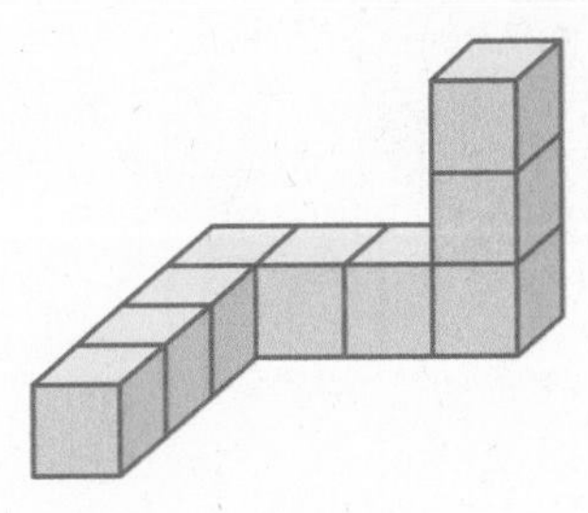

3. Volume = ____________

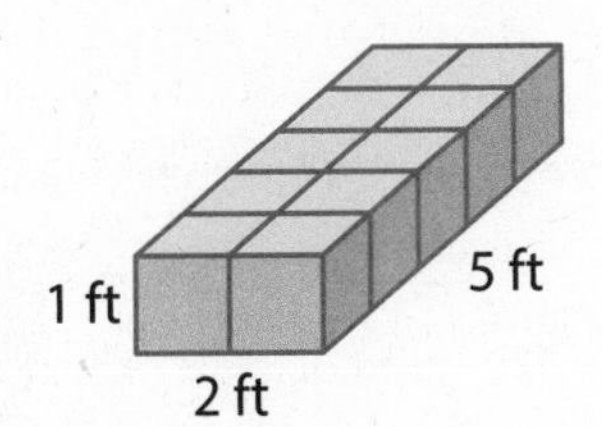

4. Volume = ____________

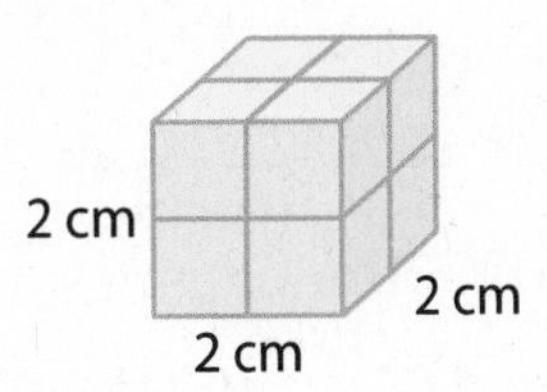

5. Cube-shaped boxes, like the one shown, are placed together for a display at a store. They form a solid without any gaps or overlaps that is 5 feet long, 4 feet wide, and 1 foot tall. What is the volume of the solid formed?

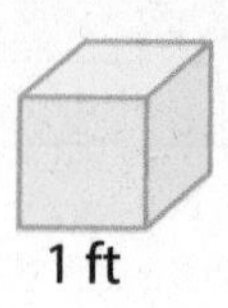

---

6. YOU BE THE TEACHER Your friend says the figures have the same volume. Is your friend correct? Explain.

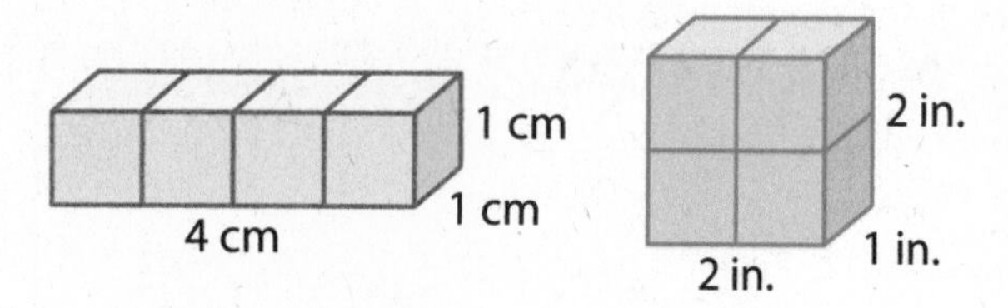

7. DIG DEEPER! Newton uses 32 centimeter cubes to build a solid. The base of the solid is shown. What is the height of the solid?

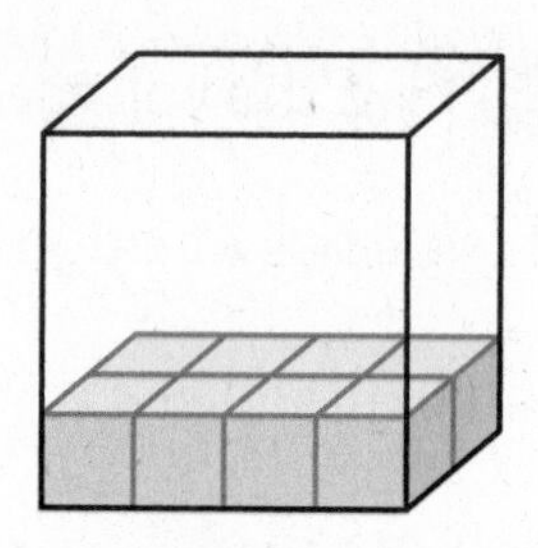

---

8. **Modeling Real Life** A container holds 27 of the spring toy boxes shown, without any gaps or overlaps. What is the volume of the container?

9. **Modeling Real Life** A store stacks boxes in an organized display. Each shaded box has a length, a width, and a height of 1 foot. What is the volume of the display?

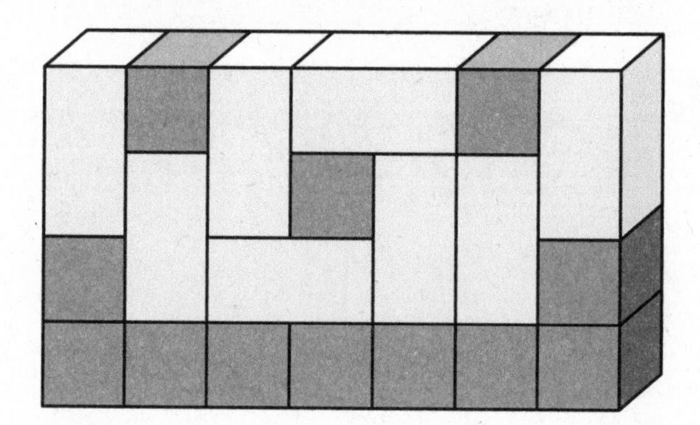

Review & Refresh

Divide.

10. $3 \div 5 =$ _____

11. $1 \div 4 =$ _____

12. $7 \div 8 =$ _____

Name ____________________

Find Volumes of Right Rectangular Prisms 13.2

Learning Target: Find volumes of right rectangular prisms.

Success Criteria:

- I can find the number of unit cubes in each layer of a rectangular prism.
- I can use the number of unit cubes in each layer to find the volume of a rectangular prism.

Explore and Grow

You can stack identical layers of unit cubes to build a rectangular prism.

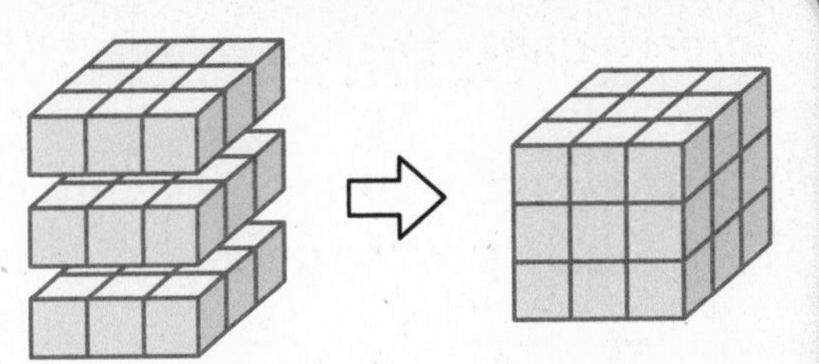

Use centimeter cubes to make one layer of a rectangular prism. Then stack identical layers and complete the table.

Number of Layers	Volume of One Layer (cubic centimeters)	Volume of Prism (cubic centimeters)
1		
2		
3		
4		

MP **Precision** Explain the relationship among the values in each row of the table.

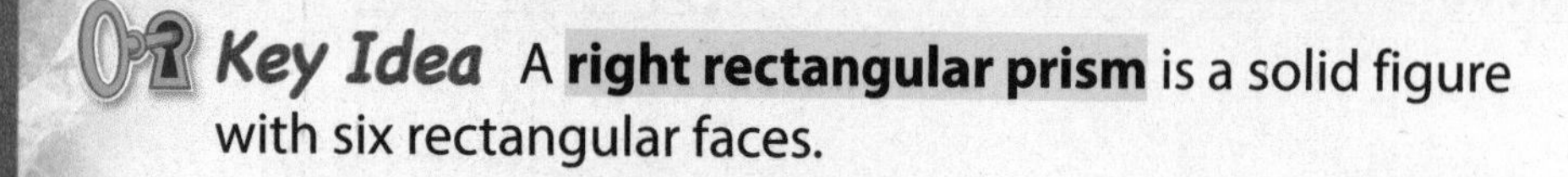

Think and Grow: Find Volumes of Right Rectangular Prisms

Key Idea A **right rectangular prism** is a solid figure with six rectangular faces.

To find the volume of a right rectangular prism, multiply the number of unit cubes that cover the **base** by the number of layers of unit cubes.

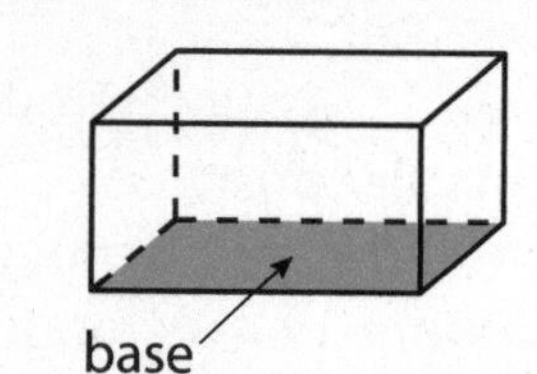

Example Find the volume of the rectangular prism.

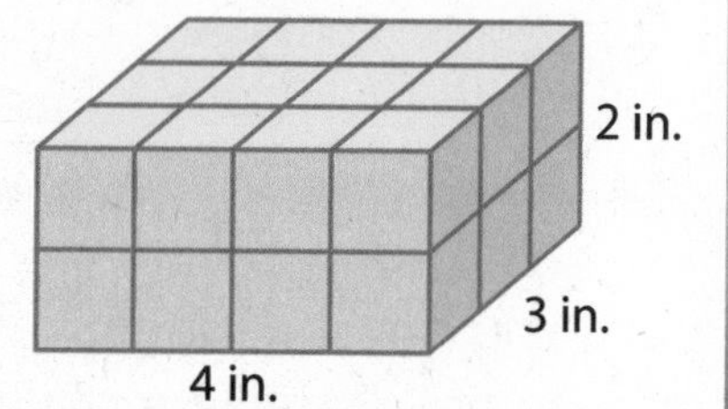

Each unit cube has an edge length of ________.

So, each unit cube has a volume of ________.

Find the number of unit cubes in a base layer.
Then multiply by the number of layers to find the volume.

A base layer is made of ____ × ____ = ____ unit cubes.

The prism is made of ____ layers of unit cubes.

So, the prism is made of ____ × ____ = ____ unit cubes.

The volume of the prism is ________.

You can also count unit cubes to find the volume of the prism.

Show and Grow *I can do it!*

Find the volume of the rectangular prism.

1. Volume = ________

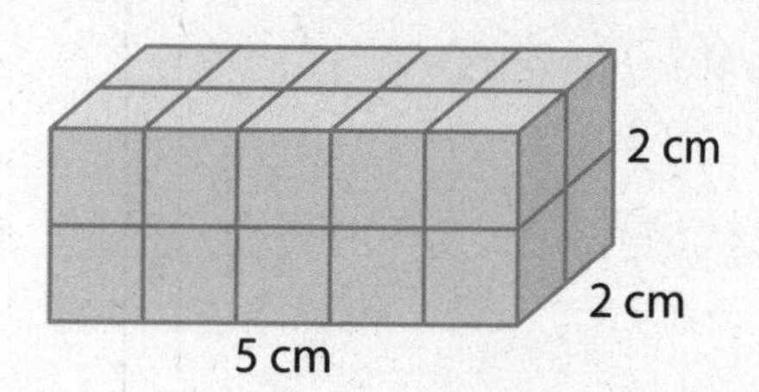

2. Volume = ________

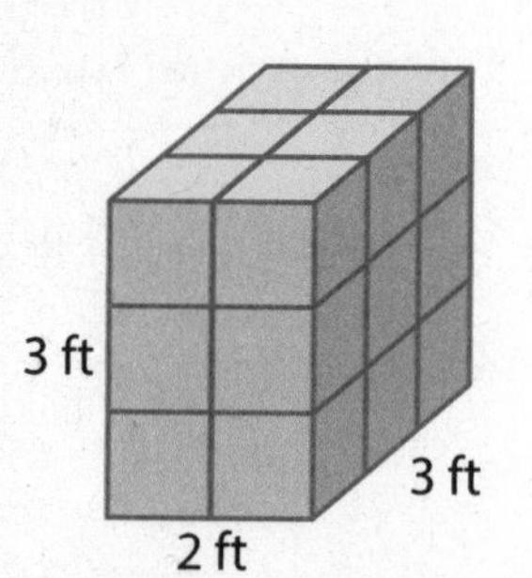

Name ______________________________

Apply and Grow: Practice

Find the volume of the rectangular prism.

3. Volume = ______________

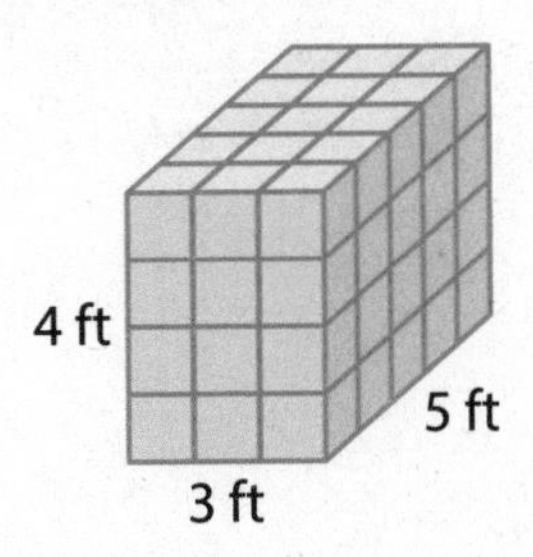

4. Volume = ______________

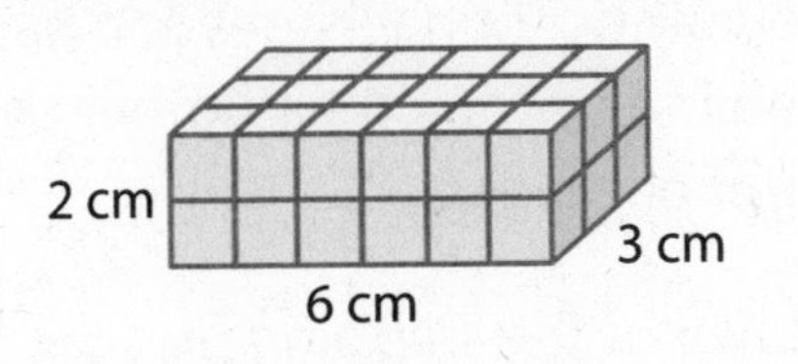

5. Volume = ______________

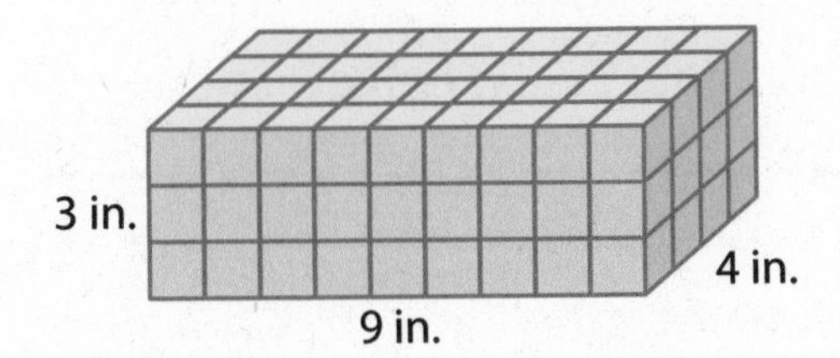

6. Volume = ______________

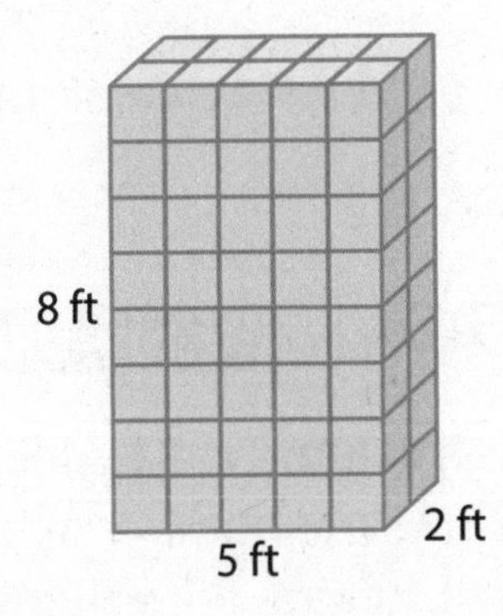

7. A rectangular prism is made of unit cubes, like the one shown. The prism is 6 centimeters long, 3 centimeters wide, and 4 centimeters tall. What is the volume of the prism?

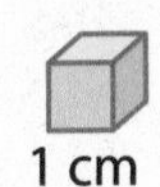

8. MP **Structure** What happens to the volume of a rectangular prism when you double its height? Justify your answer by giving an example.

9. DIG DEEPER! How many unit cubes can fit inside the rectangular prism? Explain.

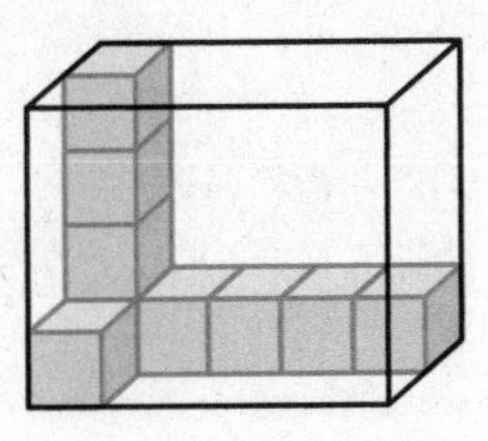

Think and Grow: Modeling Real Life

Example A package containing video games has a volume of 150 cubic inches. The diagram shows the number of packages that can fit in a shipping box. Estimate the volume of the shipping box.

Find the number of packages in a base layer of the shipping box. Then multiply by the number of layers to find the total number of packages.

A base layer of the shipping box can hold $2 \times 3 = 6$ packages. There are 4 layers. So, the shipping box holds $4 \times 6 = 24$ packages.

Multiply the volume of 1 package by the number of packages that fit in the shipping box.

$24 \times$ ______ = ______ cubic inches

So, the volume of the shipping box is about ______ cubic inches.

Show and Grow *I can think deeper!*

10. A book has a volume of 91 cubic inches. The diagram shows the number of books that can fit in a container. Estimate the volume of the container.

11. **DIG DEEPER!** Your teacher stores 12 of the crates in a supply closet. The closet is 3 feet long, 2 feet wide, and 7 feet tall. How much extra space is in the closet?

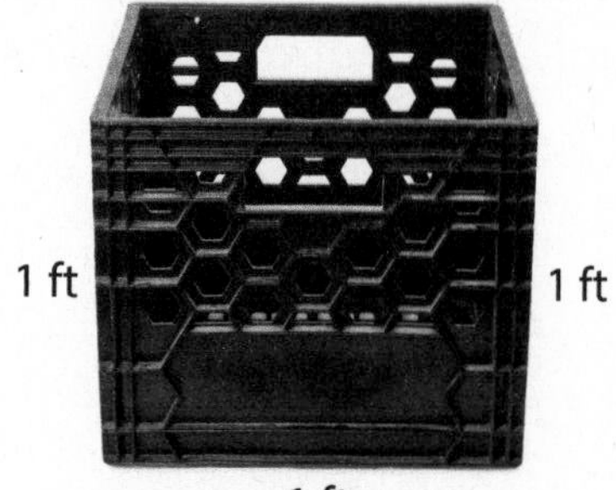

Name ______________________________

Homework & Practice 13.2

Learning Target: Find volumes of right rectangular prisms.

Example Find the volume of the rectangular prism.

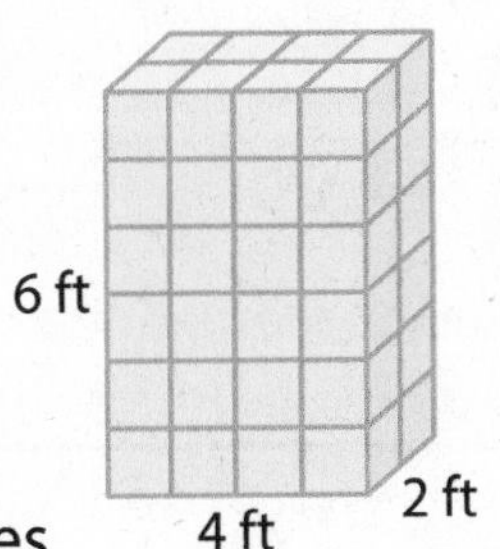

Find the number of unit cubes in a base layer. Then multiply by the number of layers to find the volume.

A base layer is made of __4__ × __2__ = __8__ unit cubes.

The prism is made of __6__ layers of unit cubes.

So, the prism is made of __8__ × __6__ = __48__ unit cubes.

The volume of the prism is __48 cubic feet__.

Find the volume of the rectangular prism.

1. Volume = ____________

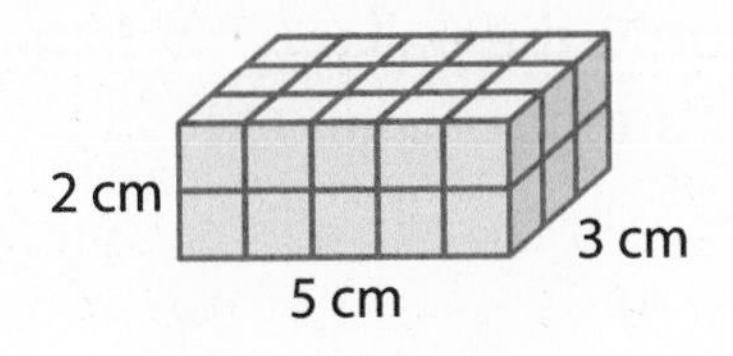

2. Volume = ____________

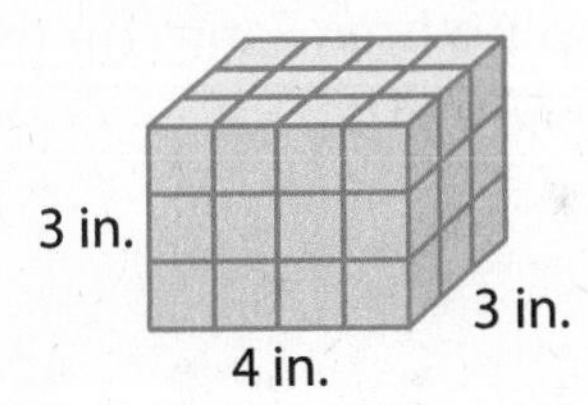

3. Volume = ____________

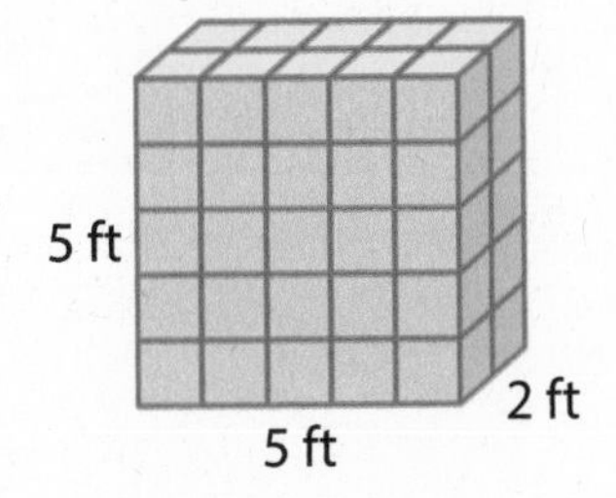

4. Volume = ____________

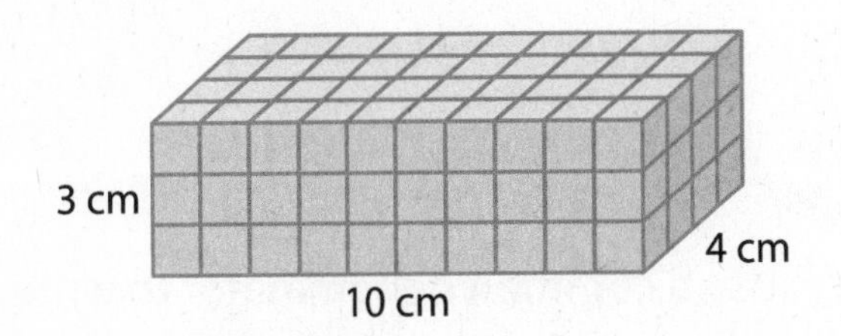

5. A rectangular prism is made of unit cubes, like the one shown. The prism is 7 inches long, 2 inches wide, and 4 inches tall. What is the volume of the prism?

6. **Reasoning** A cube has a volume of 27 cubic inches. What is the edge length of the cube?

7. **Reasoning** You, Newton, and Descartes each build the rectangular prism shown. You stack all three prisms on top of each other. What is the volume of the new prism? How many times greater is the volume of the new prism than the volume of your original prism?

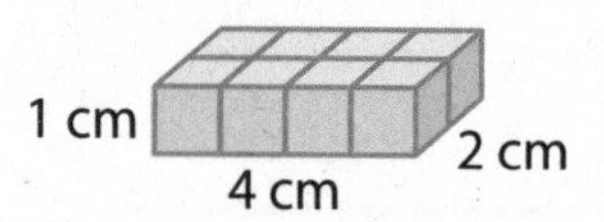

8. **Modeling Real Life** A shoe box has a volume of 336 cubic inches. The diagram shows the number of shoe boxes that can fit in a shipping box. Estimate the volume of the shipping box.

9. **Modeling Real Life** The wooden pallet can hold a maximum of 5 layers of boxes. A worker places 48 boxes, like the one shown, onto the pallet with no gaps. How many more boxes can fit?

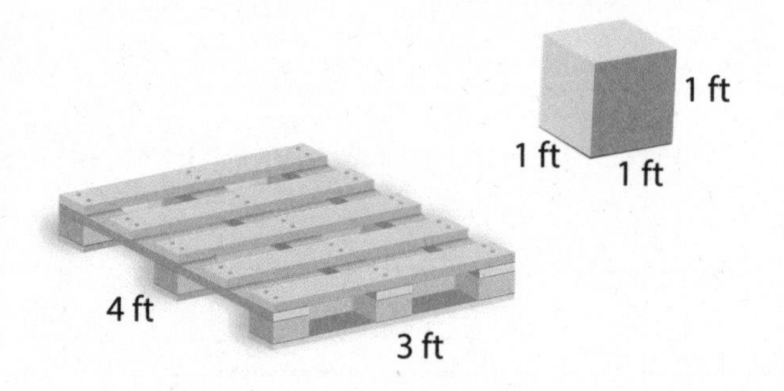

Review & Refresh

Use a common denominator to write an equivalent fraction for each fraction.

10. $\frac{2}{3}$ and $\frac{5}{9}$

11. $\frac{1}{4}$ and $\frac{3}{7}$

Name ____________________

Apply the Volume Formula 13.3

Learning Target: Use a formula to find volumes of rectangular prisms.

Success Criteria:
- I can write a formula for the volume of a rectangular prism.
- I can explain how to use the area of the base to find the volume of a rectangular prism.
- I can use a formula to find the volume of a rectangular prism.

Explore and Grow

Work with a partner. Use centimeter cubes to create several different rectangular prisms, each with a volume of 12 cubic centimeters. Record your results in the table.

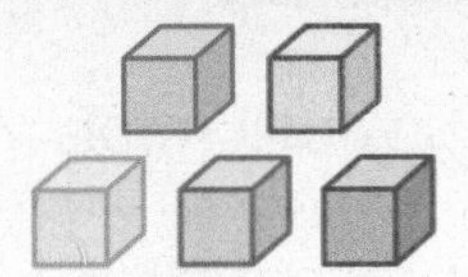

Base Length (centimeters)	Base Width (centimeters)	Height (centimeters)

What do you notice about the dimensions of each prism?

Structure How can you find the volume of a rectangular prism without using unit cubes?

Think and Grow: Use a Formula to Find Volume

Key Idea You can use the length, width, and height of a rectangular prism to find its volume.

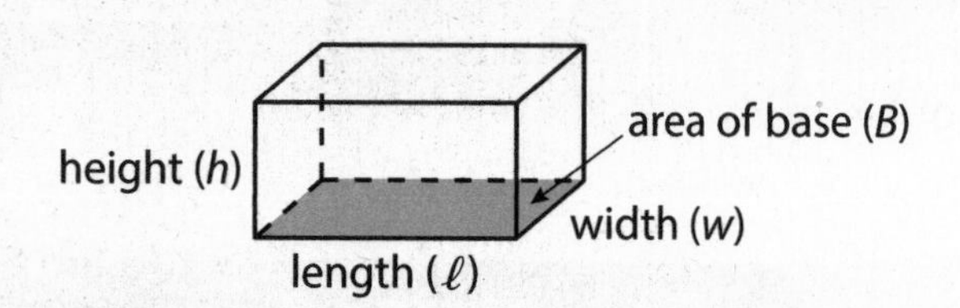

Volume of a Rectangular Prism

B ← area of base

$$V = \underbrace{\ell \times w}_{B} \times h$$

volume length width height

Example Find the volume of the rectangular prism.

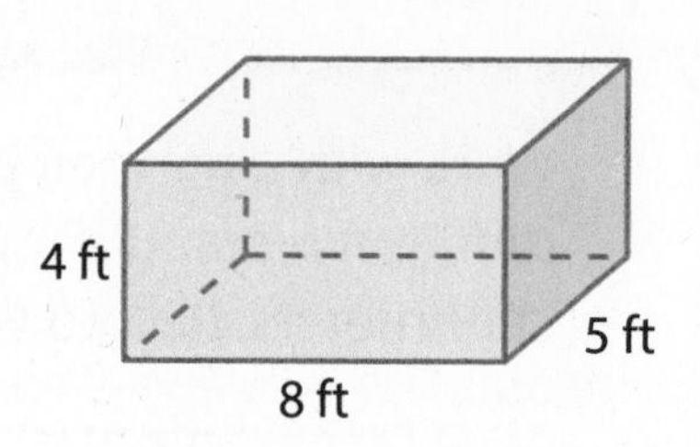

Use the formula $V = \ell \times w \times h$.

The length is _____ feet, the width is _____ feet,

and the height is _____ feet.

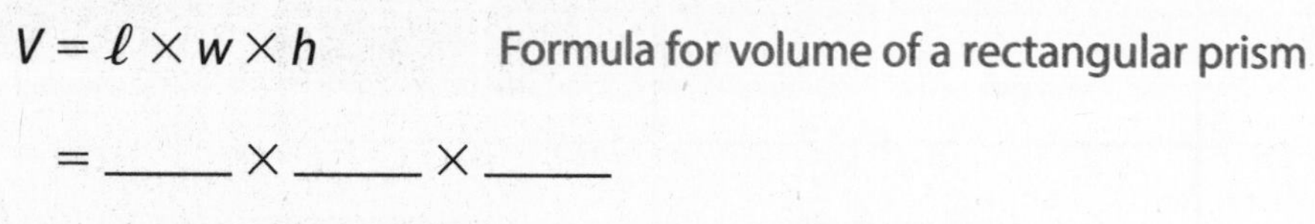

$V = \ell \times w \times h$ Formula for volume of a rectangular prism

$= \_\_\_\_ \times \_\_\_\_ \times \_\_\_\_$

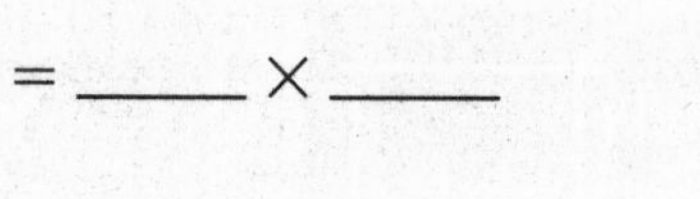

$= \_\_\_\_ \times \_\_\_\_$

$= \_\_\_\_$

The volume is ______________.

You can also use the formula $V = B \times h$, where $B = \ell \times w$.

Show and Grow I can do it!

Find the volume of the rectangular prism.

1.

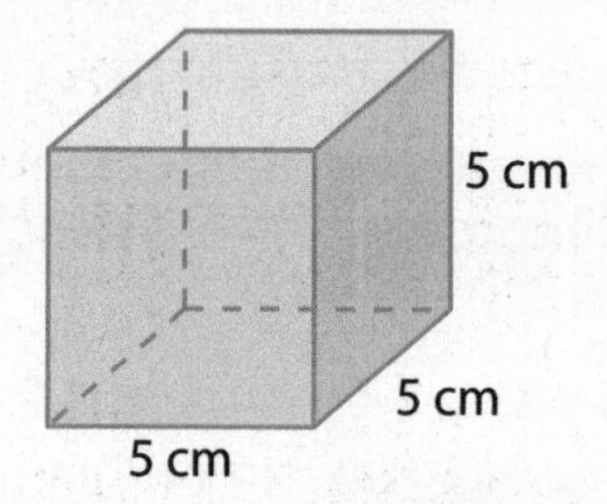

2.

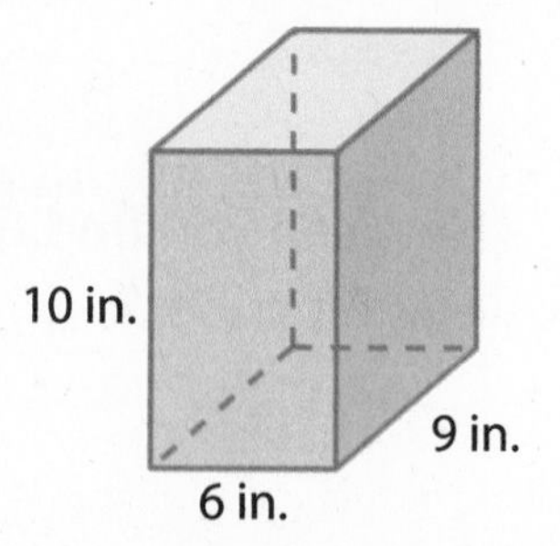

Name ___

Apply and Grow: Practice

Find the volume of the rectangular prism.

3.

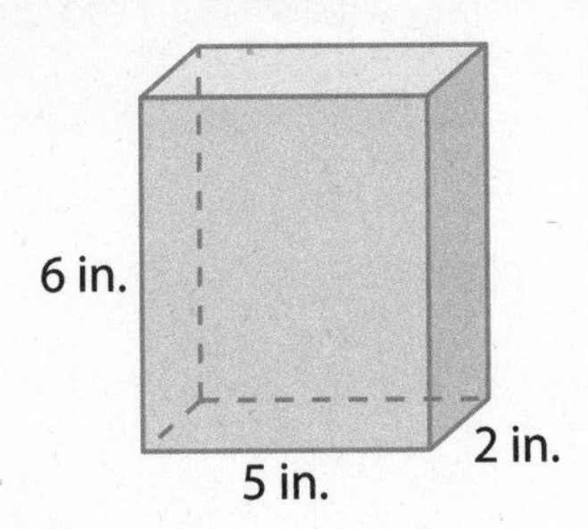

4.

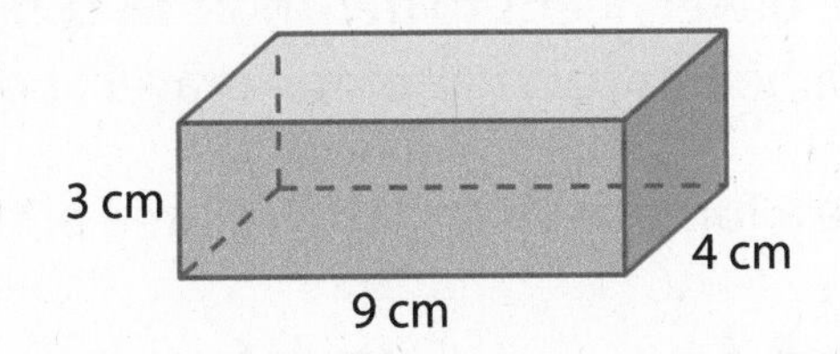

5. Newton's dog house is a rectangular prism. The house is 5 feet long, 4 feet wide, and 5 feet tall. What is the volume of the house?

6. A hole dug for an in-ground swimming pool is a rectangular prism. The hole is 20 feet long, 10 feet wide, and 6 feet tall. What is the volume of the hole?

7. MP **Structure** Compare the dimensions and volumes of the three rectangular prisms. What do you notice?

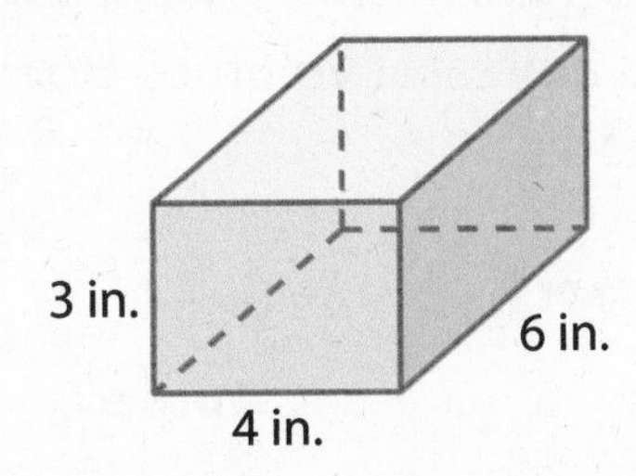

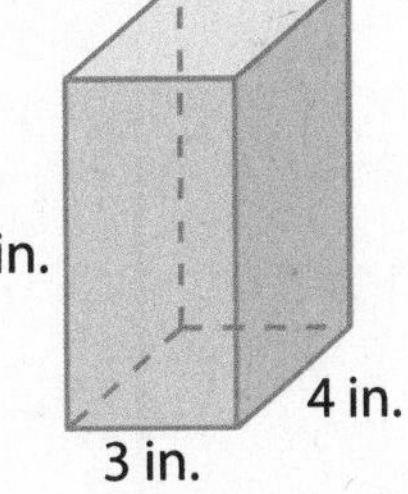

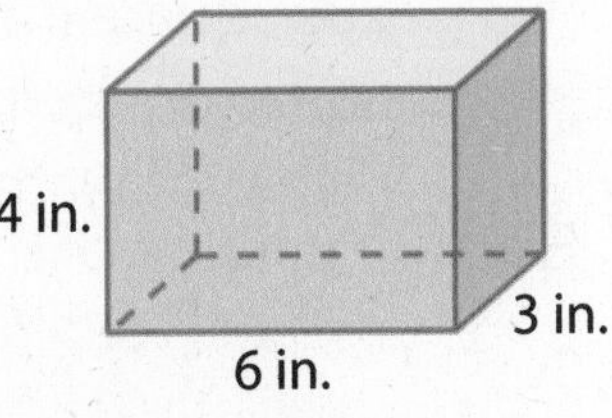

8. **DIG DEEPER!** What is the volume of Newton's rectangular prism?

Its length is 2 centimeters. Its width is twice its height. Its height is 5 times its length.

Think and Grow: Modeling Real Life

Example The dump truck bed is a rectangular prism. The area of the base is 8 square yards. The height is 2 yards. The driver needs to transport 18 cubic yards of gravel. Can he transport all of the gravel at once?

Use a formula to find the volume of the dump truck bed.

$V = B \times h$ Formula for volume of a rectangular prism

$= \_\_\_\_ \times \_\_\_\_$

$= \_\_\_\_$

The volume of the dump truck bed is ________.

Compare the volume of the bed to the amount of gravel.

____ ◯ 18

So, the driver ________ transport all of the gravel at once.

Show and Grow *I can think deeper!*

9. The aquarium is a rectangular prism. The area of the base is 288 square inches. Can the aquarium hold 4,500 cubic inches of water? Explain.

10. **DIG DEEPER!** A principal orders two types of lockers for the school. Which locker has a greater volume? How much greater?

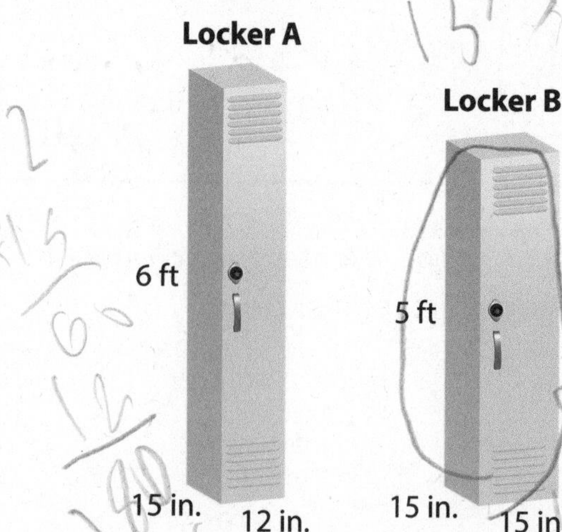

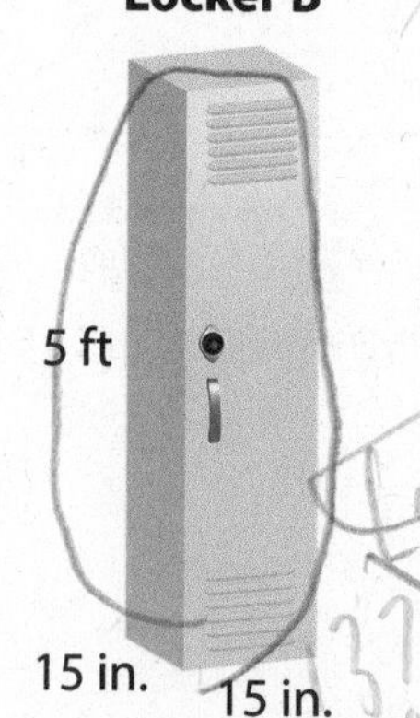

Name ______________________

Homework & Practice 13.3

Learning Target: Use a formula to find volumes of rectangular prisms.

Example Find the volume of the rectangular prism.

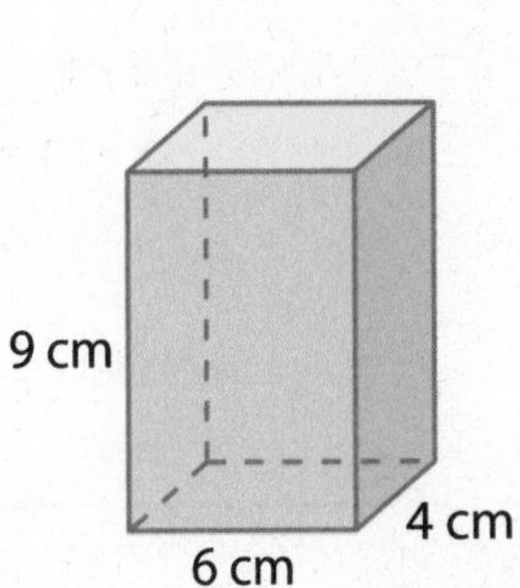

Use the formula $V = \ell \times w \times h$.

The length is __6__ centimeters,
the width is __4__ centimeters,
and the height is __9__ centimeters.

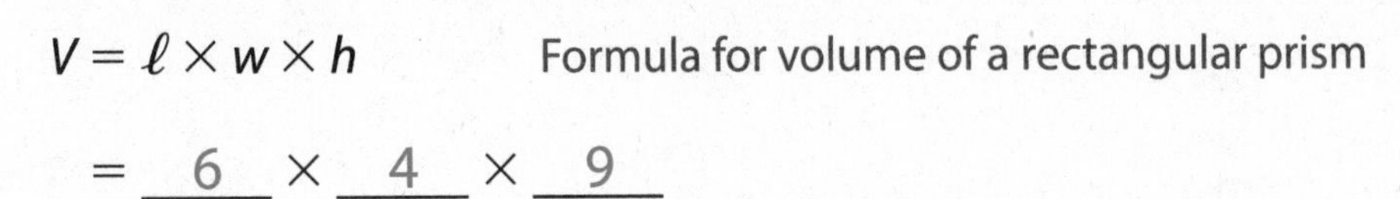

$V = \ell \times w \times h$ Formula for volume of a rectangular prism

$= \underline{6} \times \underline{4} \times \underline{9}$

$= \underline{24} \times \underline{9}$

$= \underline{216}$

The volume is 216 cubic centimeters.

You can also use the formula $V = B \times h$, where $B = \ell \times w$.

Find the volume of the rectangular prism.

1.

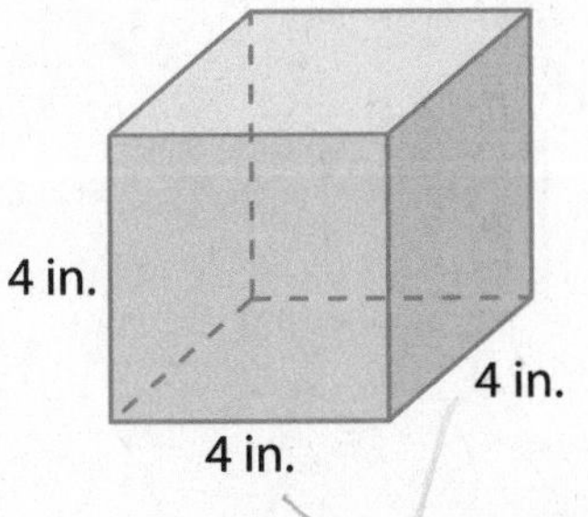

16 × 4 = 64
64 cubic inches

2.

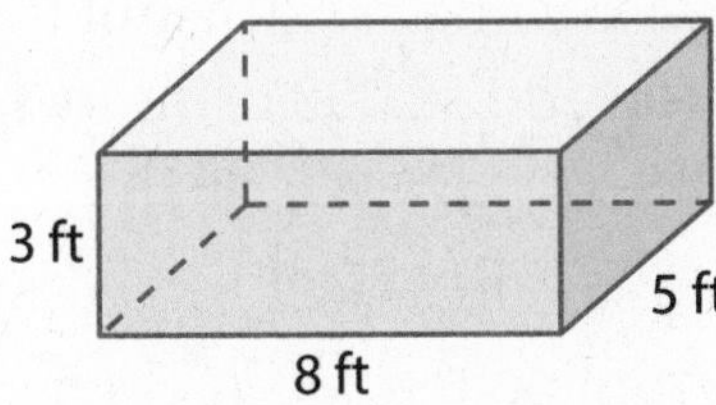

24 × 5 = 120
120 cubic feet

3.

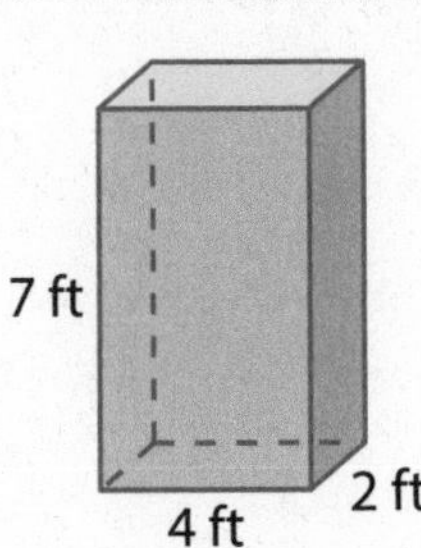

56 cubic feet

4.

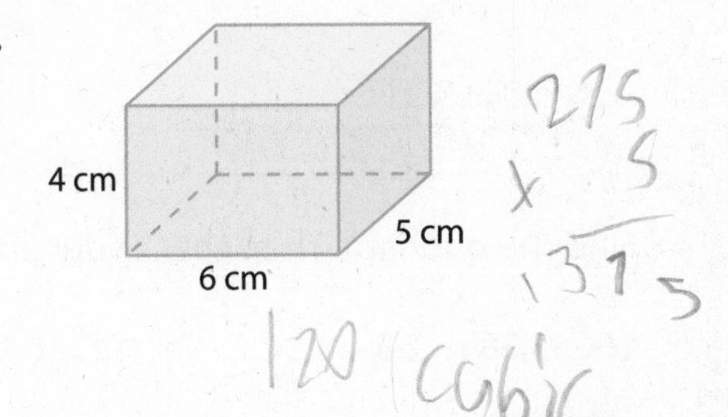

275 × 5 = 1375
120 cubic cm

5. A toy chest is a rectangular prism. The chest is 4 feet long, 2 feet wide, and 2 feet tall. What is the volume of the chest?

6. **YOU BE THE TEACHER** Your friend uses the formula $V = B \times h$ and finds $V = 10 \times 9 = 90$ cubic inches. Is your friend correct? Explain.

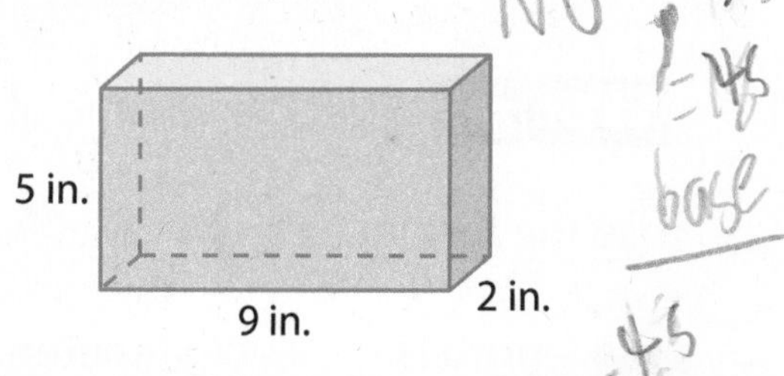

7. **MP Number Sense** Find the volume of the rectangular prism. Then double the dimensions of the prism and find the new volume. How does the new volume compare to the original volume?

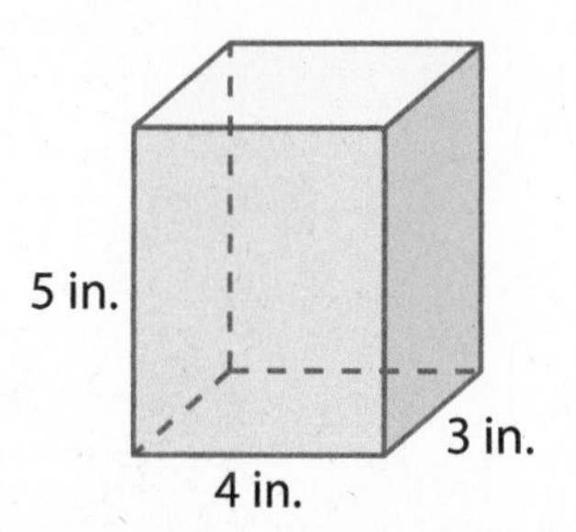

8. **Modeling Real Life** A sandbox is a rectangular prism. The area of the base is 3,600 square inches. The height is 11 inches. You add 38,000 cubic inches of sand to the box. Do you fill the sandbox to the top? Explain.

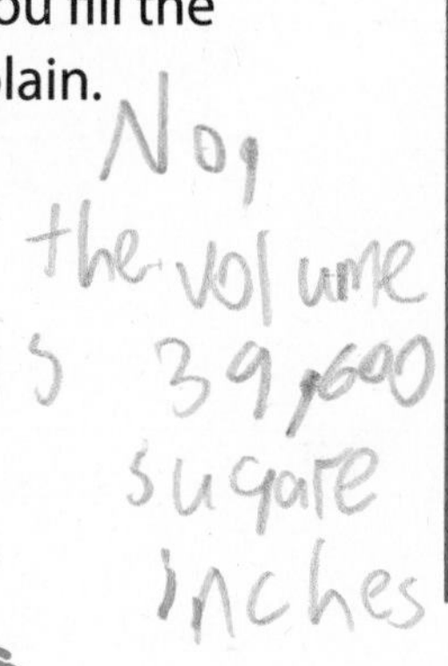

9. **DIG DEEPER!** Each piece of Descartes's clubhouse is a cube. What is the volume of the clubhouse? Explain.

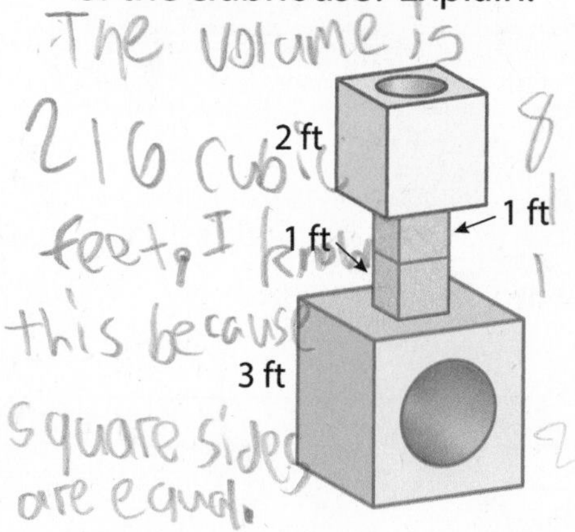

Review & Refresh

Find the quotient. Then check your answer.

10. $3.38 \div 2.6 =$ _____

11. $6.12 \div 1.53 =$ _____

12. $0.63 \div 0.9 =$ _____

Name ________________________________

Find Unknown Dimensions 13.4

Learning Target: Find unknown dimensions of rectangular prisms.

Success Criteria:

- I can find the height of a rectangular prism given the volume of the prism and the area of the base.
- I can find an unknown dimension of a rectangular prism given the volume of the prism and the other two dimensions.

Explore and Grow

For each row of the table, use centimeter cubes to create the rectangular prism described. Then complete the table.

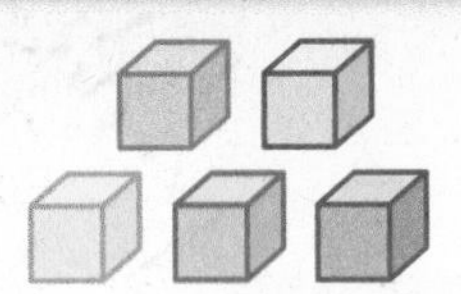

Volume (cubic centimeters)	Length (centimeters)	Width (centimeters)	Height (centimeters)
8	2	2	
15	5		1
24		2	4
36	4	3	

Construct Arguments Explain how you determined the unknown dimension for each rectangular prism.

Think and Grow: Find Unknown Dimensions

Example The volume of the rectangular prism is 56 cubic feet. Find the height.

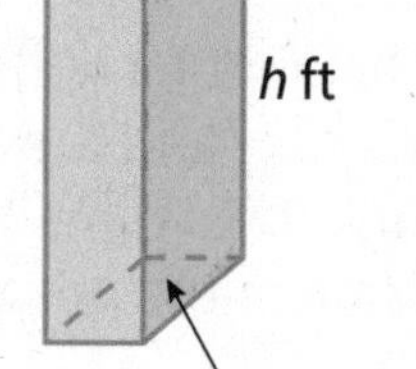

$V = B \times h$	Formula for volume of a rectangular prism
$56 = 8 \times h$	Write an equation.
$56 \div 8 = h$	Write the related division equation.
_____ $= h$	Divide 56 by 8.

The height is _____ feet.

Example The volume of the rectangular prism is 180 cubic centimeters. Find the length.

4 cm

5 cm

ℓ cm

$V = \ell \times w \times h$	Formula for volume of a rectangular prism
$180 = \ell \times 5 \times 4$	Write an equation.
$180 = \ell \times (5 \times 4)$	Associative Property of Multiplication
$180 = \ell \times 20$	Multiply the width and the height.
$180 \div 20 = \ell$	Write the related division equation.
_____ $= \ell$	Divide 180 by 20.

The length is _____ centimeters.

Show and Grow I can do it!

Find the unknown dimension of the rectangular prism.

1. Volume = 48 cubic centimeters

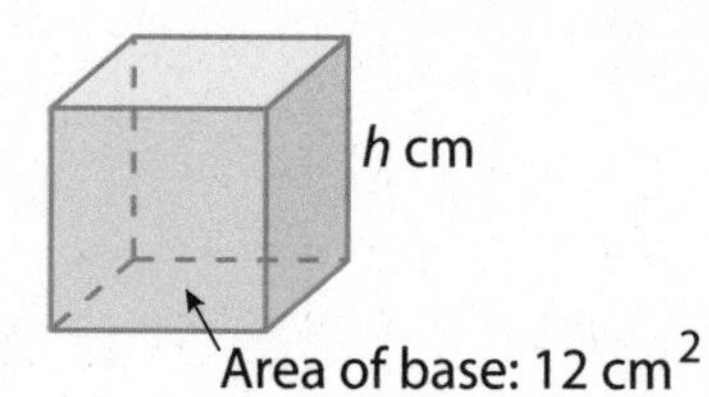

2. Volume = 81 cubic inches

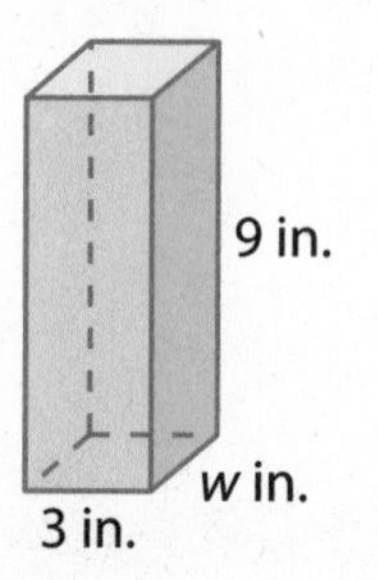

Name ____

Apply and Grow: Practice

Find the unknown dimension of the rectangular prism.

3. Volume = 400 cubic inches

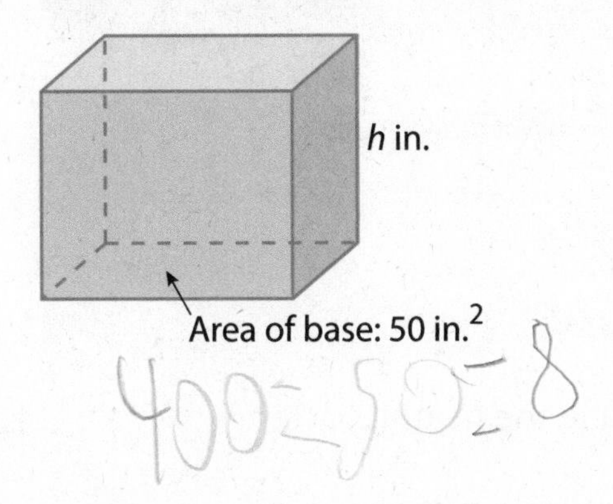

400 ÷ 50 = 8

4. Volume = 343 cubic centimeters

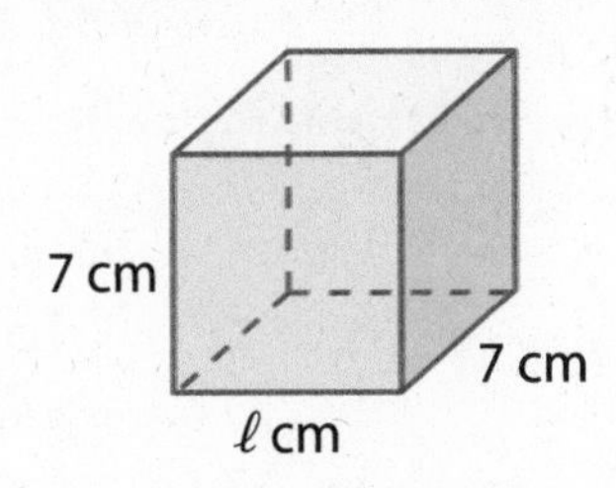

5. Volume = 132 cubic feet

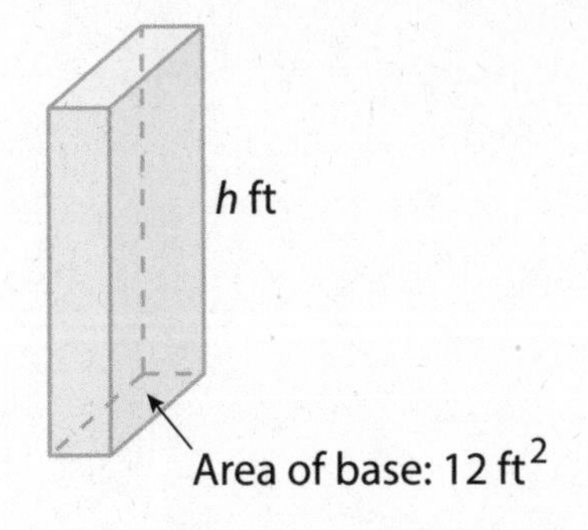

6. Volume = 512 cubic inches

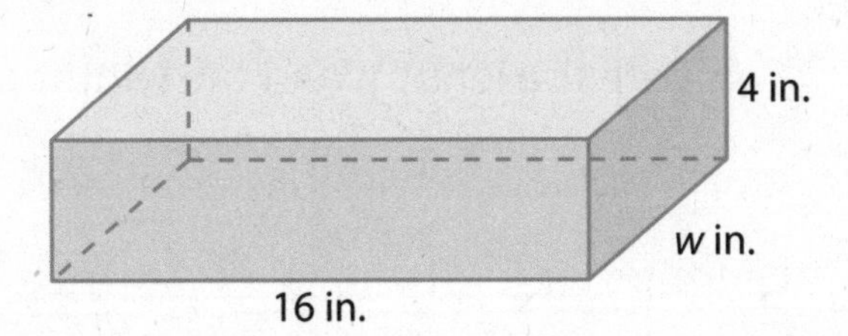

7. Writing Explain how to find an unknown dimension of a rectangular prism given its volume and the other two dimensions.

8. DIG DEEPER! A rectangular prism has a volume of 720 cubic centimeters. The height of the prism is 10 centimeters. The length is twice the width. What are the dimensions of the base?

Think and Grow: Modeling Real Life

Example A traveler buys the painting shown. The volume of the suitcase is 1,820 cubic inches. Can the painting lie flat inside the suitcase?

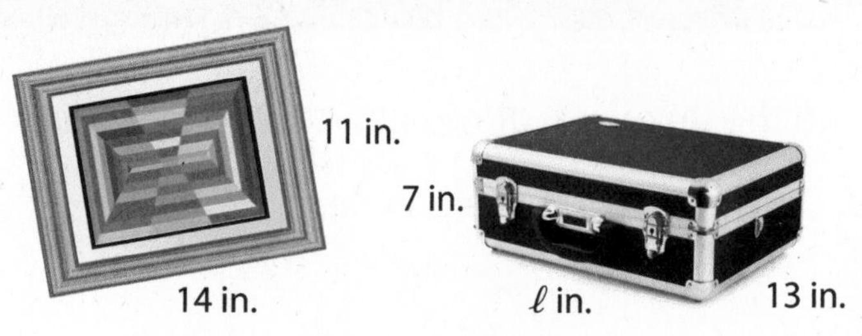

Not drawn to scale

Use a formula to find the length of the suitcase.

$V = \ell \times w \times h$	Formula for volume of a rectangular prism
$1{,}820 = \ell \times 13 \times 7$	Write an equation.
$1{,}820 = \ell \times 91$	Multiply the width and the height.
$1{,}820 \div 91 = \ell$	Write the related division equation.
$\_\_\_\_ = \ell$	Divide 1,820 by 91.

Compare the lengths and the widths of the painting and the suitcase.

	Painting		Suitcase
Length:	14 in.	◯	____
Width:	11 in.	◯	13 in.

So, the painting ____________ lie flat inside the suitcase.

Show and Grow *I can think deeper!*

9. The volume of the candle is 576 cubic centimeters. Will the candle fit completely inside the glass jar? Explain.

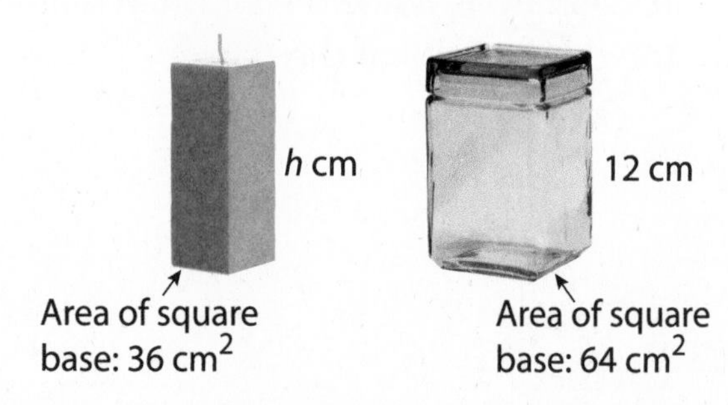

Not drawn to scale

10. **DIG DEEPER!** The popcorn boxes have the same volume, but different dimensions. What is the width of Box B?

Name ______________________________

Homework & Practice 13.4

Learning Target: Find unknown dimensions of rectangular prisms.

Example The volume of the rectangular prism is 60 cubic centimeters. Find the height.

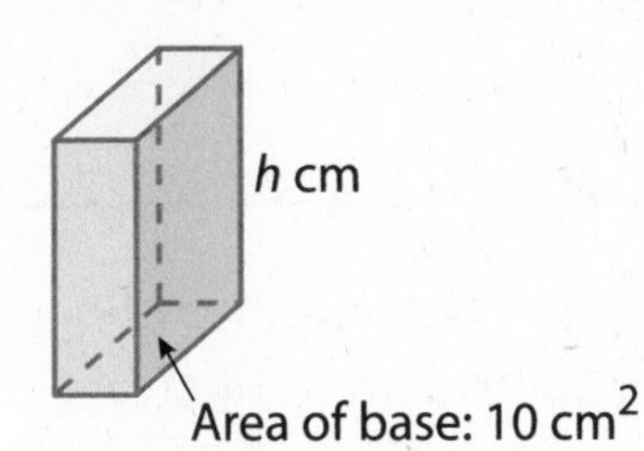

$V = B \times h$

$60 = 10 \times h$

$60 \div 10 = h$

$\underline{\ 6\ } = h$

The height is $\underline{\ 6\ }$ centimeters.

Example The volume of the rectangular prism is 280 cubic feet. Find the width.

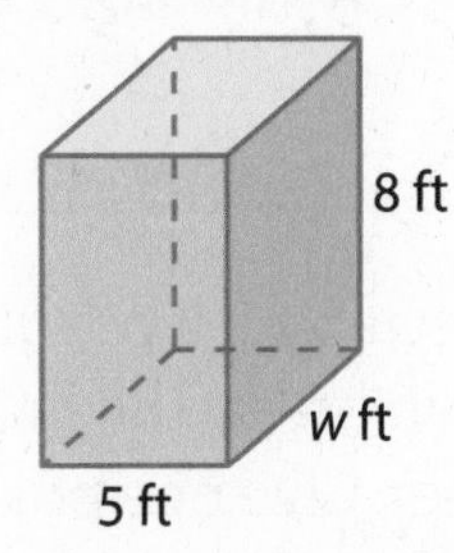

$V = \ell \times w \times h$

$280 = 5 \times w \times 8$

$280 = w \times (5 \times 8)$

$280 = w \times 40$

$280 \div 40 = w$

$\underline{\ 7\ } = w$

The width is $\underline{\ 7\ }$ feet.

Find the unknown dimension of the rectangular prism.

1. Volume = 36 cubic inches

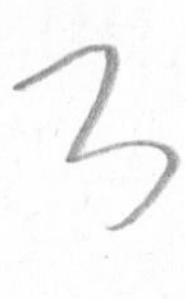

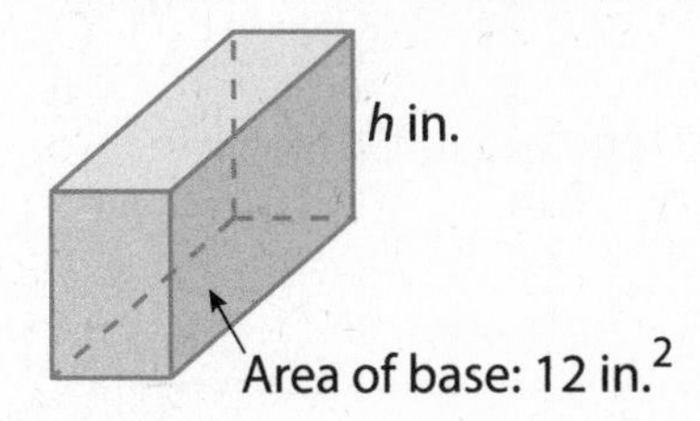

2. Volume = 210 cubic centimeters

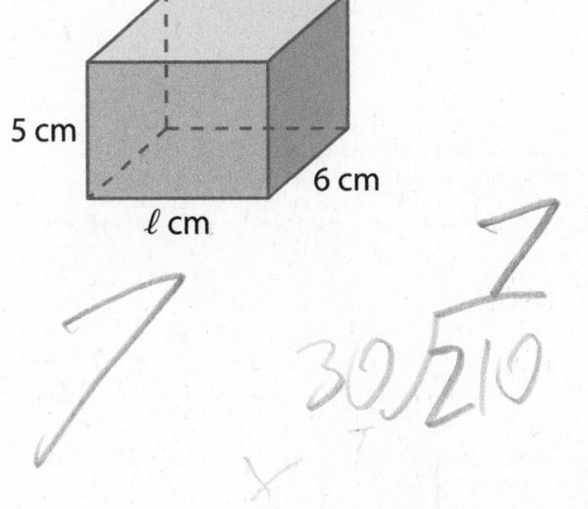

Find the unknown dimension of the rectangular prism.

3. Volume = 480 cubic feet

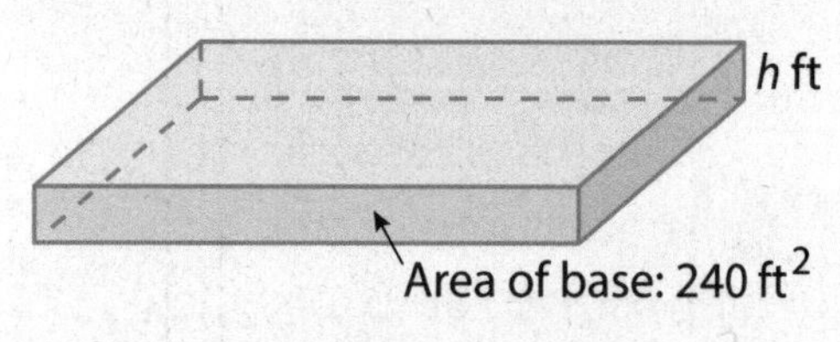

4. Volume = 351 cubic inches

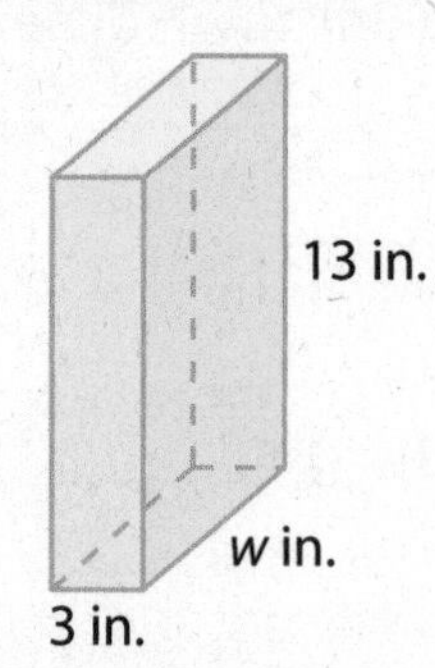

5. **Open-Ended** A rectangular prism has a volume of 144 cubic feet. The height of the prism is 6 feet. Give one possible pair of dimensions for the base.

6. **DIG DEEPER!** The volume of the rectangular prism is 64 cubic centimeters. What is the value of y?

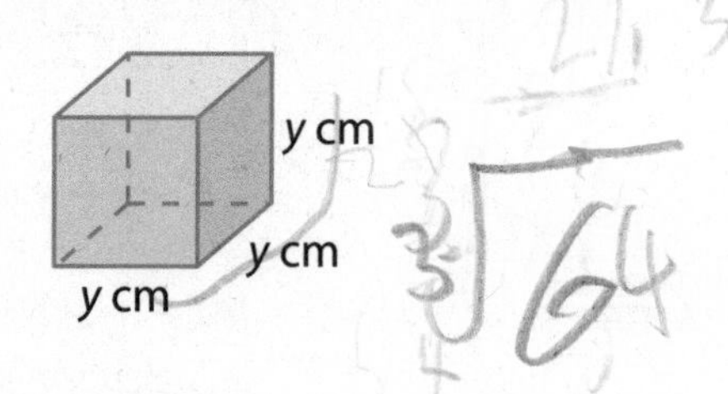

7. **Modeling Real Life** The volume of the gift is 900 cubic inches. Can the gift fit inside the shipping box? Explain.

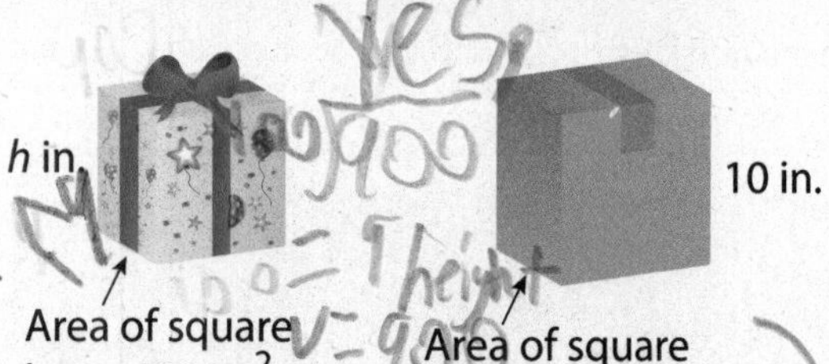

Not drawn to scale

8. **DIG DEEPER!** You fill the entire fish tank with water. Then you dump out some of the water. The volume of the water in the tank decreases by 96 cubic inches. How deep is the water now?

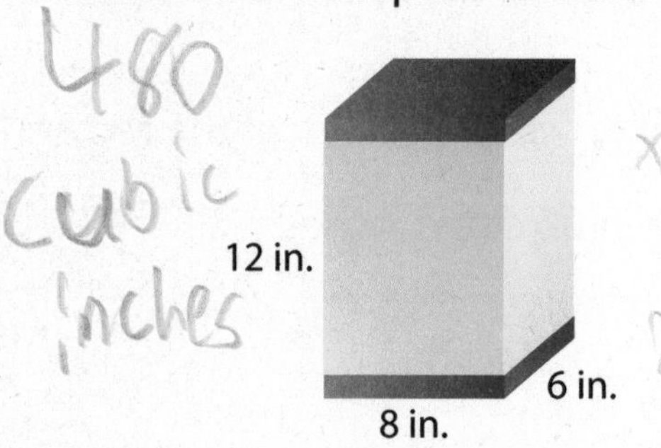

Review & Refresh

Divide.

9. $\frac{1}{3} \div 2 =$ ______

10. $\frac{1}{10} \div 6 =$ ______

11. $\frac{1}{8} \div 4 =$ ______

Name ______________________________

Find Volumes of Composite Figures 13.5

Learning Target: Find volumes of composite figures.

Success Criteria:

- I can break apart a composite figure into rectangular prisms.
- I can find an unknown dimension of a composite figure.
- I can add the volumes of rectangular prisms to find the volume of a composite figure.

Explore and Grow

Use centimeter cubes to find the volume of the figure. Explain your method.

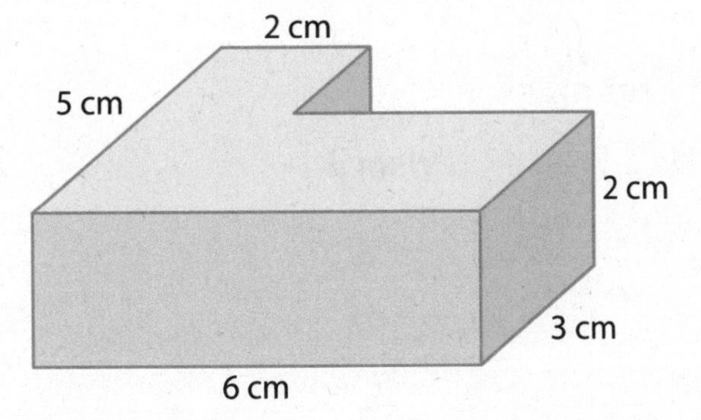

Structure Describe another way to find the volume of the figure.

Think and Grow: Find the Volume of a Composite Figure

Key Idea A **composite figure** is made of two or more solid figures. You can break apart a composite figure to find its volume.

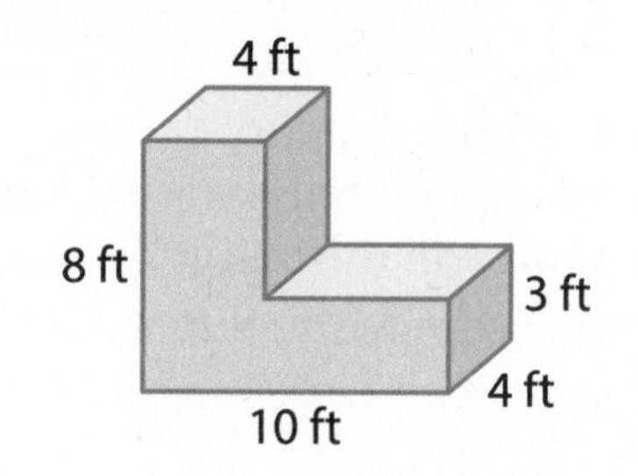

Example Find the volume of the composite figure.

Break apart the figure into rectangular prisms.
Find the dimensions of each prism.

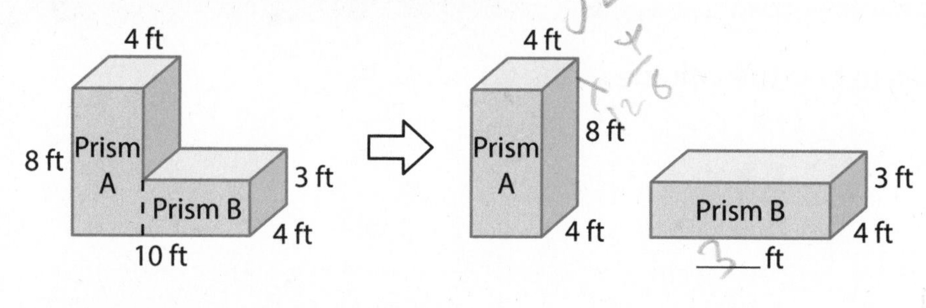

Think: How can you use the given dimensions to find the unknown length?

Find the volume of each prism.

Prism A

$V = \ell \times w \times h$

$=$ ____ $\times$ ____ $\times$ ____

$=$ ____

Prism B

$V = \ell \times w \times h$

$=$ ____ $\times$ ____ $\times$ ____

$=$ ____

Add the volumes of the prisms to find the total volume.

____ + ____ = ____

So, the volume of the composite figure is ____ cubic feet.

Show and Grow I can do it!

1. Find the volume of the composite figure.

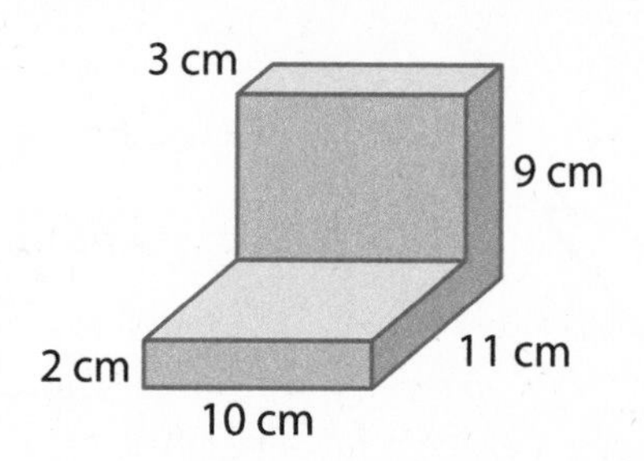

Name ______________________________

Apply and Grow: Practice

Find the volume of the composite figure.

2.

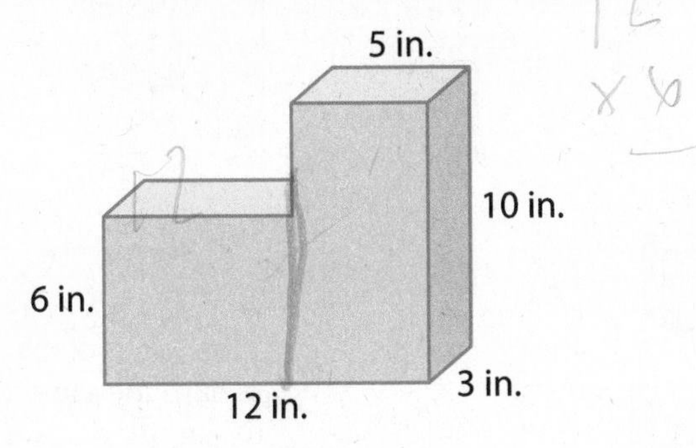

3.

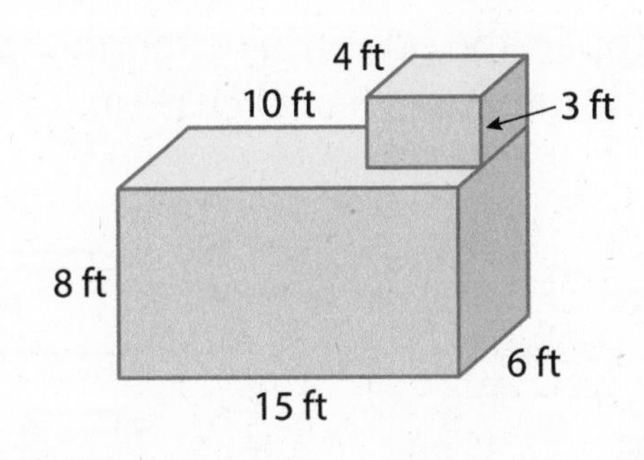

4.

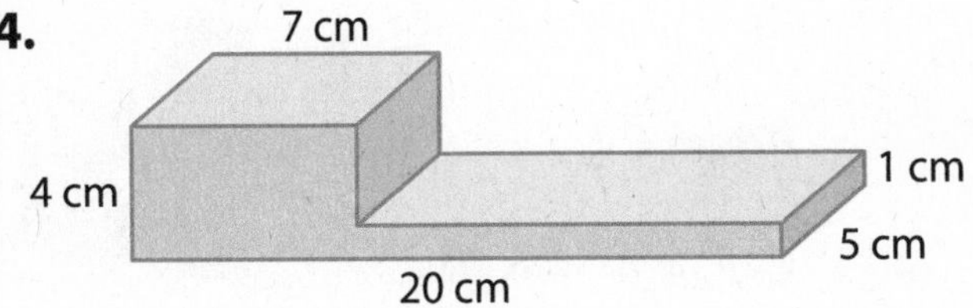

5.

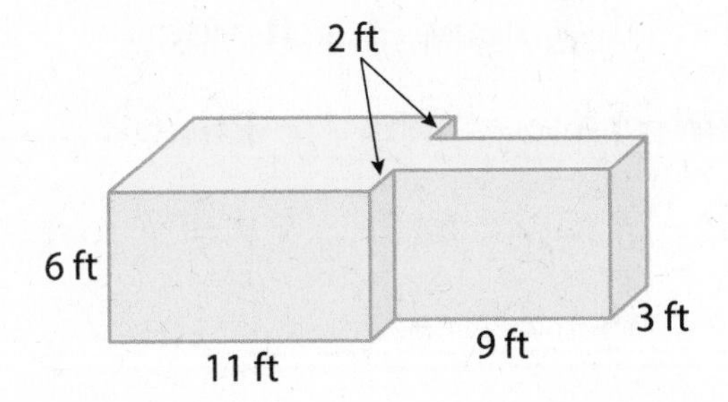

6. **MP Number Sense** Write one possible equation for the volume of the composite figure.

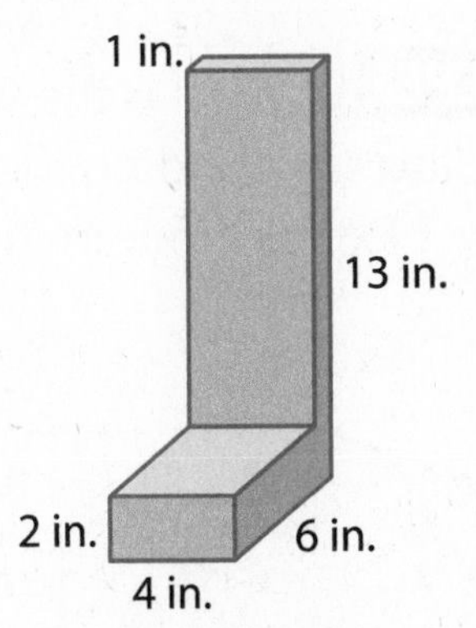

7. **YOU BE THE TEACHER** Your friend finds the volume of the composite figure. Is your friend correct? Explain.

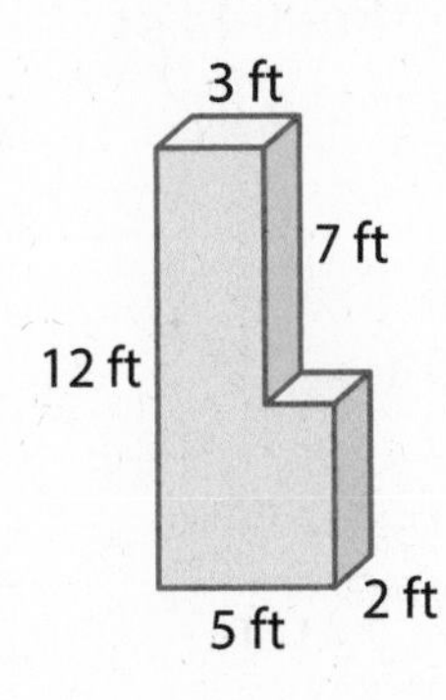

$12 \times 2 \times 3 = 72$

$5 \times 2 \times 5 = 50$

$72 + 50 = 122$ cubic feet

Think and Grow: Modeling Real Life

Example The depth of the swimming pool is 5 feet. How many cubic feet of water can the pool hold?

40 ft
30 ft
20 ft
12 ft

Break apart the pool into rectangular prisms. Find the dimensions of each prism.

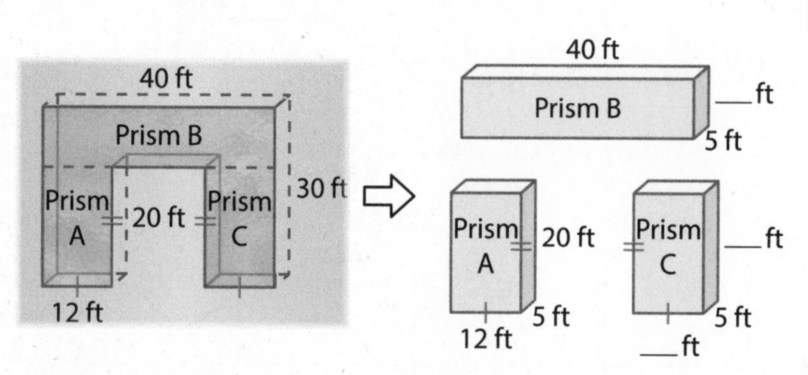

Remember, red tick marks mean the edges of the figure have the same length.

Find the volume of each prism.

Prism A	**Prism B**	**Prism C**
$V = \ell \times w \times h$	$V = \ell \times w \times h$	$V = \ell \times w \times h$
$= 12 \times 5 \times 20$	$= 40 \times 5 \times$ ____	$=$ ____ $\times 5 \times$ ____
$=$ ____	$=$ ____	$=$ ____

Add the volumes of the prisms to find the total volume. ____ + ____ + ____ = ____

So, the pool can hold ____ cubic feet of water.

Show and Grow *I can think deeper!*

8. How many cubic centimeters of wood are needed to make the game piece?

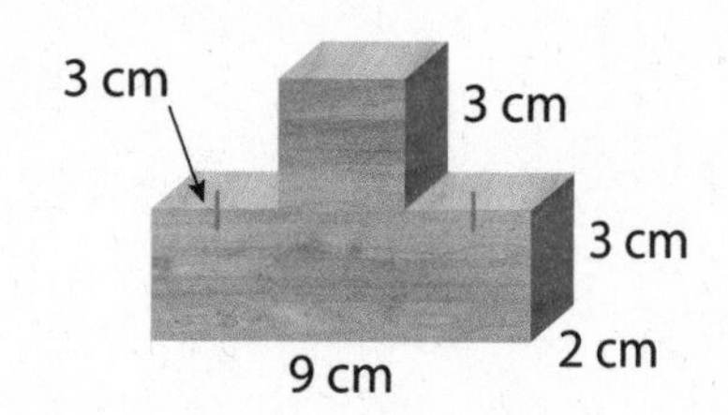

9. **DIG DEEPER!** What is the volume of the building surrounding the outdoor courtyard? Explain your method.

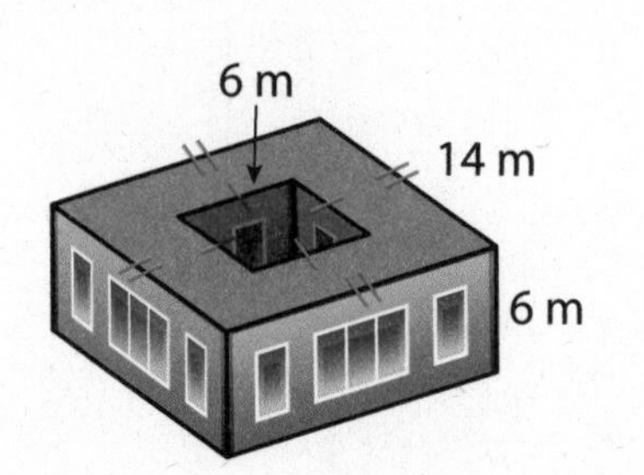

Name ____________________

Homework & Practice 13.5

Learning Target: Find volumes of composite figures.

Example Find the volume of the composite figure.

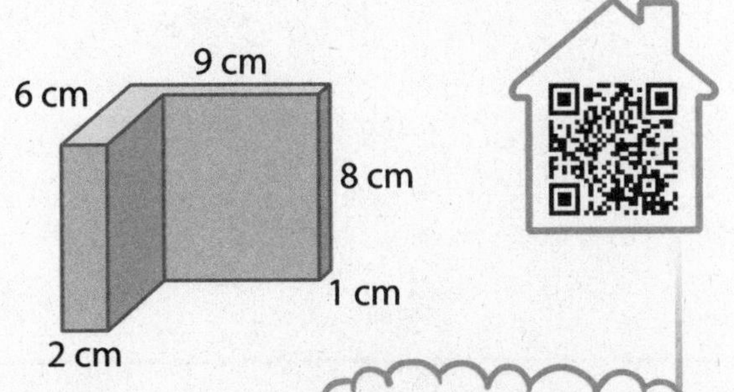

Break apart the figure into rectangular prisms. Find the dimensions of each prism.

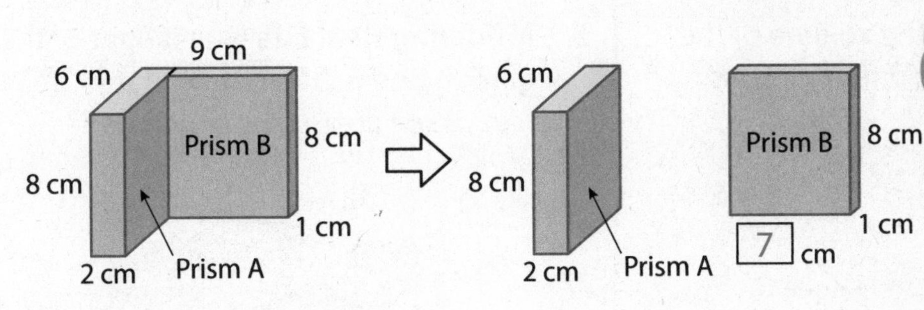

Think: How can you use the given dimensions to find the unknown length?

Find the volume of each prism.

Prism A

$V = \ell \times w \times h$

$= \underline{2} \times \underline{6} \times \underline{8}$

$= \underline{96}$

Prism B

$V = \ell \times w \times h$

$= \underline{7} \times \underline{1} \times \underline{8}$

$= \underline{56}$

Add the volumes of the prisms to find the total volume.

$\underline{96} + \underline{56} = \underline{152}$

So, the volume of the composite figure is $\underline{152}$ cubic centimeters.

1. Find the volume of the composite figure.

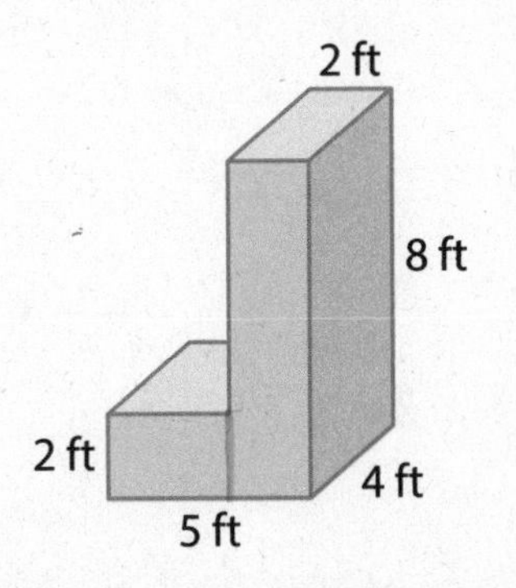

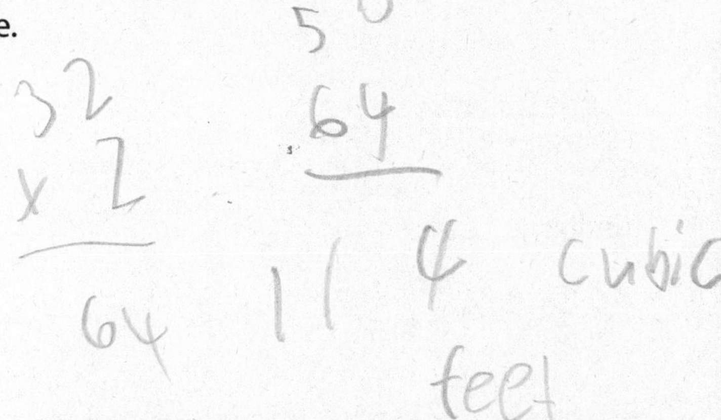

Find the volume of the composite figure.

2.

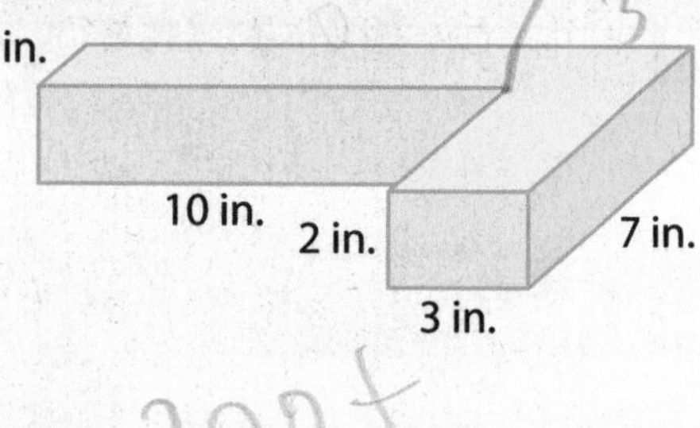

3.

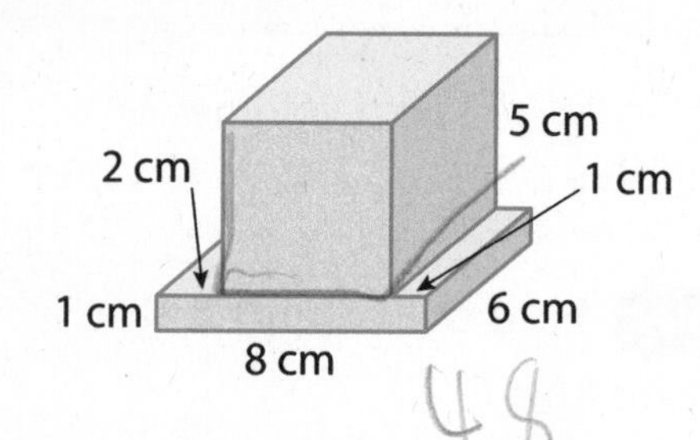

4. MP **Structure** Find the volume of the composite figure two different ways.

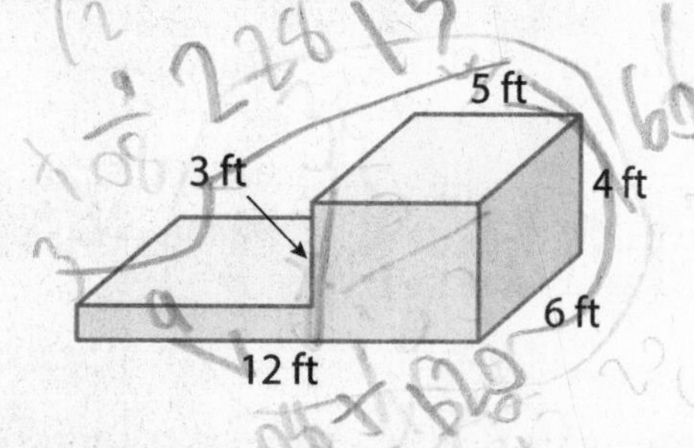

5. **Modeling Real Life** The height of the garden bed is 2 feet. How many cubic feet of soil are in the garden bed?

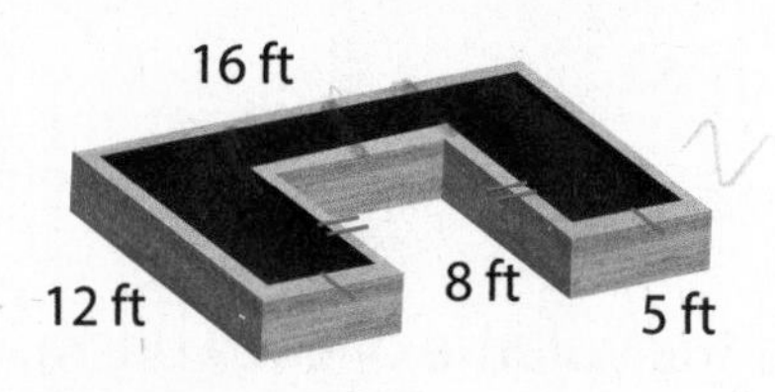

6. **DIG DEEPER!** The dimensions of each step are the same. How many cubic inches of concrete does a worker need to make the staircase?

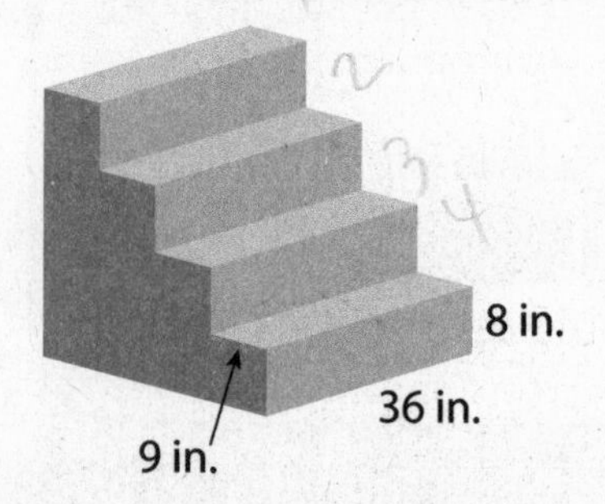

Review & Refresh

Multiply.

7. $\frac{4}{5} \times 5 =$ ______

8. $\frac{1}{2} \times 6 =$ ______

9. $\frac{7}{8} \times 3 =$ ______

Name ______________________________

Performance Task 13

Mechanical engineers follow guidelines when designing and installing elevators to make sure passengers are safe and the elevators are wheelchair accessible. Many elevators have heating and air conditioning units attached to them, so it is important to know the volume inside the elevator to keep the air a comfortable temperature.

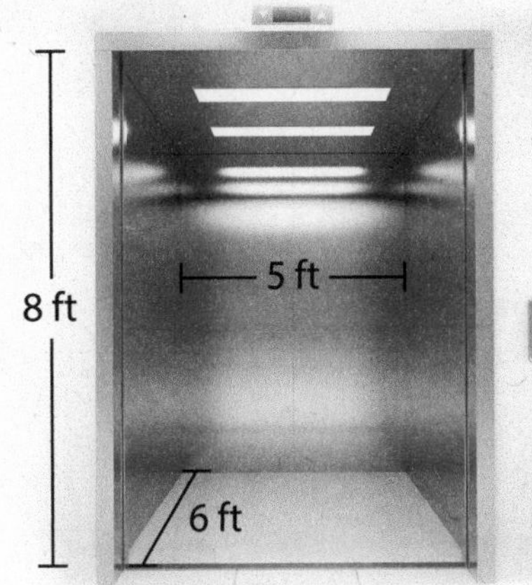

Maximum Capacity: 2,500 pounds

1. The inside of the elevator shown is in the shape of a rectangular prism.

   a. What is the volume of the inside of the elevator?

   b. Another elevator has the same height and width. The length of the inside is doubled. How does this affect the volume?

   36 × 3 = 108; 108 + 120 = 228

   c. Five people ride with their luggage on the elevator. Each person weighs about 200 pounds. The total combined weight of their luggage is about 425 pounds. About how many more pounds can the elevator hold?

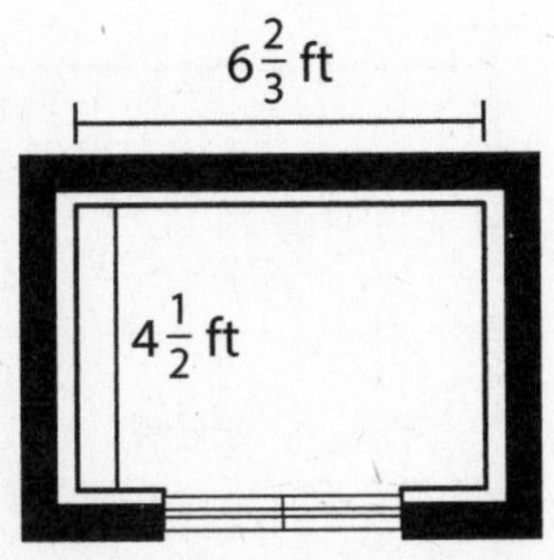

Minimum requirements for a wheelchair accessible elevator.

2. The inside of all new elevators must be wheelchair accessible. The minimum requirements are shown. The inside volume of an older elevator is 144 cubic feet. The base is a square and the height is 9 feet.

   a. Explain why the older elevator does *not* meet the wheelchair requirements.

   b. A new wheelchair-accessible elevator is added to the building. Its height is 9 feet. Find the minimum volume of the new elevator. How does this compare to the volume of the old elevator?

Volume Solve and Connect

Directions:

1. Players take turns rolling a die.
2. On your turn, move your piece the number of spaces shown on the die.
3. Find the volume of the rectangular prism.
4. Cover the volume with a counter. Your turn is over.
5. The first player to get three in a row, horizontally, vertically, or diagonally, wins!

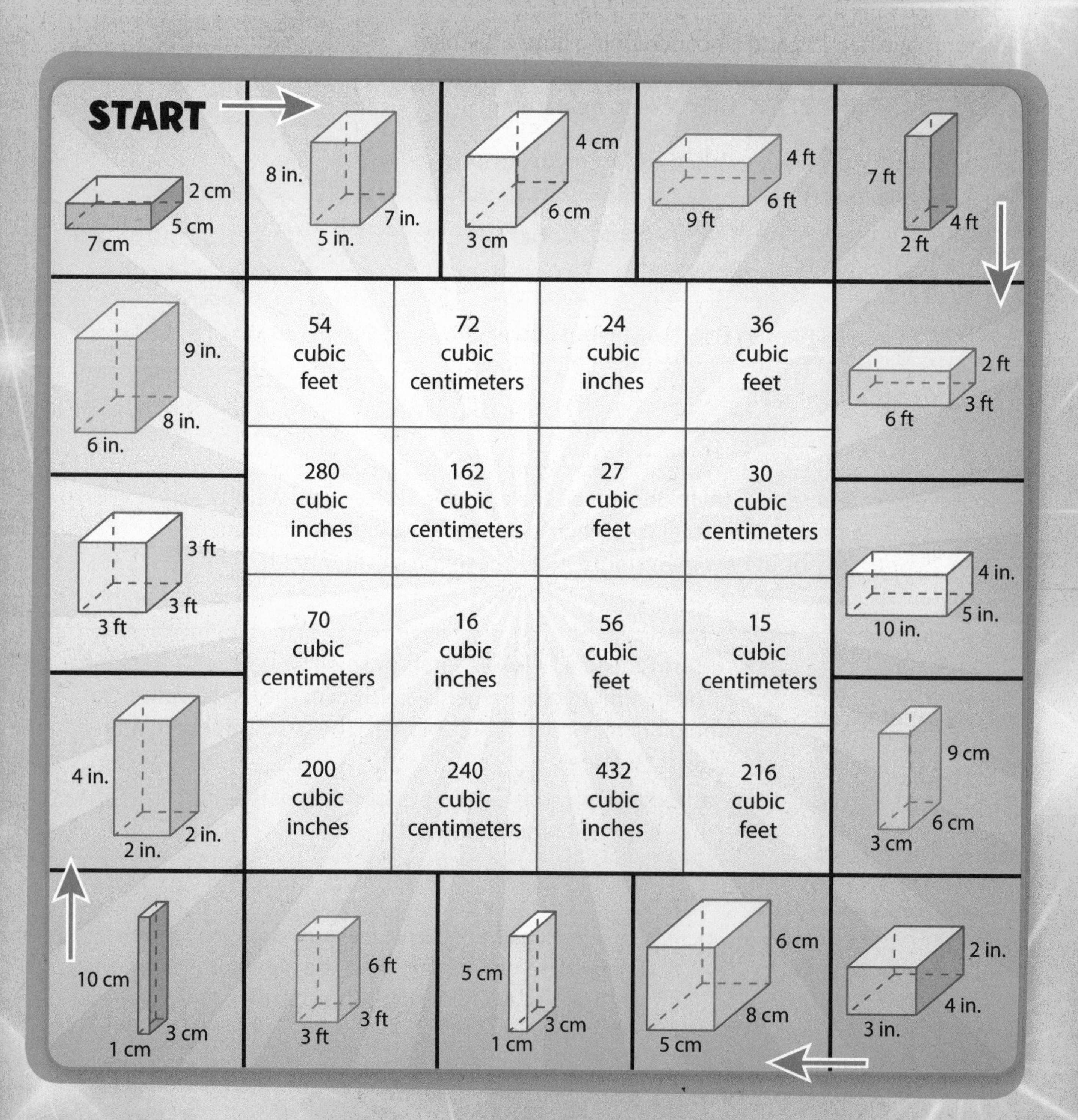

Name ______________________

Chapter Practice 13

13.1 Understand the Concept of Volume

Find the volume of the figure.

1. Volume = 10

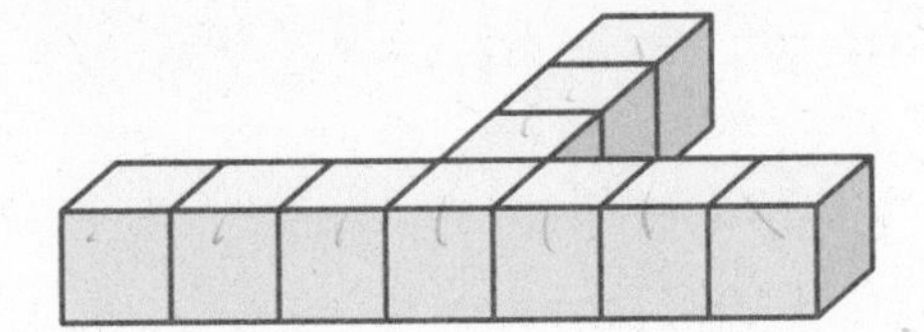

2. Volume = 5

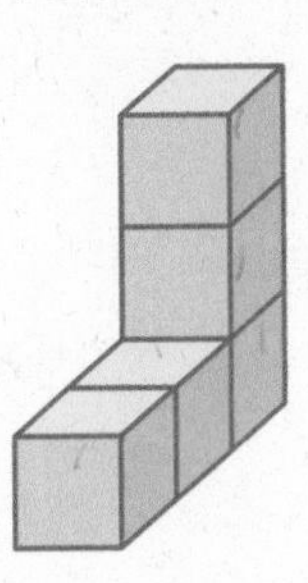

3. Volume = 8

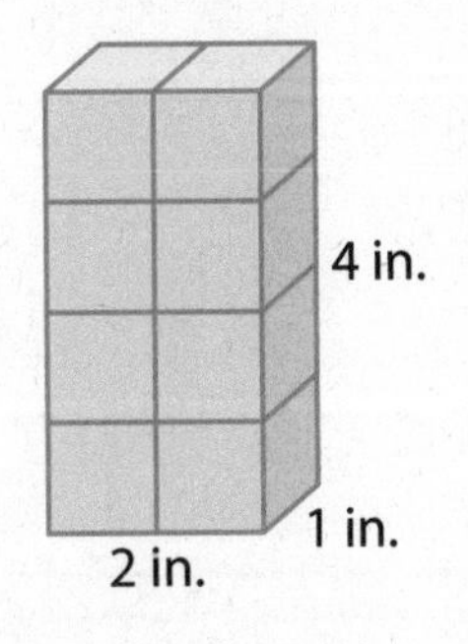

4. Volume = 18

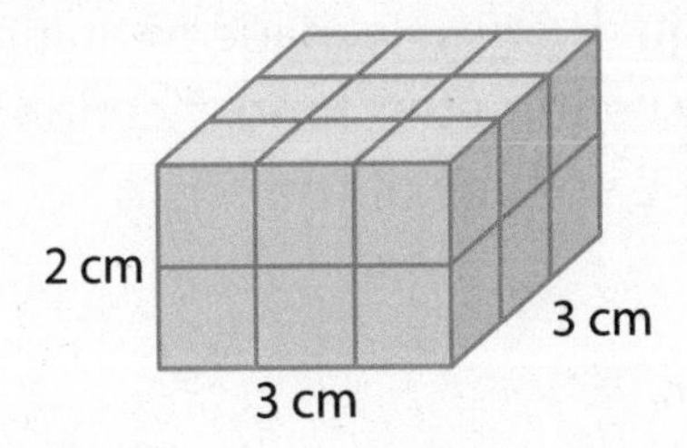

5. Volume = 16

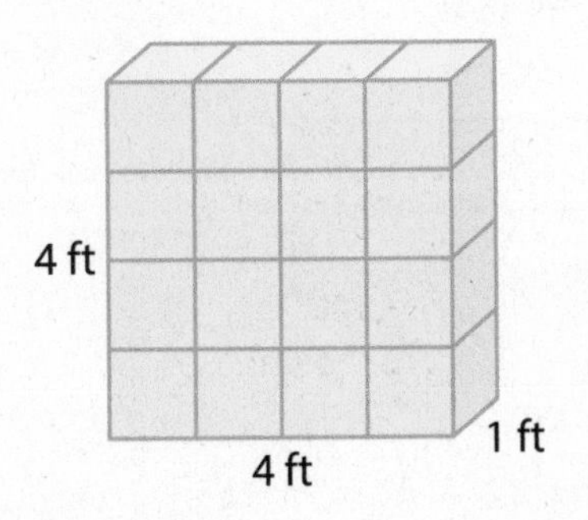

6. Volume = 16

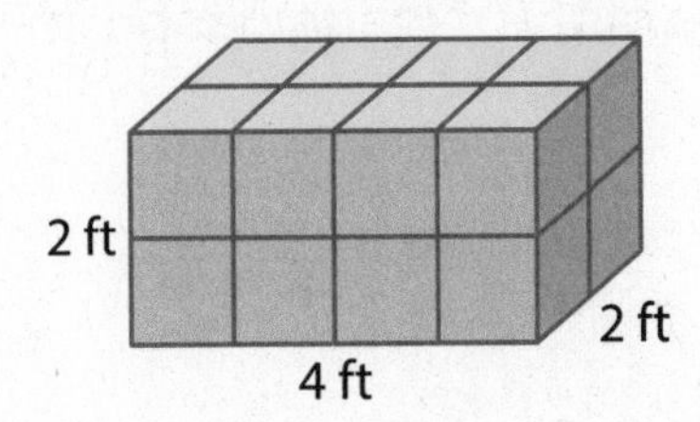

7. Modeling Real Life A container holds 30 of the phone boxes shown, without any gaps or overlaps. What is the volume of the container?

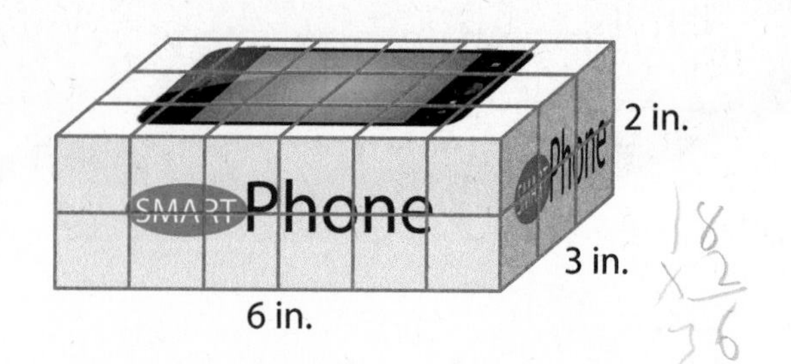

36

13.2 Find Volumes of Right Rectangular Prisms

Find the volume of the rectangular prism.

8. Volume = 54

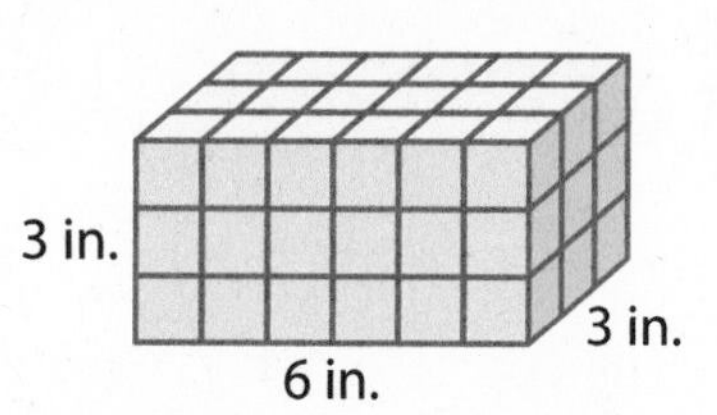

9. Volume = 84

28 × 3 = 84

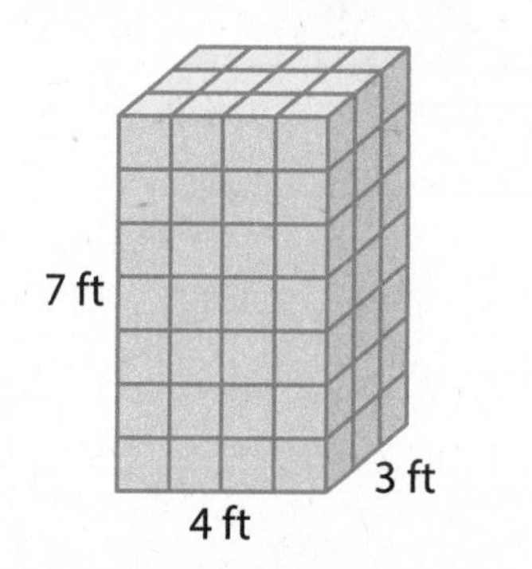

10. A rectangular prism is made of unit cubes, like the one shown. The prism is 9 centimeters long, 3 centimeters wide, and 5 centimeters tall. What is the volume of the prism?

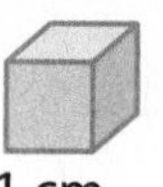

135

13.3 Apply the Volume Formula

Find the volume of the rectangular prism.

11.

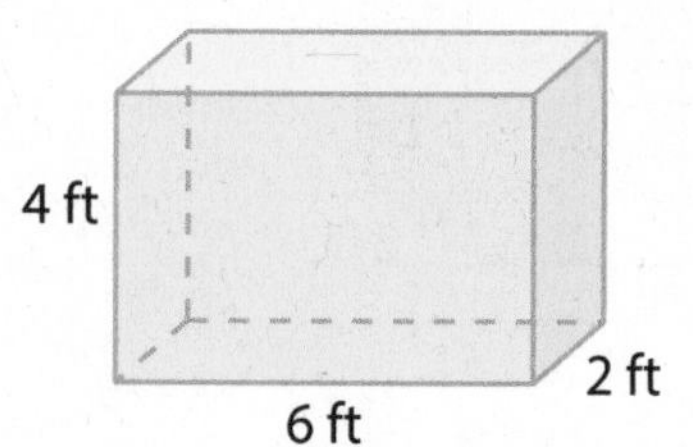

12.

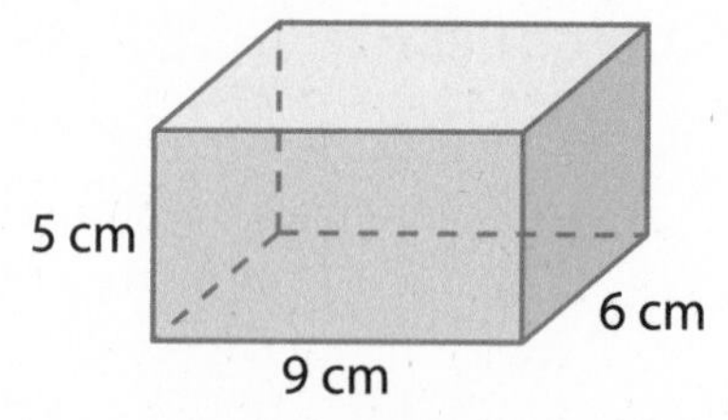

270

13. Descartes's cat cave is a rectangular prism. The cave is 14 inches long, 13 inches wide, and 12 inches tall. What is the volume of the cave?

13.4 Find Unknown Dimensions

Find the unknown dimension of the rectangular prism.

14. Volume = 81 cubic feet

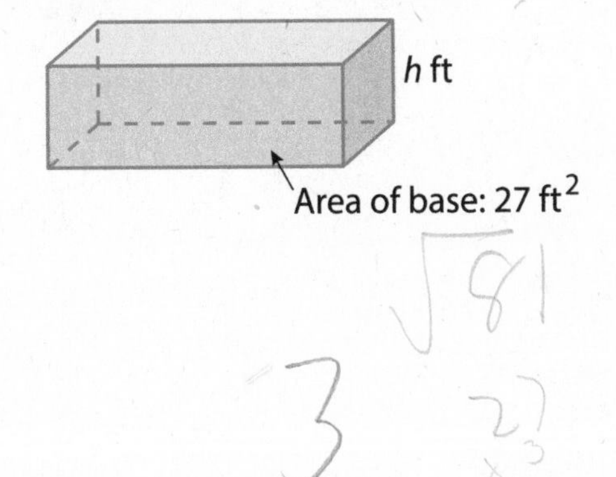

15. Volume = 288 cubic centimeters

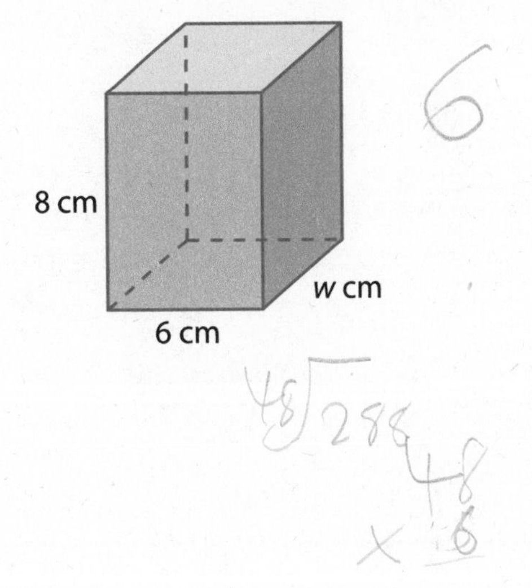

16. Volume = 729 cubic inches

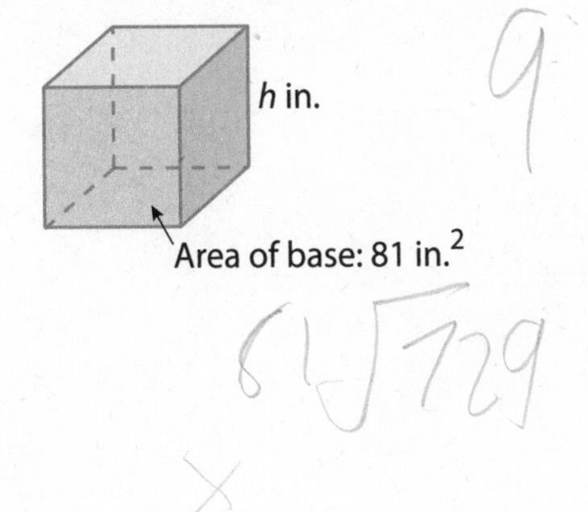

17. Volume = 216 cubic feet

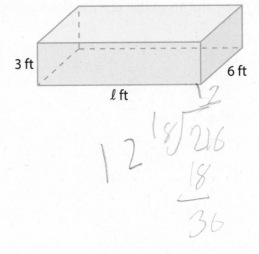

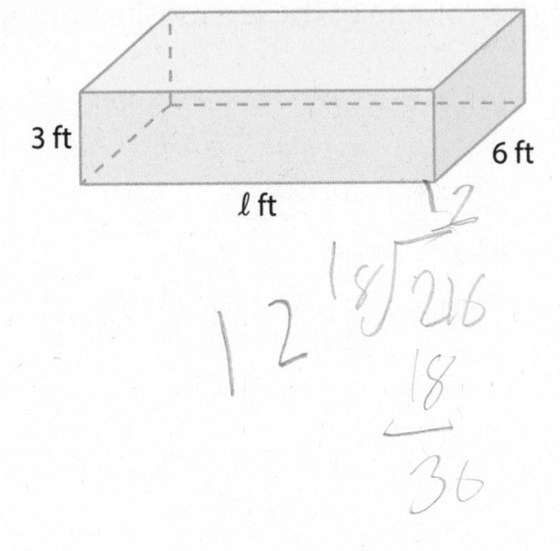

18. **Open-Ended** A rectangular prism has a volume of 128 cubic feet. The height of the prism is 8 feet. Give one possible set of dimensions for the base.

19. **Modeling Real Life** You buy the drawing pad shown. The volume of the drawer is 800 cubic inches. Can the drawing pad lie flat inside the drawer?

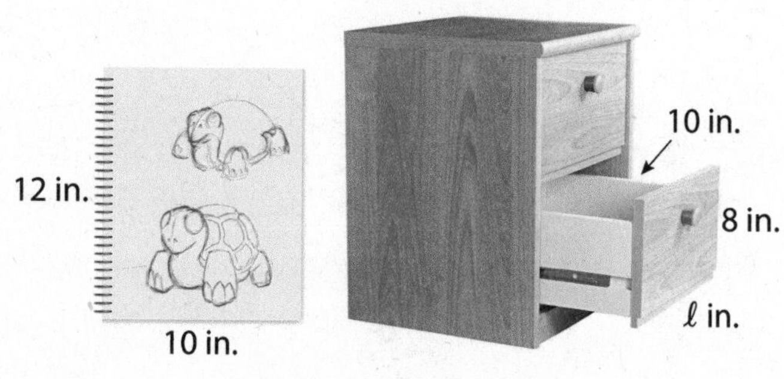

Not drawn to scale

13.5 Find Volumes of Composite Figures

Find the volume of the composite figure.

20.

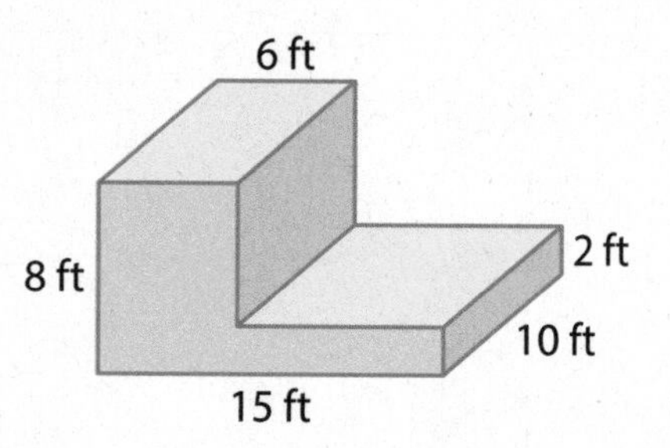

21.

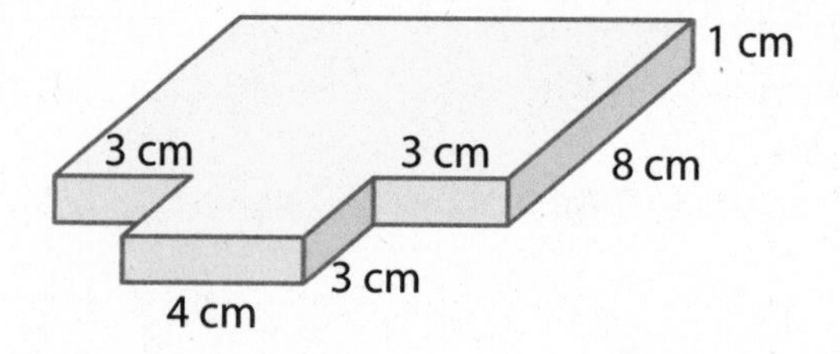

14 Classify Two-Dimensional Shapes

Chapter Learning Target:
Understand two-dimensional shapes.

Chapter Success Criteria:
- I can define two-dimensional shapes.
- I can explain different shapes and their features.
- I can compare shapes.
- I can draw shapes.

- Solar panels absorb energy from the Sun to generate electricity. Where are some places you might see solar panels?
- Why is it important to know the shape and dimensions of a roof before installing solar panels?

Name ______________________________

14 Vocabulary

Review Words

acute angle
obtuse angle
right angle
straight angle

Organize It

Use the review words to complete the graphic organizer.

Angles

______ • An angle that is open less than a right angle • An angle that measures less than 90°	______ • An L-shaped angle • An angle that measures 90°
______ • An angle that is open more than a right angle and less than a straight angle • An angle that measures between 90° and 180°	______ • An angle that forms a straight line • An angle that measures 180°

Define It

Use your vocabulary cards to match.

1. rectangle

2. rhombus

3. trapezoid

A quadrilateral that has exactly one pair of parallel sides

A parallelogram that has four sides with the same length

A parallelogram that has four right angles

Chapter 14 Vocabulary Cards

acute triangle

equiangular triangle

equilateral triangle

isosceles triangle

obtuse triangle

parallelogram

rectangle

rhombus

A triangle that has three angles with the same measure

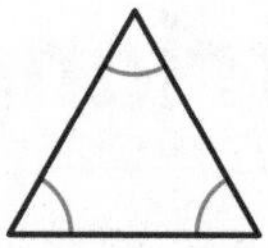

A triangle that has three acute angles

A triangle that has two sides with the same length

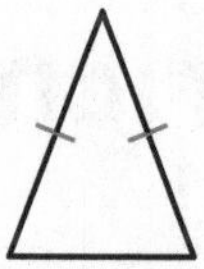

A triangle that has three sides with the same length

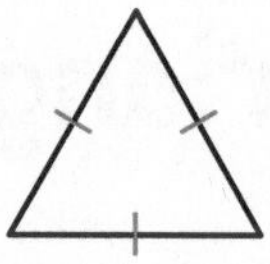

A quadrilateral that has two pairs of parallel sides

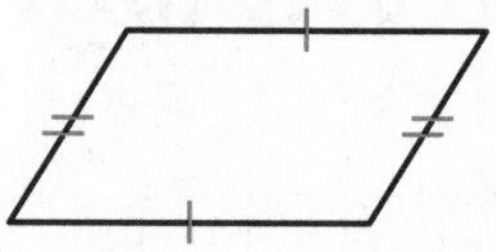

A triangle that has one obtuse angle

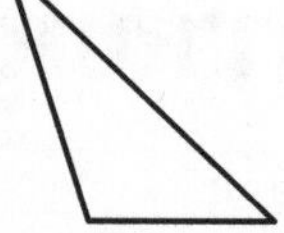

A parallelogram that has four sides with the same length

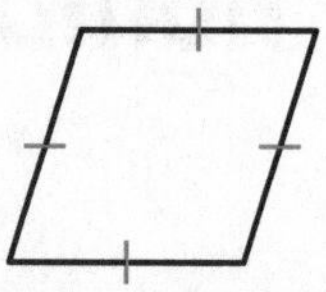

A parallelogram that has four right angles

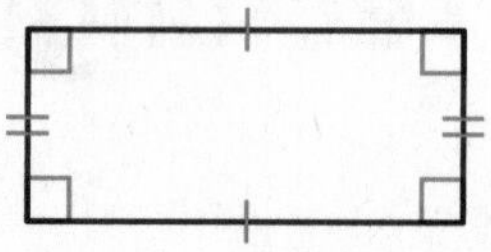

Chapter 14 Vocabulary Cards

right triangle

scalene triangle

square

trapezoid

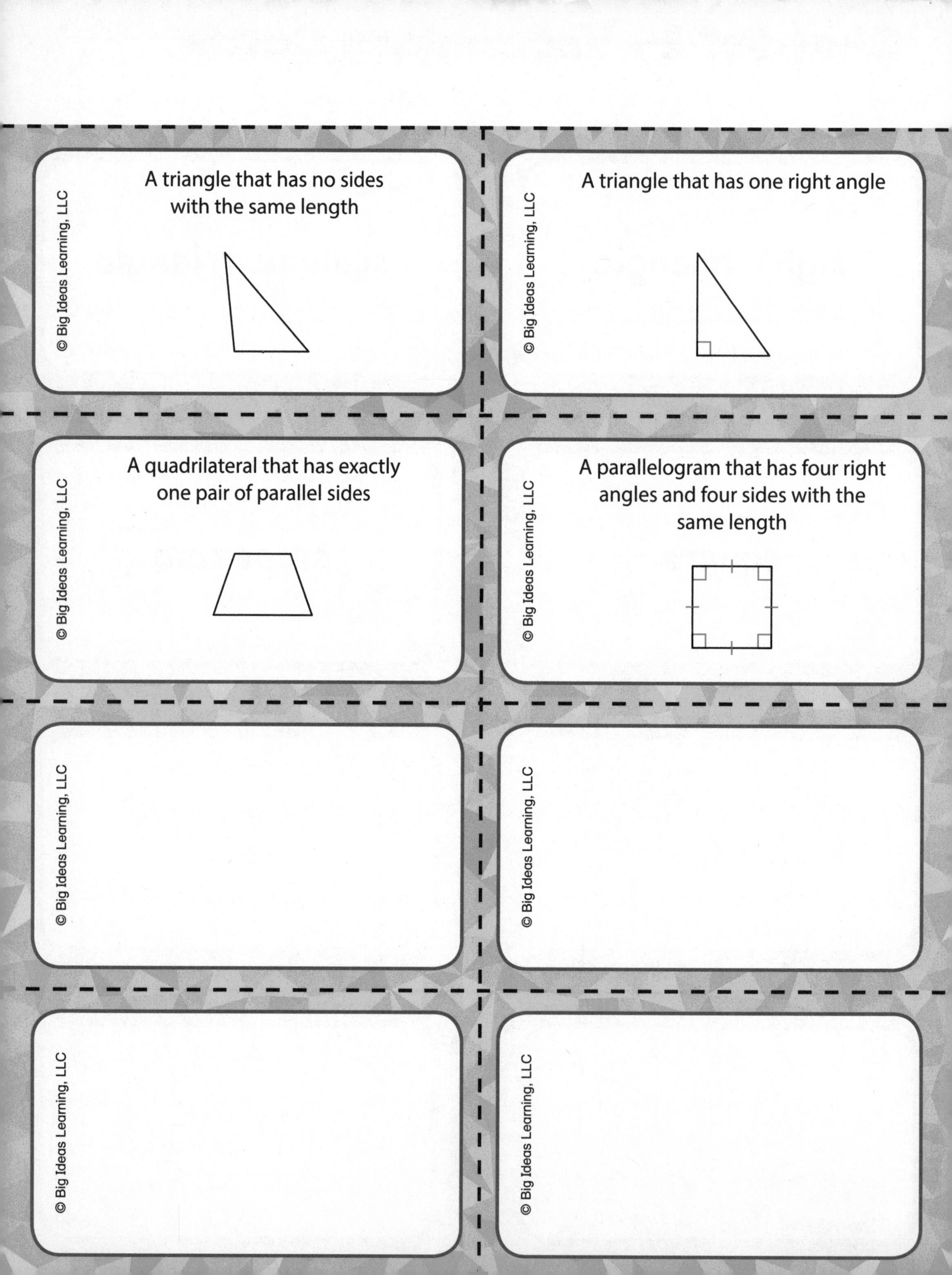
© Big Ideas Learning, LLC
A triangle that has no sides with the same length
© Big Ideas Learning, LLC
A triangle that has one right angle
© Big Ideas Learning, LLC
A quadrilateral that has exactly one pair of parallel sides
© Big Ideas Learning, LLC
A parallelogram that has four right angles and four sides with the same length
© Big Ideas Learning, LLC
© Big Ideas Learning, LLC
© Big Ideas Learning, LLC
© Big Ideas Learning, LLC

Name ______________________________

Classify Triangles 14.1

Learning Target: Classify triangles by their angles and their sides.

Success Criteria:
- I can identify an angle of a triangle as right, acute, or obtuse.
- I can determine whether sides of a triangle have the same length.
- I can use angles and sides to classify a triangle.

Explore and Grow

Draw and label a triangle for each description. If a triangle *cannot* be drawn, explain why.

triangle with three acute angles	triangle with two obtuse angles
triangle with two sides that have the same length	triangle with no sides that have the same length

Precision Draw a triangle that meets two of the descriptions above.

Think and Grow: Classify Triangles

Key Idea Triangles can be classified by their sides.

Remember, red tick marks indicate sides with the same length, and red arcs indicate angles with the same measure.

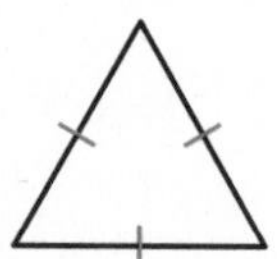

An **equilateral triangle** has three sides with the same length.

An **isosceles triangle** has two sides with the same length.

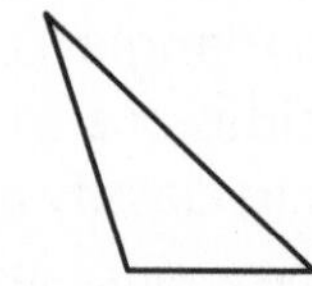

A **scalene triangle** has no sides with the same length.

Key Idea Triangles can be classified by their angles.

An **acute triangle** has three acute angles.

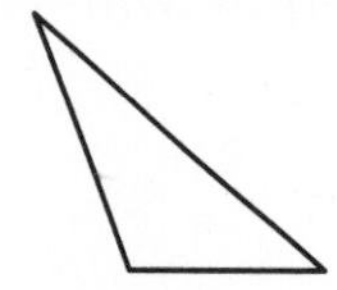

An **obtuse triangle** has one obtuse angle.

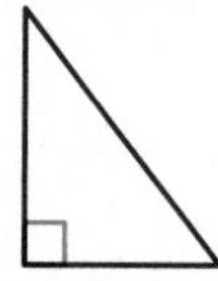

A **right triangle** has one right angle.

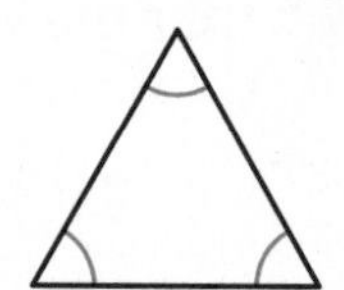

An **equiangular triangle** has three angles with the same measure.

Example Classify the triangle by its angles and its sides.

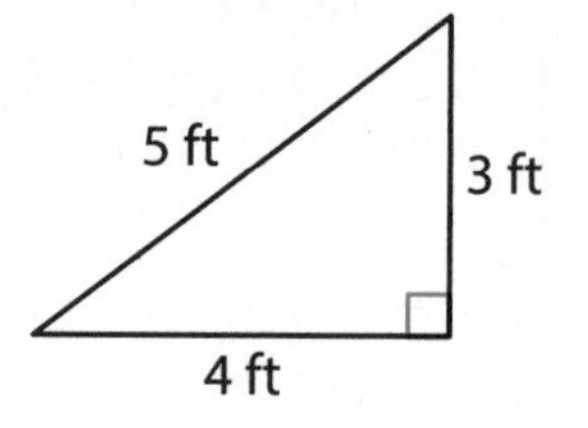

The triangle has one ________ angle

and ______ sides with the same length.

So, it is a ________ ________ triangle.

Show and Grow *I can do it!*

Classify the triangle by its angles and its sides.

1.

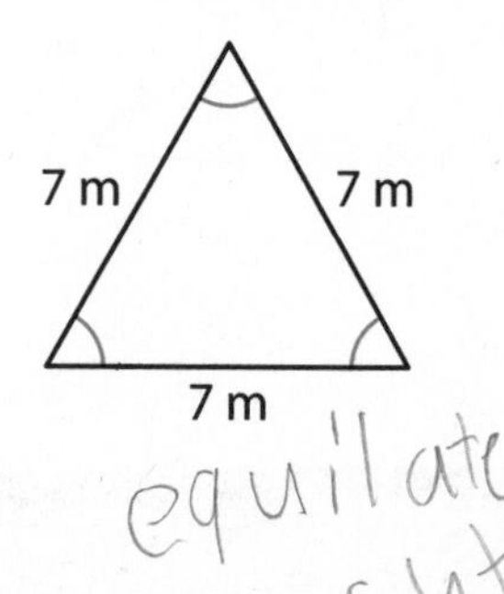

equilateral acute

2.

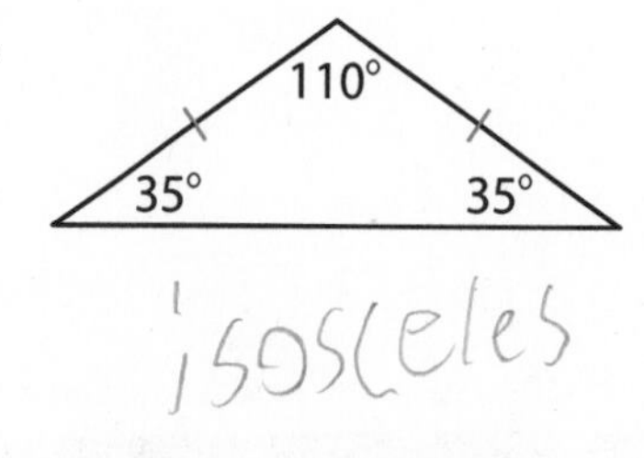

isosceles acute

3.

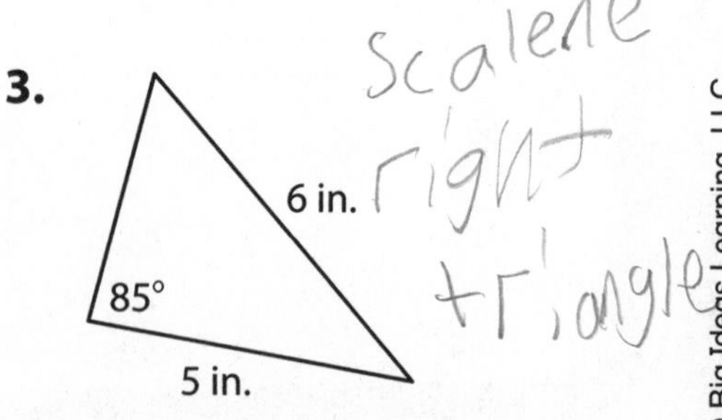

Scalene right triangle

Name ________________________________

Apply and Grow: Practice

Classify the triangle by its angles and its sides.

4.

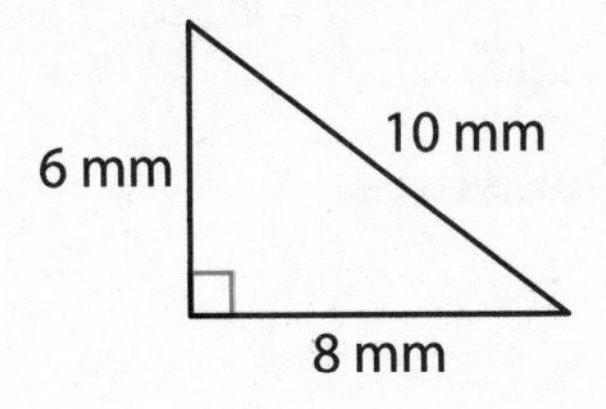

5.

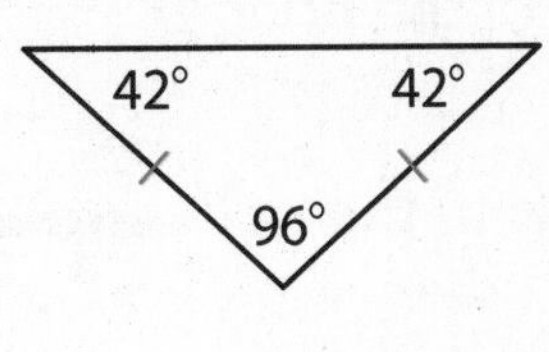

6.

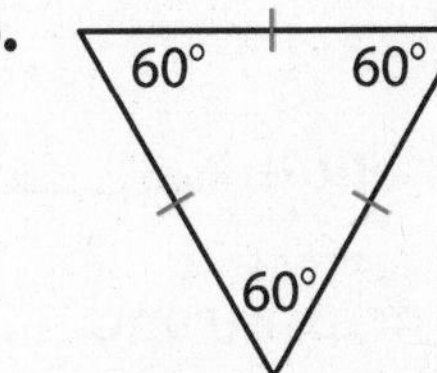

7.

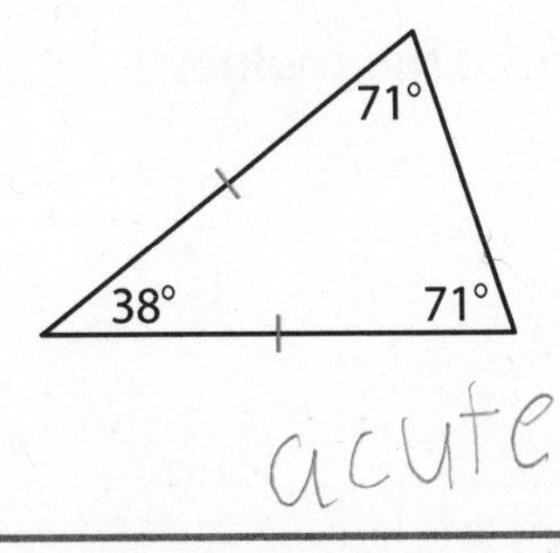

8.

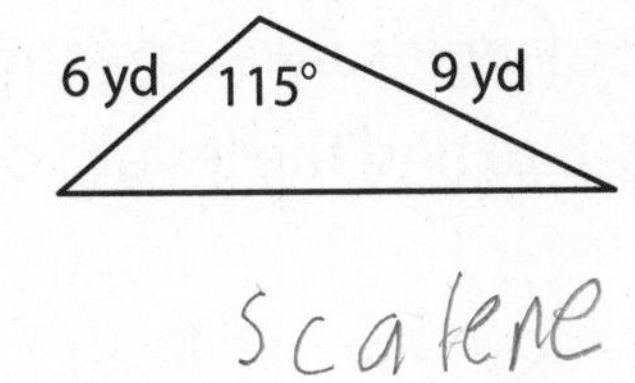

9.

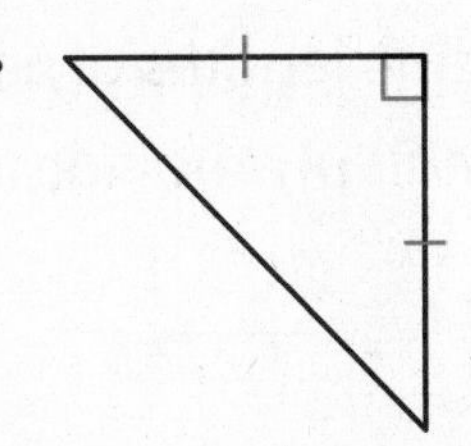

10. A triangular sign has a 40° angle, a 55° angle, and an 85° angle. None of its sides have the same length. Classify the triangle by its angles and its sides.

11. **YOU BE THE TEACHER** Your friend says the triangle is an acute triangle because it has two acute angles. Is your friend correct? Explain.

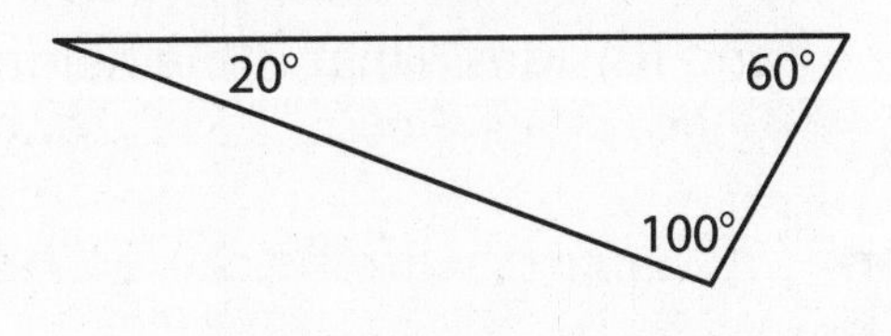

12. **DIG DEEPER!** Draw one triangle for each category. Which is the appropriate category for an equiangular triangle? Explain your reasoning.

Acute Triangles

Obtuse Triangles

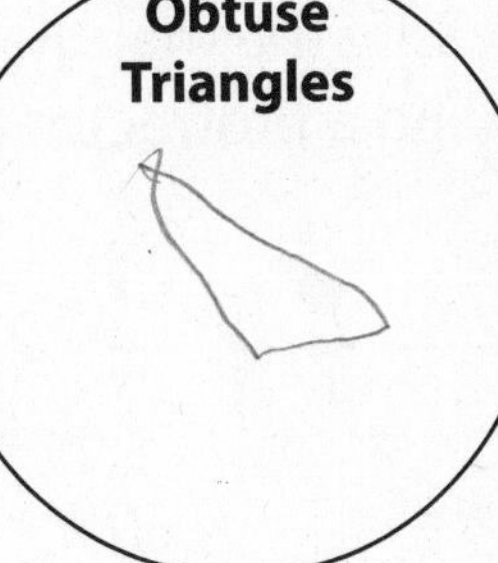

Right Triangles

Think and Grow: Modeling Real Life

Example A bridge contains several identical triangles. Classify each triangle by its angles and its sides. What is the length of the bridge?

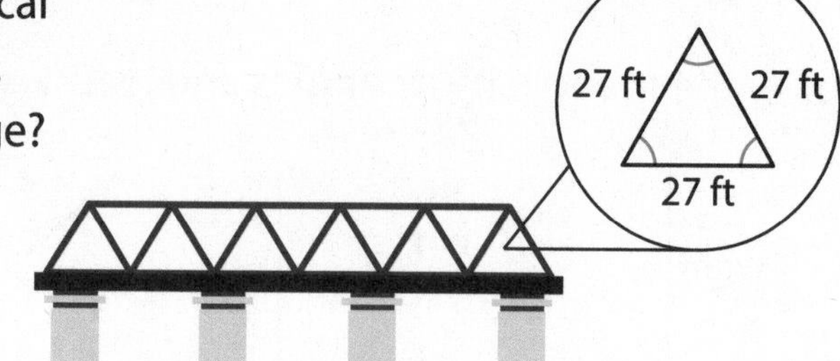

Each triangle has ______ angles with the same measure and ______ sides with the same length.

So, each triangle is ________ and ________.

The side lengths of 6 identical triangles meet to form the length of the bridge. So, multiply the side length by 6 to find the length of the bridge.

$27 \times 6 =$ ______

So, the bridge is ________ long.

Show and Grow *I can think deeper!*

13. The window is made using identical triangular panes of glass. Classify each triangle by its angles and its sides. What is the height of the window?

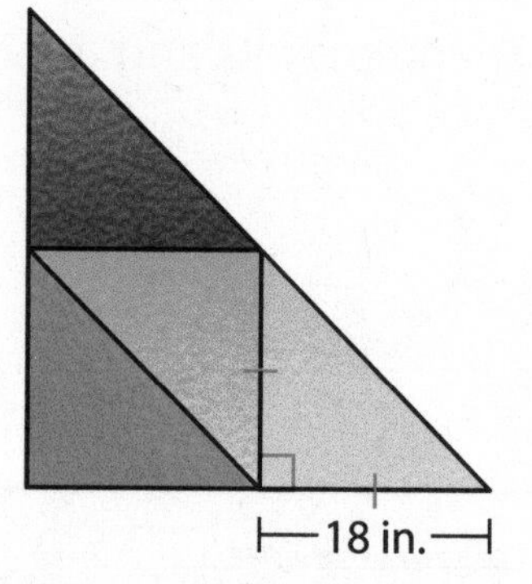

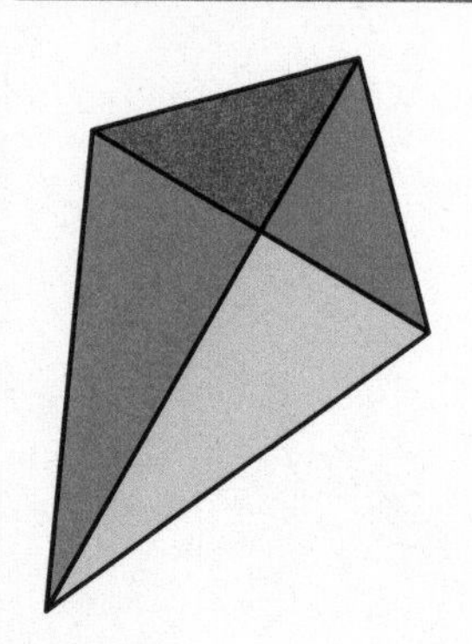

14. **DIG DEEPER!** You connect four triangular pieces of fabric to make the kite. Classify the triangles by their angles and their sides. Use a ruler and a protractor to verify your answer.

Name ______________________________

Homework & Practice 14.1

Learning Target: Classify triangles by their angles and their sides.

Example Classify the triangle by its angles and its sides.

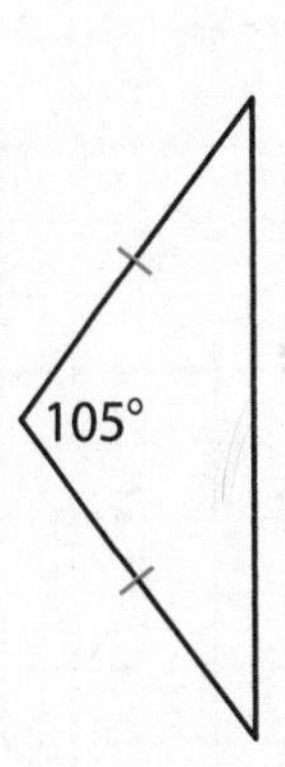

The triangle has one __obtuse__ angle

and __2__ sides with the same length.

So, it is an __obtuse__ __isosceles__ triangle.

Classify the triangle by its angles and its sides.

1.

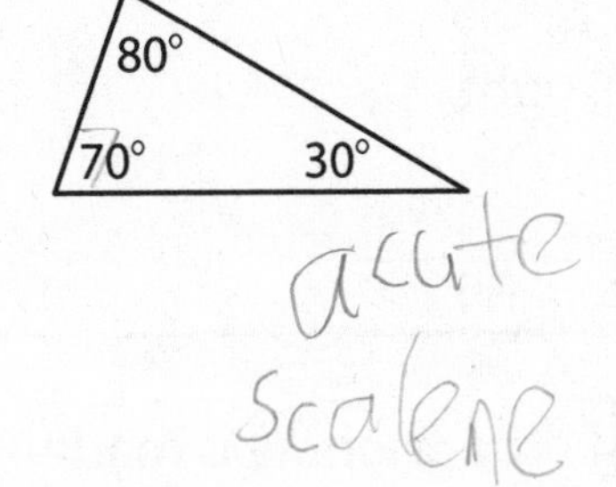

acute
scalene

2.

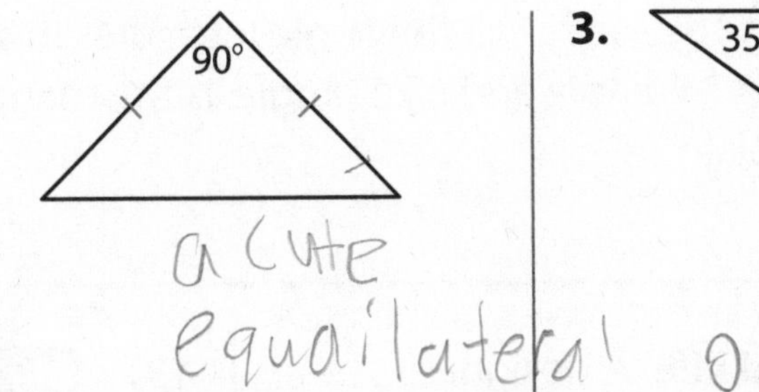

acute
equailateral

3.

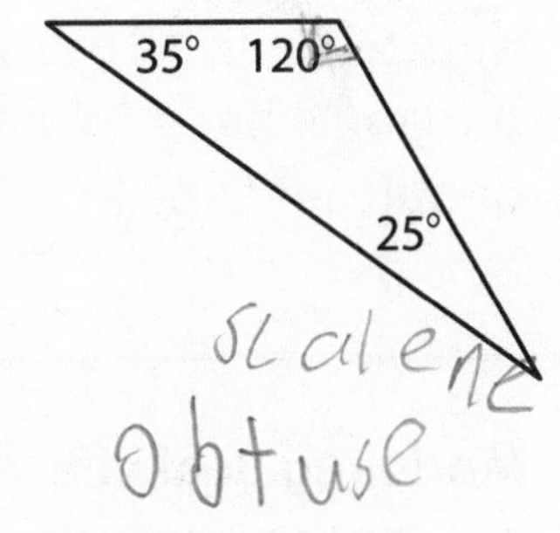

scalene
obtuse

4.

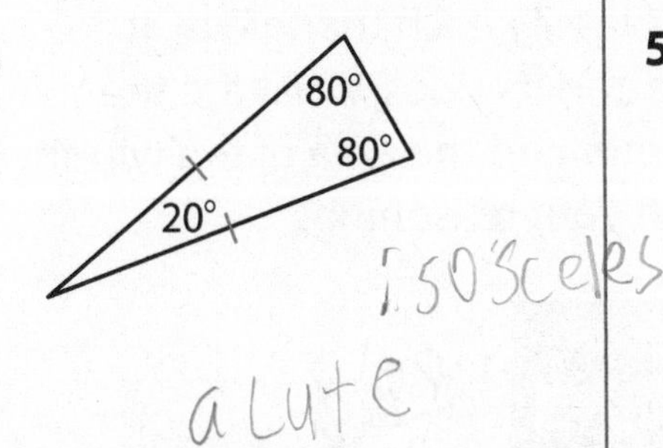

isosceles
alute

5.

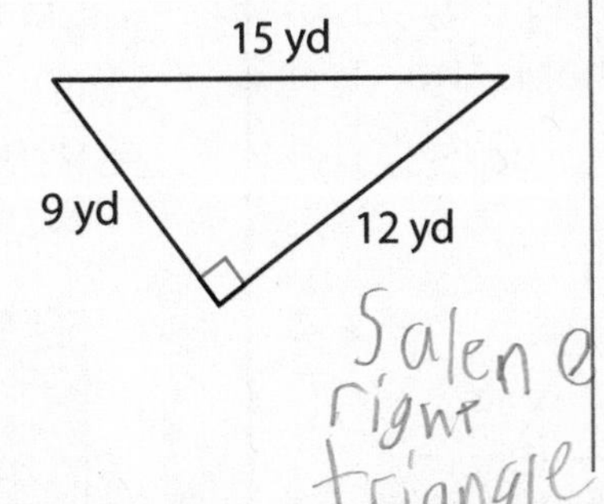

Salene
right
triangle

6.

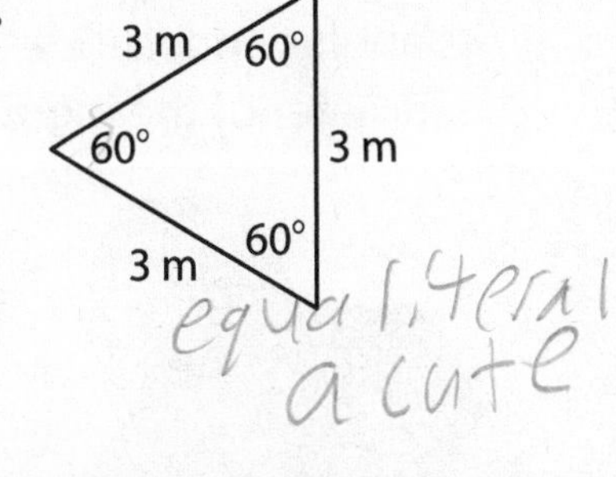

equalitera l
acute

7. A triangular race flag has two 65° angles and a 50° angle. Two of its sides have the same length. Classify the triangle by its angles and its sides.

isosceles
acute

8. A triangular measuring tool has a 90° angle and no sides of the same length. Classify the triangle by its angles and its sides.

right triangle
scalene

9. MP **Structure** Draw a triangle with vertices $A(2, 2)$, $B(2, 6)$, and $C(6, 2)$ in the coordinate plane. Classify the triangle by its angles and its sides. Explain your reasoning.

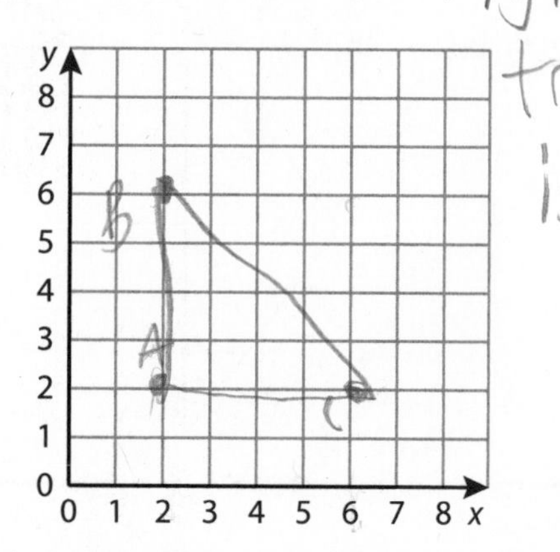

10. YOU BE THE TEACHER Your friend says that both Newton and Descartes are correct. Is your friend correct? Explain.

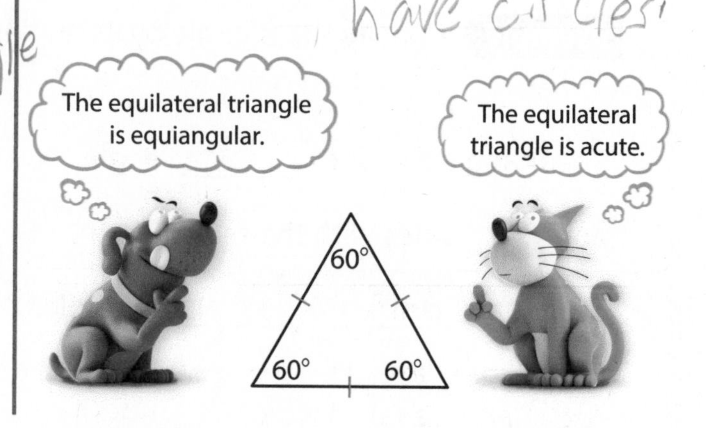

The equilateral triangle is equiangular.
The equilateral triangle is acute.
60°
60°
60°

11. DIG DEEPER! The sum of all the angle measures in a triangle is 180°. A triangle has a 34° angle and a 26° angle. Is the triangle acute, right, or obtuse? Explain.

12. Modeling Real Life A designer creates the logo using identical triangles. Classify each triangle by its angles and its sides. What is the perimeter of the logo?

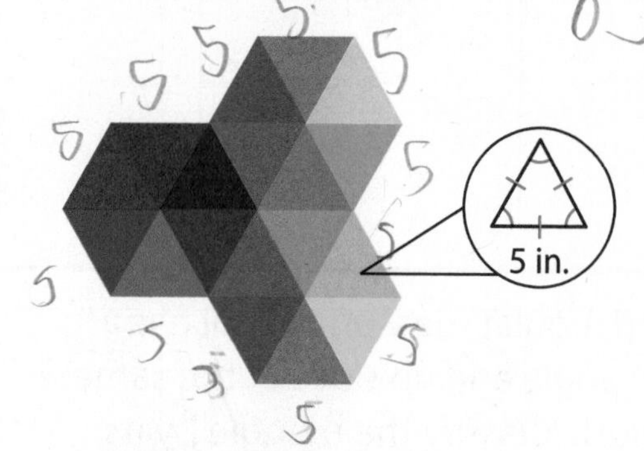

13. DIG DEEPER! The window is made using identical triangular panes of glass. Classify each triangle by its angles and its sides. What are the perimeter and the area of the window? Explain your reasoning.

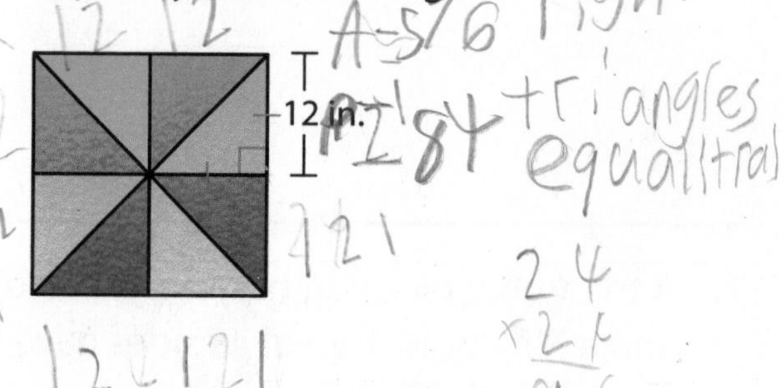

Review & Refresh

Multiply.

14. $\frac{1}{8} \times 2 =$ ______

15. $\frac{3}{4} \times 5 =$ ______

16. $3 \times \frac{5}{12} =$ ______

Name ______________________________

Classify Quadrilaterals 14.2

Learning Target: Classify quadrilaterals by their angles and their sides.

Success Criteria:

- I can identify parallel sides and sides with the same length in a quadrilateral.
- I can identify right angles in a quadrilateral.
- I can use angles and sides to classify a quadrilateral.

Explore and Grow

Draw and label a quadrilateral for each description. If a quadrilateral cannot be drawn, explain why.

quadrilateral with exactly one pair of parallel sides	quadrilateral with two pairs of parallel sides
quadrilateral with four right angles	quadrilateral with four sides that have the same length

180 40
−60
120

MP Precision Draw a quadrilateral that meets three of the descriptions above.

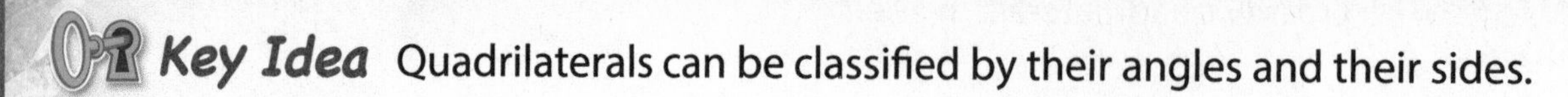

Think and Grow: Classify Quadrilaterals

Key Idea Quadrilaterals can be classified by their angles and their sides.

A **trapezoid** is a quadrilateral that has exactly one pair of parallel sides.

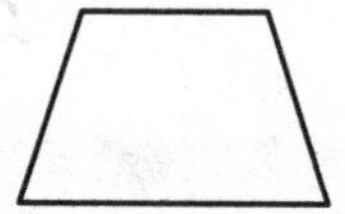

A **parallelogram** is a quadrilateral that has two pairs of parallel sides. Opposite sides have the same length.

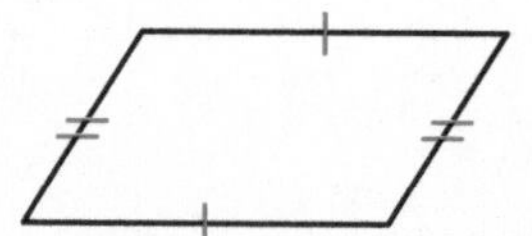

A **rectangle** is a parallelogram that has four right angles.

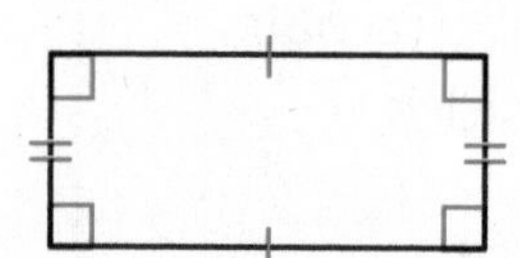

A **rhombus** is a parallelogram that has four sides with the same length.

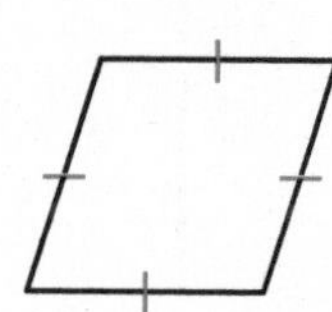

A **square** is a parallelogram that has four right angles and four sides with the same length.

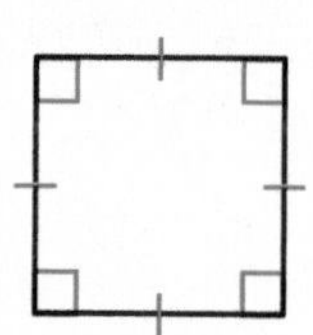

Example Classify the quadrilateral in as many ways as possible.

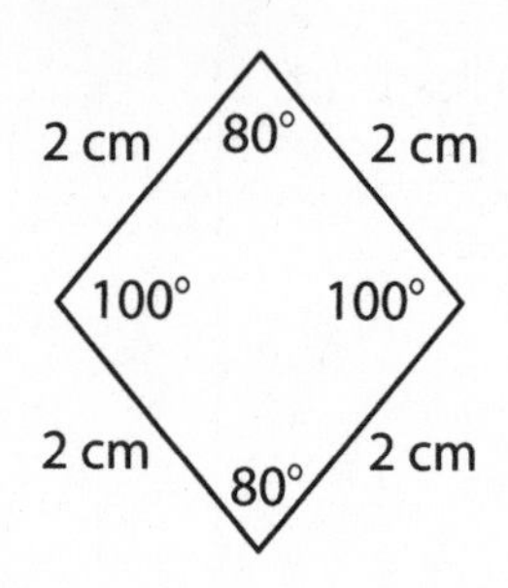

The quadrilateral has 0 right angles,

2 pairs of parallel sides, and

4 sides with the same length.

So, it is a ________ and a ________.

Show and Grow I can do it!

Classify the quadrilateral in as many ways as possible.

1.

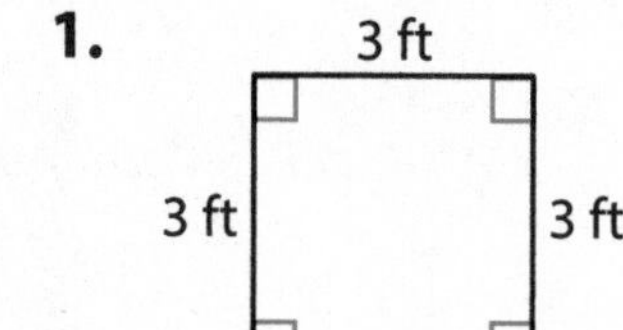

2.

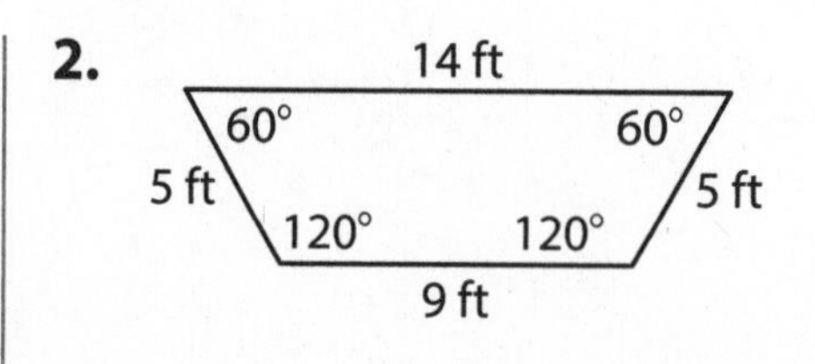

Name ____________________

Apply and Grow: Practice

Classify the quadrilateral in as many ways as possible.

3. 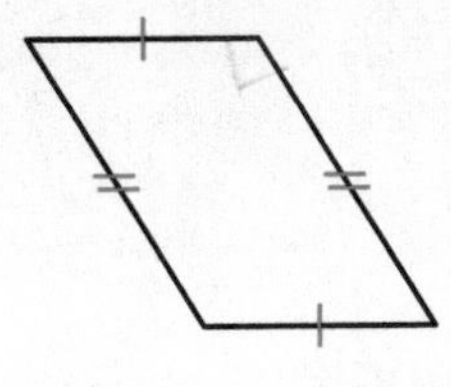

Rhombus and a Parallelogram

4. 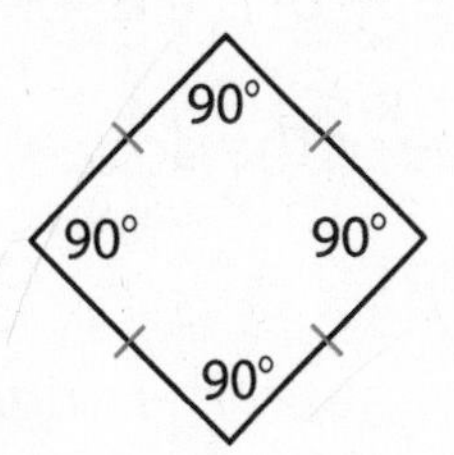

A rhombus, parallelog

5. 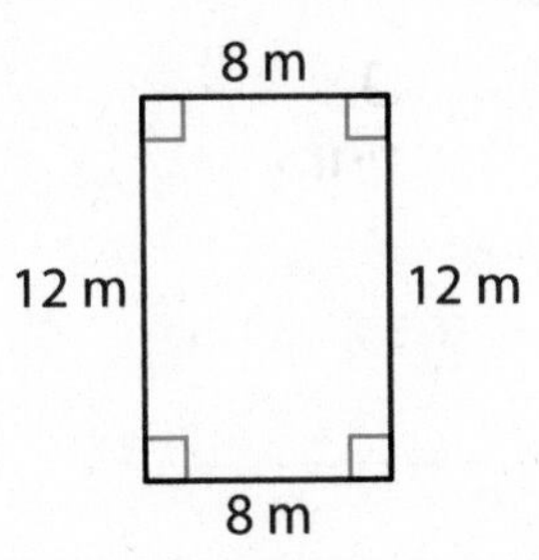

6. 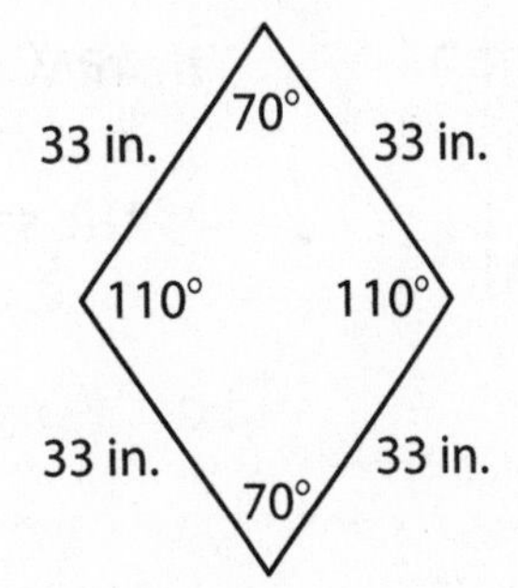

7. A sign has the shape of a quadrilateral that has two pairs of parallel sides, four sides with the same length, and no right angles.

8. A tabletop has the shape of a quadrilateral with exactly one pair of parallel sides.

9. **YOU BE THE TEACHER** Your friend says that a quadrilateral with at least two right angles must be a parallelogram. Is your friend correct? Explain.

10. **Which One Doesn't Belong?** Which set of lengths *cannot* be the side lengths of a parallelogram?

4 m, 2 m, 4 m, 2 m

5 in., 5 in., 5 in., 5 in.

7 ft, 8 ft, 8 ft, 7 ft

9 yd, 5 yd, 5 yd, 3 yd

Think and Grow: Modeling Real Life

Example The dashed line shows how you cut the bottom of a rectangular door so it opens more easily. Classify the new shape of the door.

Draw the new shape of the door.

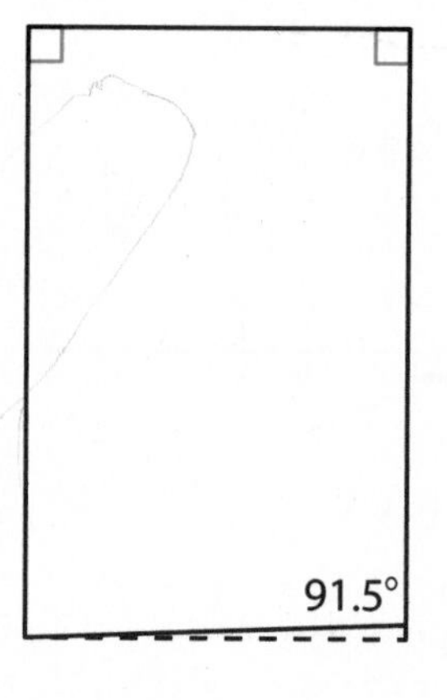

The original shape of the door was a rectangle,

so it had ______ pairs of parallel sides.

The new shape of the door has

exactly ______ pair of parallel sides.

So, the new shape of the door is a ______________.

Show and Grow *I can think deeper!*

11. The dashed line shows how you cut the corner of the trapezoidal piece of fabric. The line you cut is parallel to the opposite side. Classify the new shape of the four-sided piece of fabric.

12. **DIG DEEPER!** A farmer encloses a section of land using the four pieces of fencing. Name all of the four-sided shapes that the farmer can enclose with the fencing.

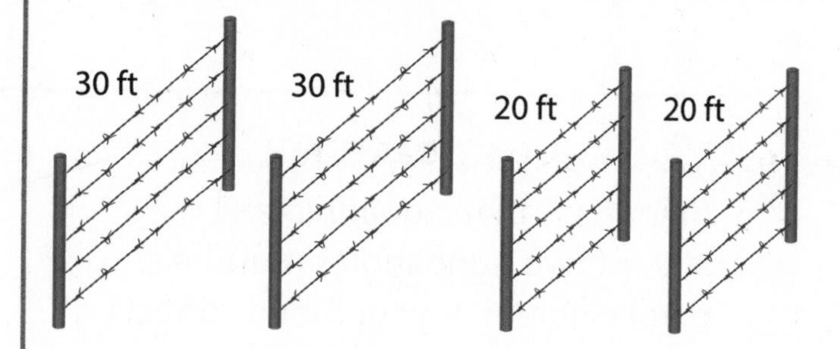

Name ______________________________

Homework & Practice 14.2

Learning Target: Classify quadrilaterals by their angles and their sides.

Example Classify the quadrilateral in as many ways as possible.

The quadrilateral has __4__ right angles,

__2__ pairs of parallel sides, and

__opposite__ sides with the same length.

So, it is a __parallelogram__ and a __rectangle__.

Classify the quadrilateral in as many ways as possible.

1. 5 yd, 6 yd, 12 yd, 5 yd

Trapozoied and a Parallelogram

2.

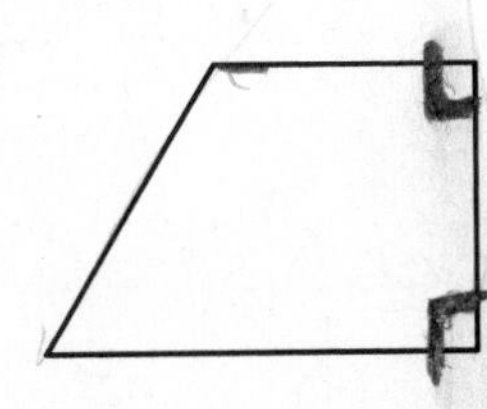

Trapoziod parallelogram

3.

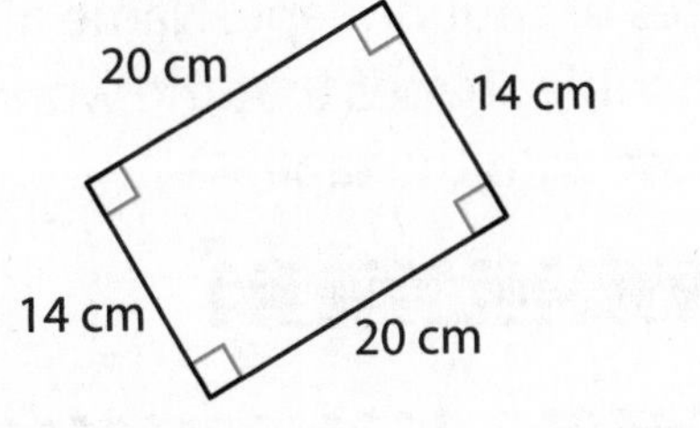

Rectangle Parallelogram

4.

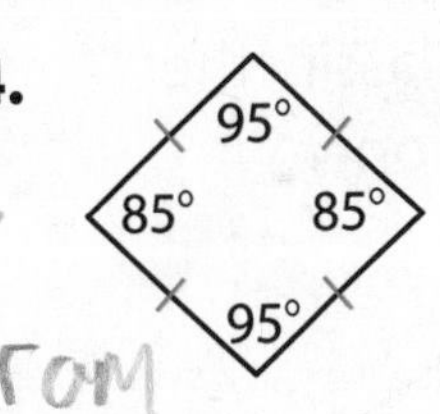

Trapoziod parallelogram

5. A name tag has the shape of a quadrilateral that has two pairs of parallel sides and four right angles. Opposite sides are the same length, but not all four sides are the same length.

Parallelogam rect.

6. A napkin has the shape of a quadrilateral that has two pairs of parallel sides, four sides with the same length, and four right angles.

parallelogram square

7. **MP Reasoning** Can you draw a quadrilateral that is *not* a square, but has four right angles? Explain.

8. **MP Structure** Plot two more points in the coordinate plane to form a square. What two points can you plot to form a parallelogram? What two points can you plot to form a trapezoid? Do not use the same pair of points twice.

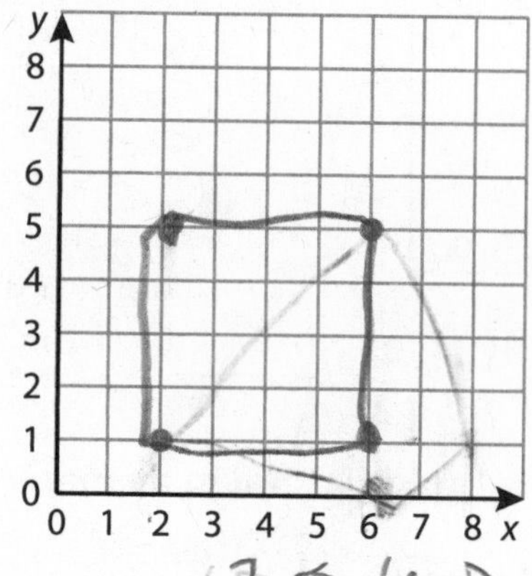

Parallelogram: ______

Trapezoid: ______

9. **DIG DEEPER!** Which quadrilateral can be classified as a parallelogram, rectangle, square, *and* rhombus? Explain.

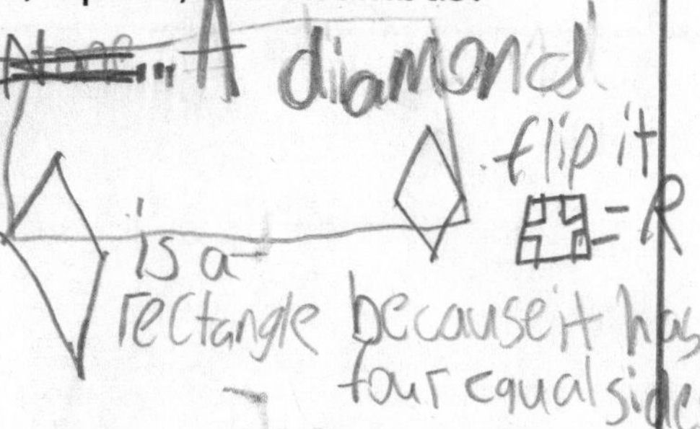

10. **Modeling Real Life** The dashed line shows how you fold the flap of the envelope so it closes. Classify the new shape of the envelope.

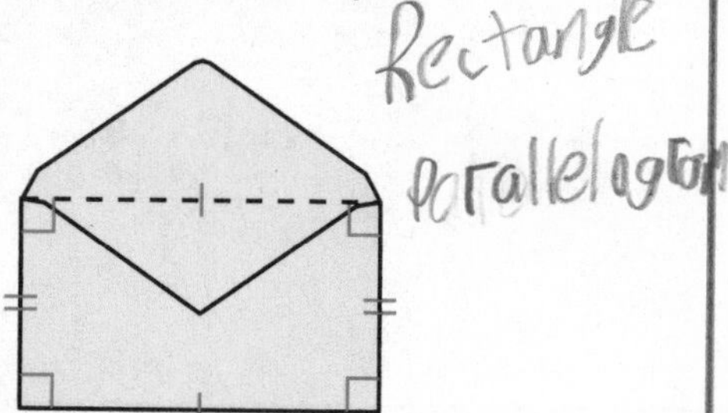

11. **DIG DEEPER!** A construction worker tapes off a section of land using the four pieces of caution tape. Name all of the possible shapes that the worker can enclose with the tape.

CAUTION CAUTION CAUTION CAUTIO
6 yd

CAUTION CAUTION CAUTION CAUTIO
18 ft

CAUTION CAUTION CAUTION CAUTIO
6 yd

CAUTION CAUTION CAUTION CAUTIO
6 yd

Subtract.

12. $\frac{2}{3} - \frac{1}{6} = $ ______

13. $\frac{1}{2} - \frac{7}{18} = $ ______

14. $\frac{2}{5} - \frac{1}{9} = $ ______

Name ______________________________

Relate Quadrilaterals 14.3

Learning Target: Understand the hierarchy of quadrilaterals.

Success Criteria:

- I can arrange quadrilaterals in a Venn diagram based on their properties.
- I can use a Venn diagram to make statements about the relationships among quadrilaterals.

Explore and Grow

Label the Venn diagram to show the relationships among quadrilaterals. The first one has been done for you.

Quadrilaterals
- trapezoids
- parallelograms
- rectangles
- rhombuses
- squares

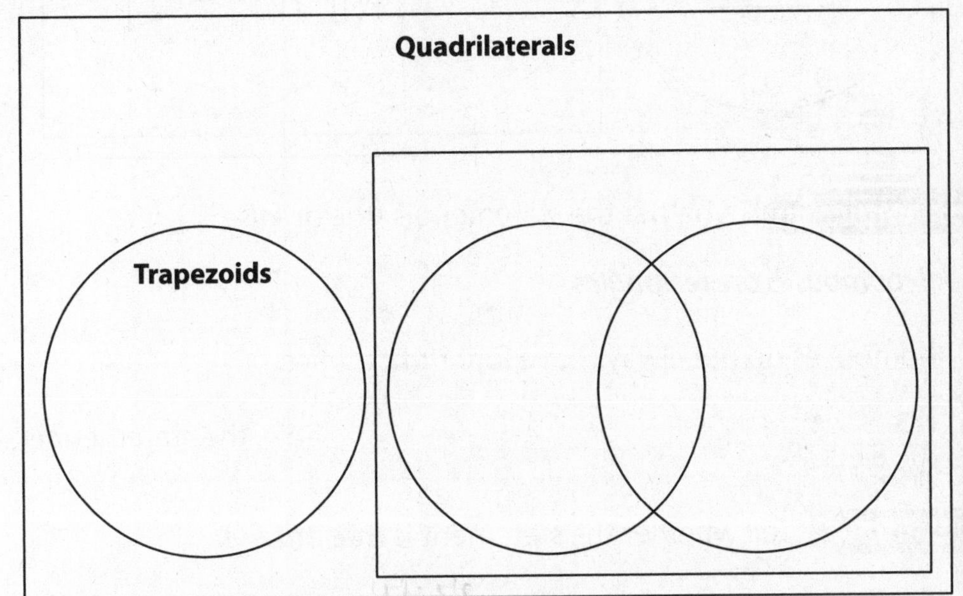

Reasoning Explain how you decided where to place each quadrilateral.

Think and Grow: Relate Quadrilaterals

Key Idea The Venn diagram shows the relationships among quadrilaterals.

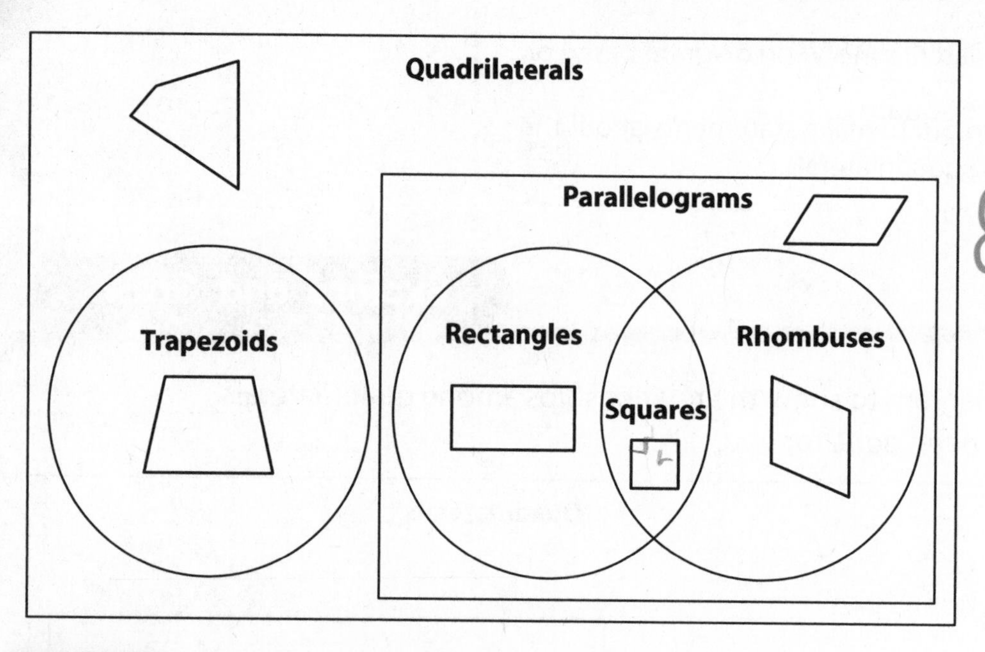

Because rhombuses are a subcategory of parallelograms, a property of parallelograms is also a property of rhombuses.

Example Tell whether the statement is *true* or *false*.

All rhombuses are rectangles.

Rhombuses do not always have four right angles.

So, the statement is false.

Example Tell whether the statement is *true* or *false*.

All rectangles are parallelograms.

All rectangles have two pairs of parallel sides.

So, the statement is __________.

Show and Grow *I can do it!*

Tell whether the statement is *true* or *false*. Explain.

1. Some rhombuses are squares.

2. All parallelograms are rectangles.

Name ____

Apply and Grow: Practice

Tell whether the statement is *true* or *false*. Explain.

3. All rectangles are squares.

4. Some parallelograms are trapezoids.

5. Some rhombuses are rectangles.

6. All trapezoids are quadrilaterals.

7. All squares are rhombuses.

8. Some trapezoids are squares.

9. MP **Reasoning** Use the word cards to complete the graphic organizer.

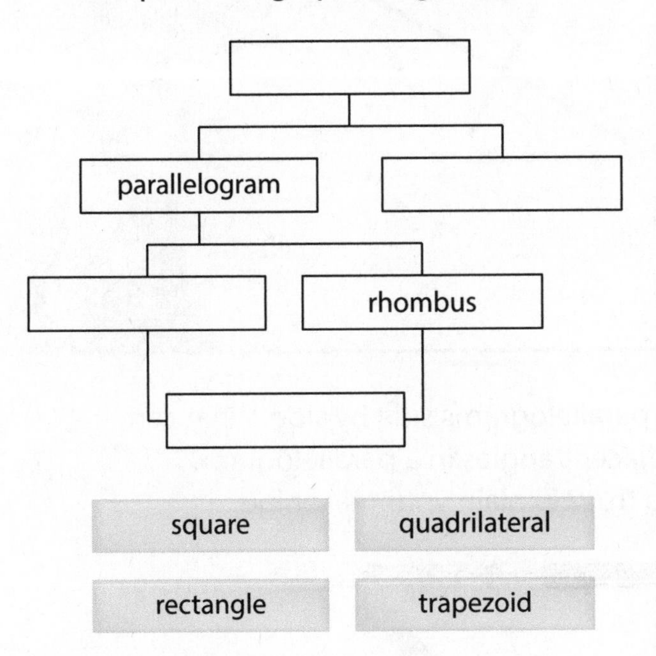

10. MP **Reasoning** All rectangles are parallelograms. Are all parallelograms rectangles? Explain.

11. MP **Precision** Newton says the figure is a square. Descartes says the figure is a parallelogram. Your friend says the figure is a rhombus. Are all three correct? Explain.

Think and Grow: Modeling Real Life

Example You use toothpicks to create several parallelograms. You notice that opposite angles of parallelograms have the same measure. For what other quadrilaterals is this also true?

Parallelograms have the property that opposite angles have the same measure. Subcategories of parallelograms must also have this property.

__________, __________, and __________ are subcategories of parallelograms.

Rectangle

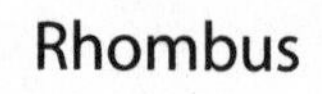

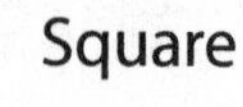

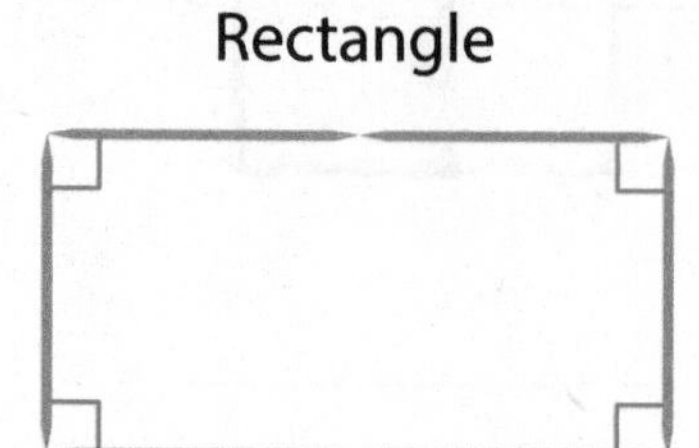

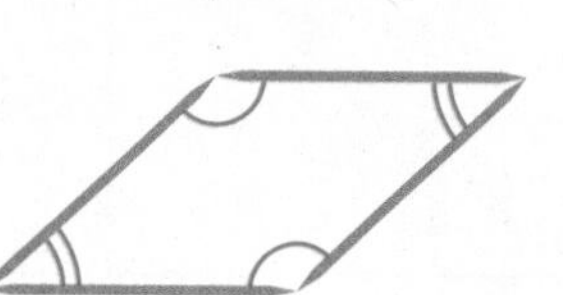

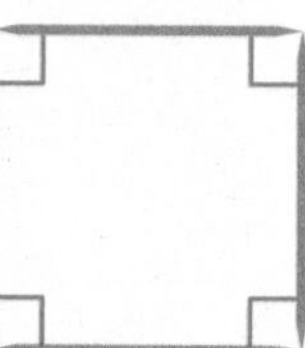

So, __________, __________, and __________ also have opposite angles with the same measure.

Show and Grow *I can think deeper!*

12. You use pencils to create several rhombuses. You notice that diagonals of rhombuses are perpendicular and divide each other into two equal parts. For what other quadrilateral is this also true? Explain your reasoning.

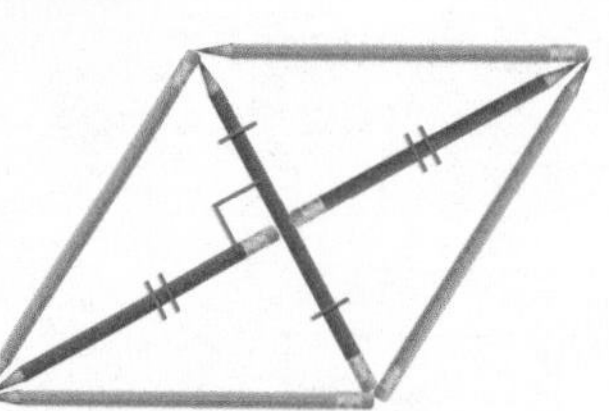

A *diagonal* is a line segment that connects opposite vertices of a quadrilateral.

13. **DIG DEEPER!** You place two identical parallelograms side by side. What can you conclude about the measures of adjacent angles in a parallelogram? For what other quadrilaterals is this also true? Explain your reasoning.

Name ______________________________

Performance Task 14

A homeowner wants to install solar panels on her roof to generate electricity for her house. A solar panel is 65 inches long and 39 inches wide.

1. a. The shape of the panel has 4 right angles. Sketch and classify the shape of the solar panel.

b. There are 60 identical solar cells in a solar panel, arranged in an array. Ten cells meet to form the length of the panel, and six cells meet to form the width. Classify the shape of each solar cell. Explain your reasoning.

2. The homeowner measures three sections of her roof.

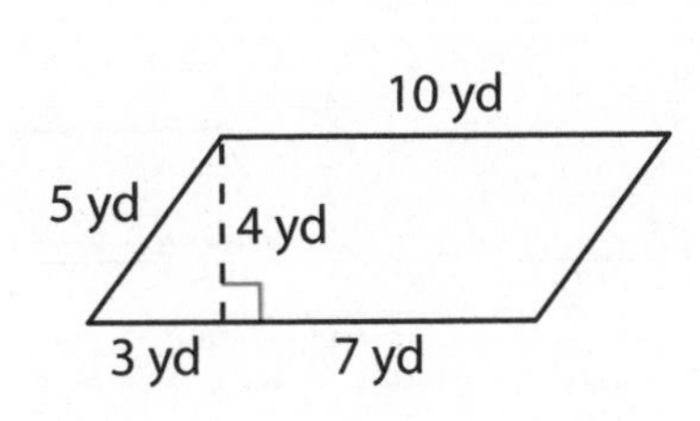

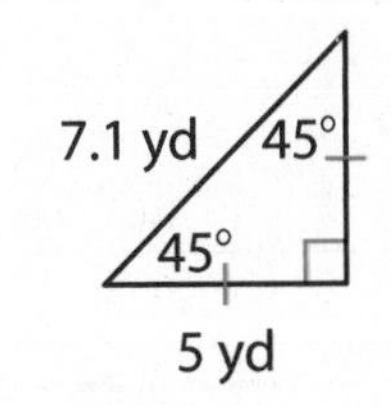

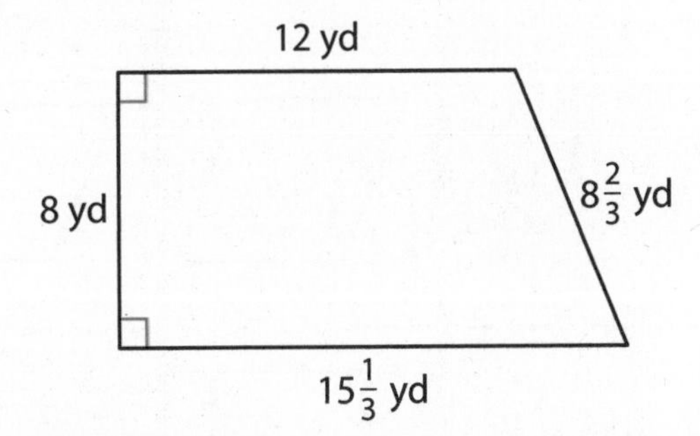

a. Classify the shape of each section in as many ways as possible.

b. About how many solar panels can fit on the measured sections of the roof? Explain your reasoning.

3. One solar panel can produce about 30 kilowatt-hours of electricity each month. The homeowner uses her electric bills to determine that she uses about 1,200 kilowatt-hours of electricity each month.

a. How many solar panels should the homeowner install on her roof?

b. Will all of the solar panels fit on the measured sections of the roof? Explain.

Quadrilateral Lineup

Directions:

1. Players take turns spinning the spinner.
2. On your turn, cover a quadrilateral that matches your spin.
3. If you land on *Lose a Turn*, then do not cover a quadrilateral.
4. The first player to get four in a row twice, horizontally, vertically, or diagonally, wins!

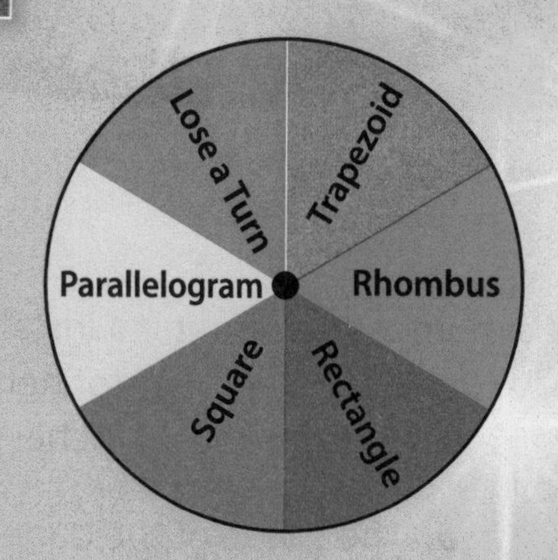

Name ______________________________

Chapter Practice

14.1 Classify Triangles

Classify the triangle by its angles and its sides.

1. 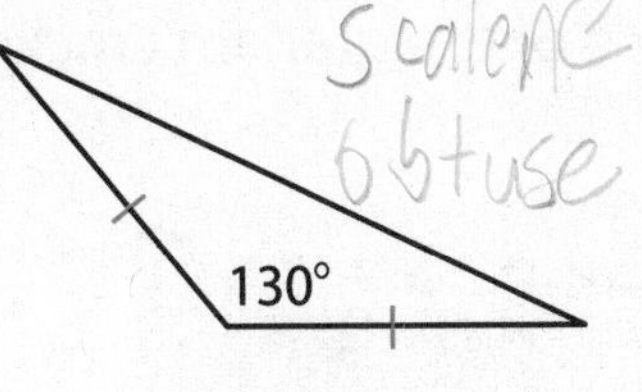

2. 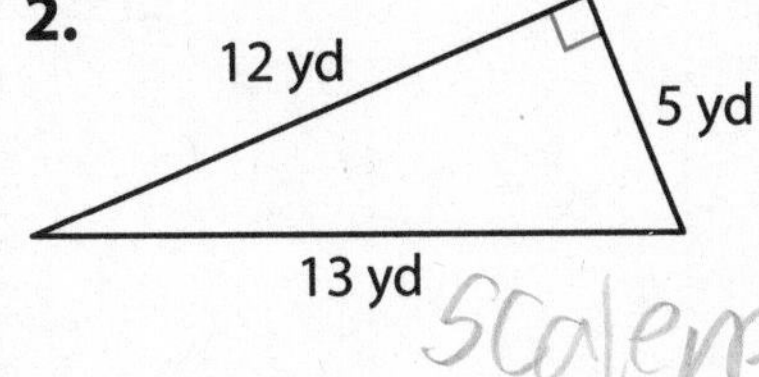

3. 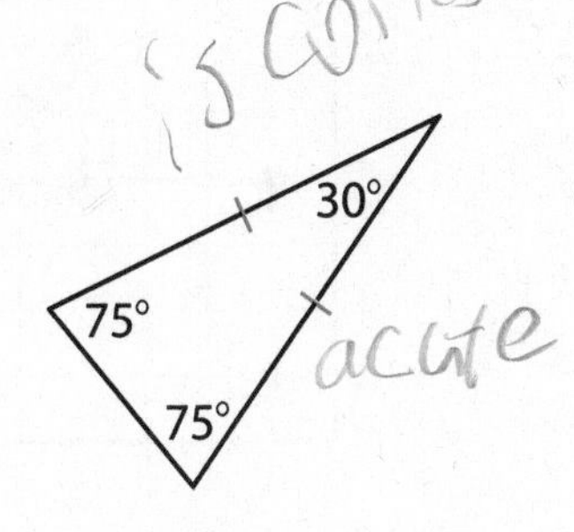

4. 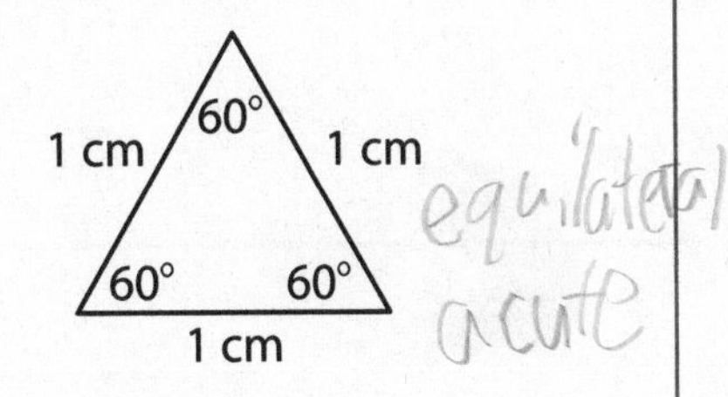

5. 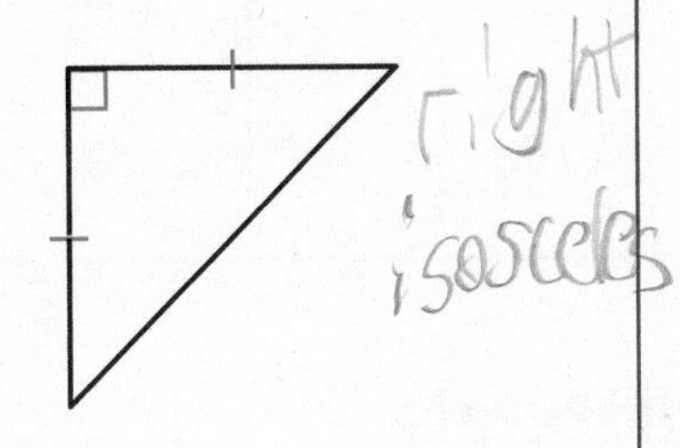

6. 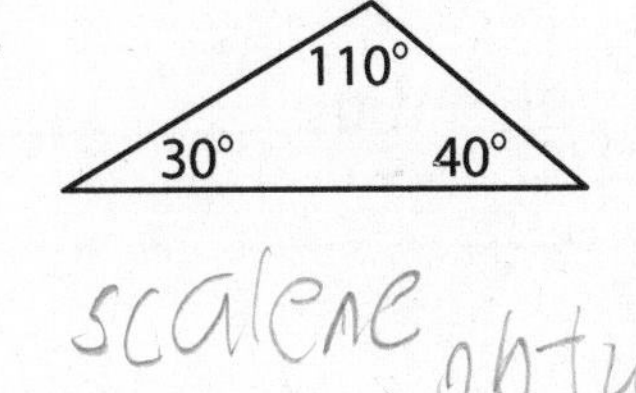

14.2 Classify Quadrilaterals

Classify the quadrilateral in as many ways as possible.

7. 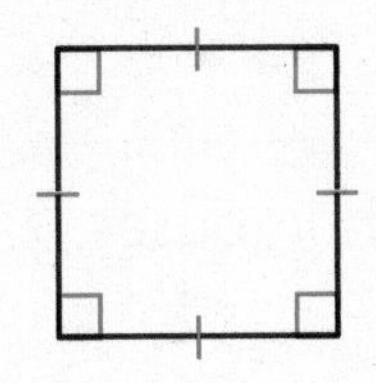

8. 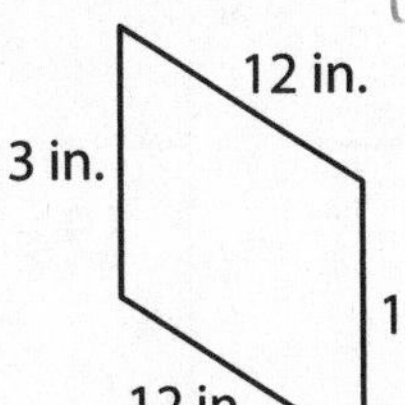

9. 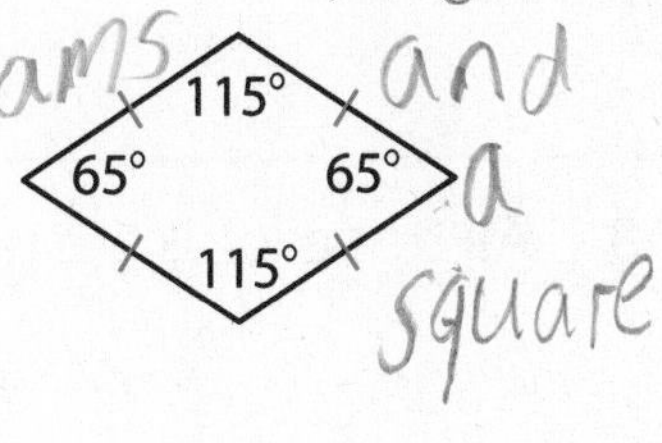

10. 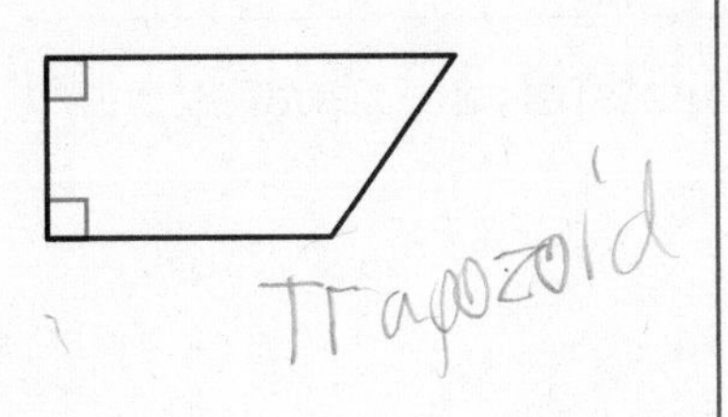

11. 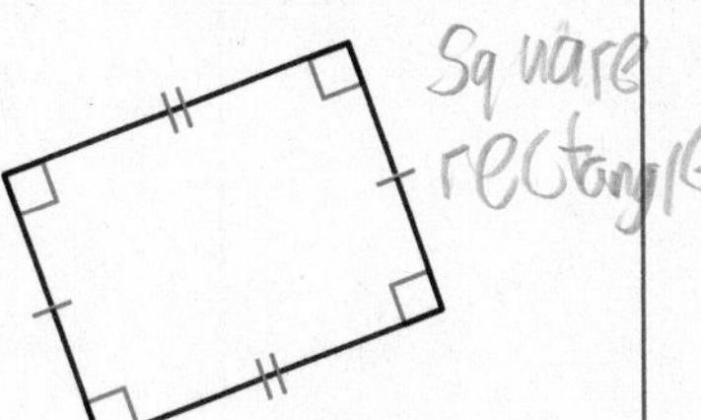

12. 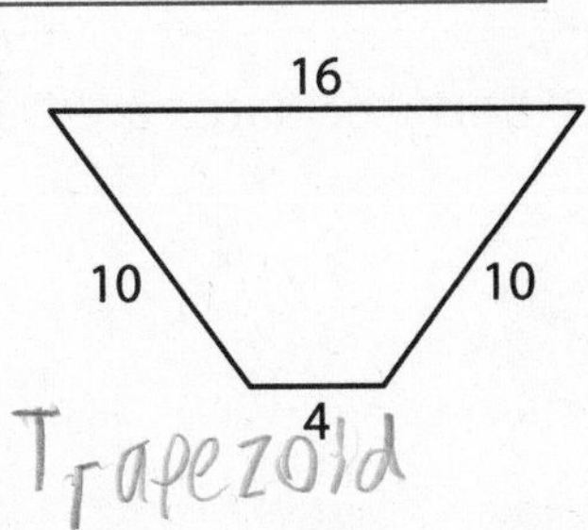

13. **Structure** Plot two more points in the coordinate plane to form a rectangle. What two points can you plot to form a trapezoid? What two points can you plot to form a rhombus? Do not use the same pair of points twice.

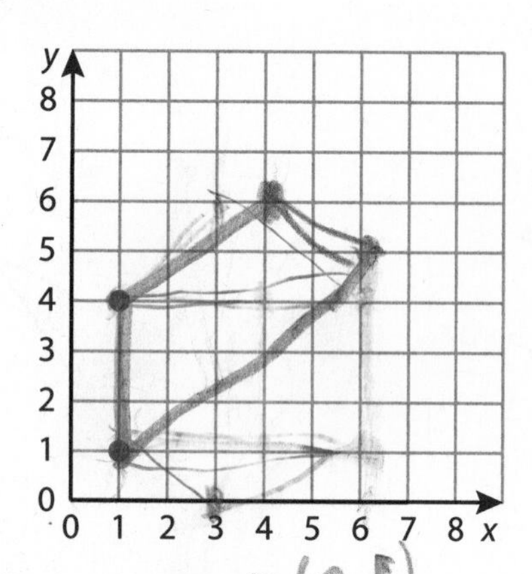

Trapezoid: (4,6)(6,5)

Rhombus: (4,4)(4,1)

14. **Reasoning** Can you draw a quadrilateral that has exactly two right angles? Explain.

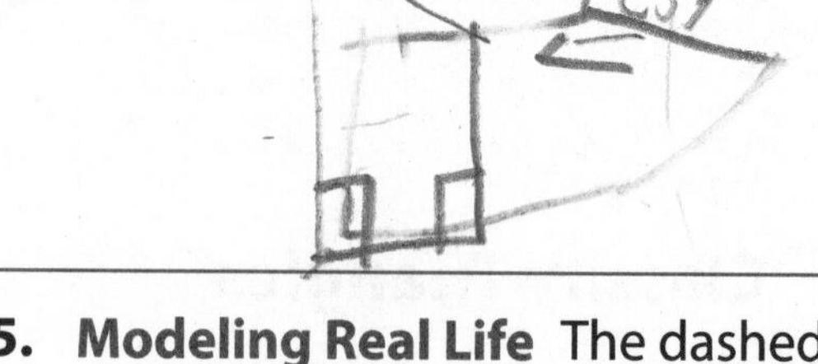

15. **Modeling Real Life** The dashed line shows how you break apart the graham cracker. Classify the new shape of each piece of the graham cracker.

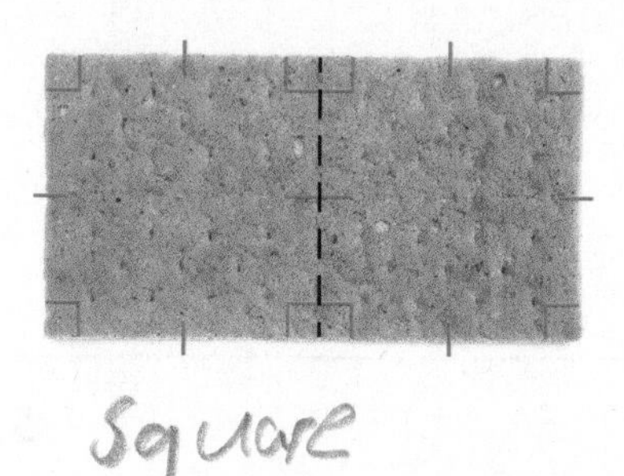

Square

14.3 Relate Quadrilaterals

Tell whether the statement is *true* or *false*. Explain.

16. All rectangles are quadrilaterals.

True

17. Some parallelograms are squares.

True

18. All trapezoids are rectangles.

False

19. Some rectangles are rhombuses.

True

20. Some squares are trapezoids.

False

21. All quadrilaterals are squares.

False

Name ____

Cumulative Practice 1–14

1. Which model shows 0.4×0.2?

Ⓐ
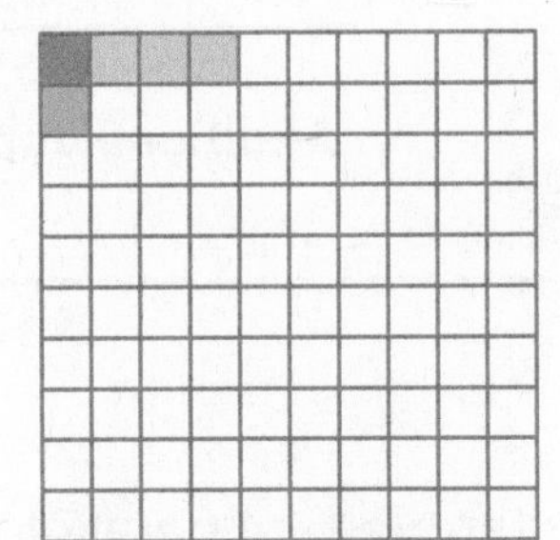

Ⓑ
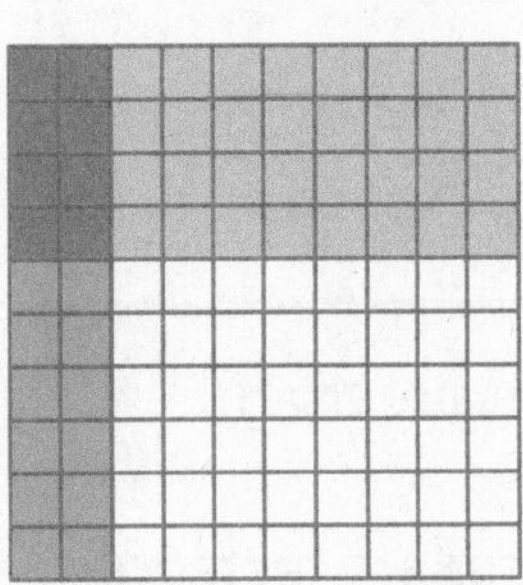

Ⓒ
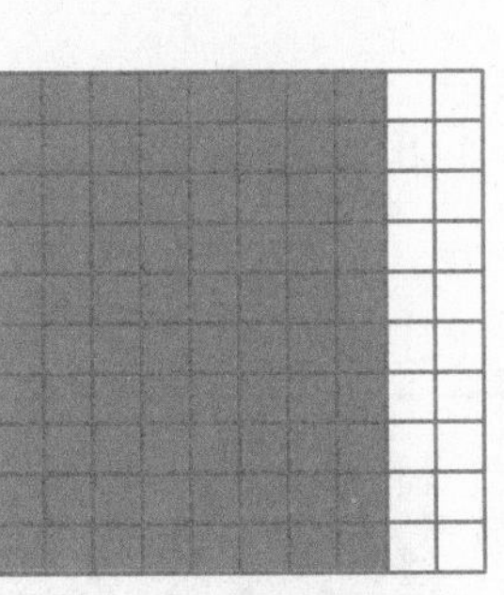

Ⓓ
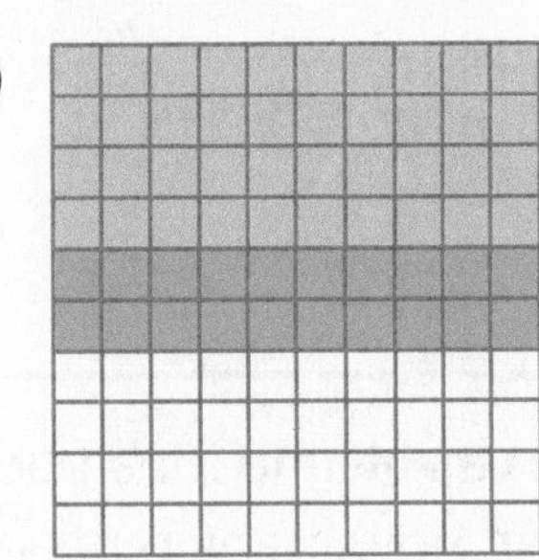

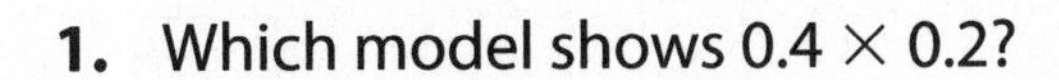

---

2. A triangle has angle measures of 82°, 53°, and 45°. Classify the triangle by its angles.

Ⓐ acute

Ⓑ equiangular

Ⓒ obtuse

Ⓓ right

---

3. Which expressions have an estimated difference of $\frac{1}{2}$?

☐ $\frac{11}{12} - \frac{7}{8}$

☐ $\frac{9}{10} - \frac{1}{5}$

☐ $\frac{4}{3} - \frac{1}{2}$

☐ $\frac{15}{16} - \frac{4}{9}$

☐ $\frac{5}{4} - \frac{11}{20}$

☐ $\frac{1}{8} - \frac{1}{6}$

---

4. A rectangular prism has a volume of 288 cubic centimeters. The height of the prism is 8 centimeters. The base is a square. What is a side length of the base?

Ⓐ 6 cm

Ⓑ 9 cm

Ⓒ 18 cm

Ⓓ 36 cm

5. A sandwich at a food stand costs \$3.00. Each additional topping costs the same extra amount. The coordinate plane shows the costs, in dollars, of sandwiches with different numbers of additional toppings. What is the cost of a sandwich with 3 additional toppings?

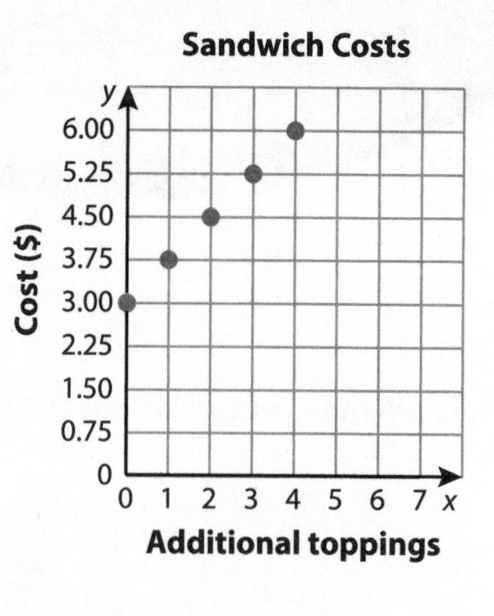

Ⓐ \$0.00

Ⓑ \$3.00

Ⓒ \$5.25

Ⓓ \$6.00

---

6. Which statements are *true*?

☐ All rhombuses are rectangles.

☐ All squares and rectangles are parallelograms.

☐ All squares are rhombuses.

☐ Every trapezoid is a quadrilateral.

---

7. Your friend makes a volcano for a science project. She uses 10 cups of vinegar. How many pints of vinegar does he use?

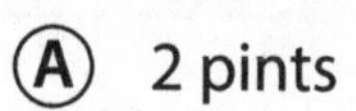

Ⓐ 2 pints

Ⓑ 5 pints

Ⓒ 20 pints

Ⓓ 40 pints

---

8. The volume of the rectangular prism is 432 cubic centimeters. What is the length of the prism?

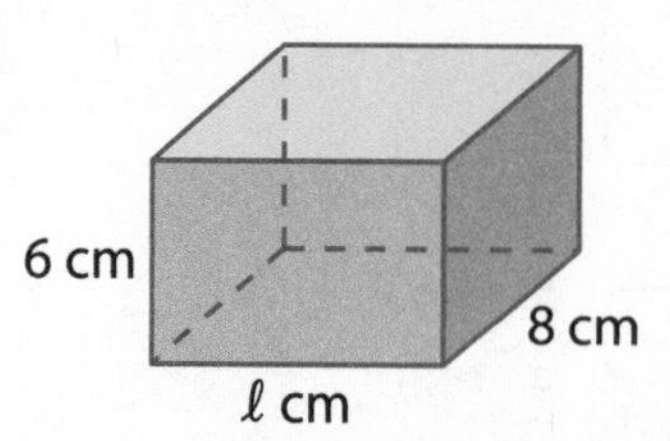

Ⓐ 9 cm

Ⓑ 18 cm

Ⓒ 384 cm

Ⓓ 418 cm

9. Descartes draws a pentagon by plotting another point in the coordinate plane and connecting the points. Which coordinates could he use?

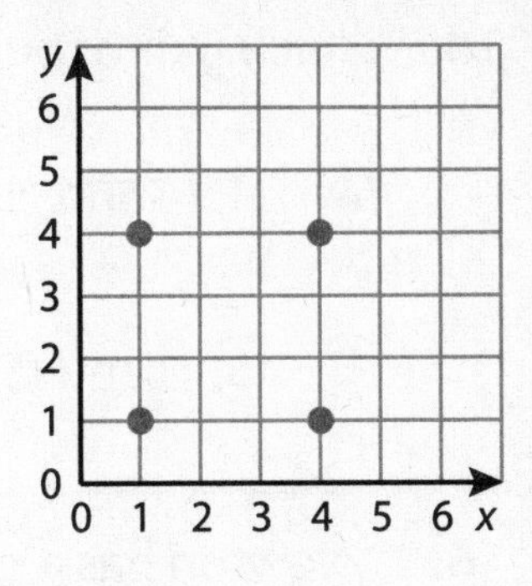

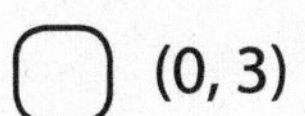

◯ (0, 3) ◯ (2, 4)

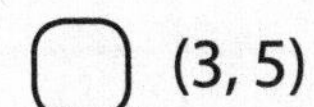

◯ (3, 5) ◯ (6, 2)

---

10. Newton rides to the dog park in a taxi. He owes the driver $12. He calculates the driver's tip by multiplying $12 by 0.15. How much does he pay the driver, including the tip?

Ⓐ $1.80 Ⓑ $10.20

Ⓒ $13.80 Ⓓ $30.00

---

11. A quadrilateral has four sides with the same length, two pairs of parallel sides, and four 90° angles. Classify the quadrilateral in as many ways as possible.

◯ square ◯ trapezoid

◯ parallelogram ◯ rhombus

---

12. Which ordered pair represents the location of a point shown in the coordinate plane?

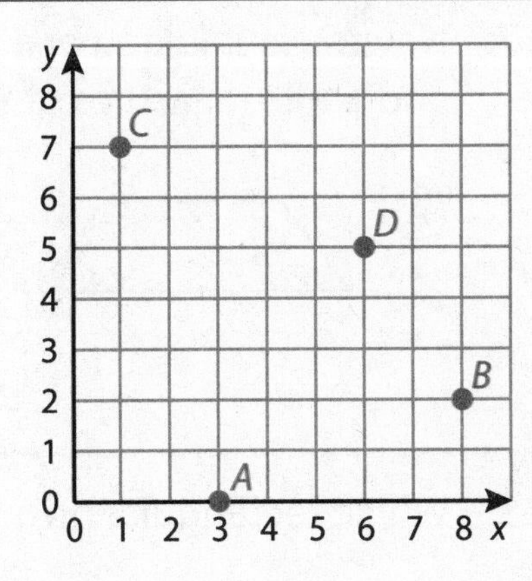

Ⓐ (0, 3)  Ⓑ (1, 7)

Ⓒ (5, 6) Ⓓ (2, 8)

---

13. What is the product of 5,602 and 17?

14. Which pair of points do *not* lie on a line that is perpendicular to the *x*-axis?

Ⓐ (1, 6) and (1, 8)

Ⓑ (4, 0) and (4, 4)

Ⓒ (5, 1) and (5, 3)

Ⓓ (7, 2) and (6, 2)

---

15. Newton has a gift in the shape of a rectangular prism that has a volume of 10,500 cubic inches. The box he uses to ship the gift is shown.

Think Solve Explain

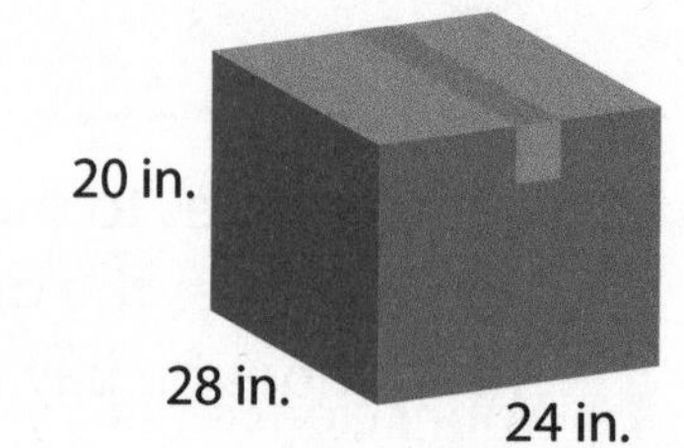

Part A What is the volume of the box?

Part B What is the volume, in cubic inches, of the space inside the box that is *not* taken up by the gift? Explain.

---

16. Which expressions have a product greater than $\frac{5}{6}$?

- ▢ $\frac{5}{6} \times \frac{7}{8}$
- ▢ $\frac{5}{6} \times \frac{1}{3}$
- ▢ $\frac{5}{6} \times \frac{9}{9}$
- ▢ $\frac{5}{6} \times \frac{3}{2}$
- ▢ $\frac{5}{6} \times \frac{4}{5}$
- ▢ $\frac{5}{6} \times 1\frac{3}{4}$

---

17. Newton is thinking of a shape that has 4 sides, only one pair of parallel sides, and angle measures of 90°, 40°, 140°, and 90°. Which is Newton's shape?

 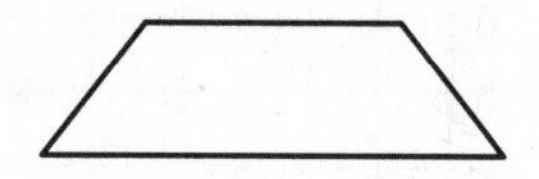

Ⓑ 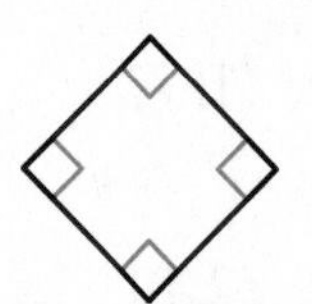

Ⓒ 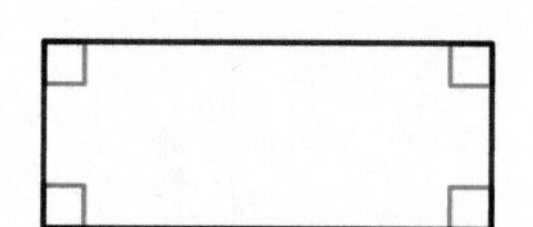

Ⓓ 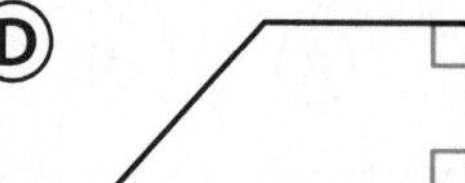

---

18. Which rectangular prisms have a volume of 150 cubic feet?

- ▢ length = 2 ft, height = 25 ft, width = 3 ft
- ▢ length = 50 ft, height = 50 ft, width = 50 ft
- ▢ length = 3 ft, height = 10 ft, width = 5 ft
- ▢ length = 6 ft, height = 5 ft, width = 5 ft

Name ______________________________

STEAM Performance Task 1–14

Each student in your grade makes a constellation display by making holes for the stars of a constellation on each side of the display. Each display is a rectangular prism with a square base.

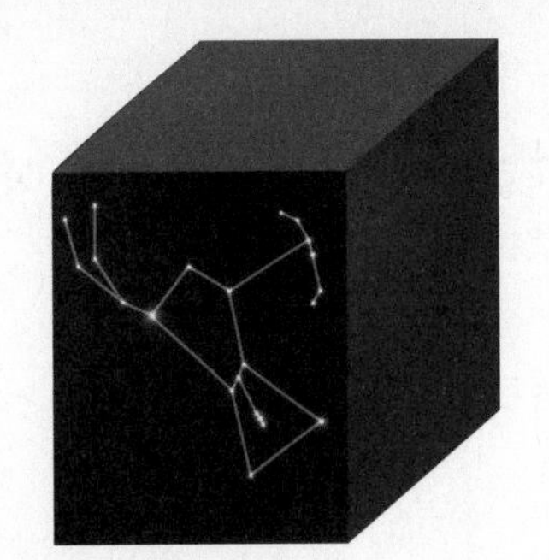

1. Your science teacher orders a display for each student. The diagram shows the number of packages that can fit in a shipping box.

   a. How many displays come in one box?

   b. There are 108 students in your grade. How many boxes of displays does your teacher order? Explain.

   c. The volume of the shipping box is 48,000 cubic inches. What is the volume of each display?

   d. The height of each display is 15 inches. What are the dimensions of the square base?

   e. Estimate the dimensions of the shipping box.

   f. You paint every side of the display except the bottom. What is the total area you will paint?

   g. You need a lantern to light up your display. Does the lantern fit inside of your display? Explain.

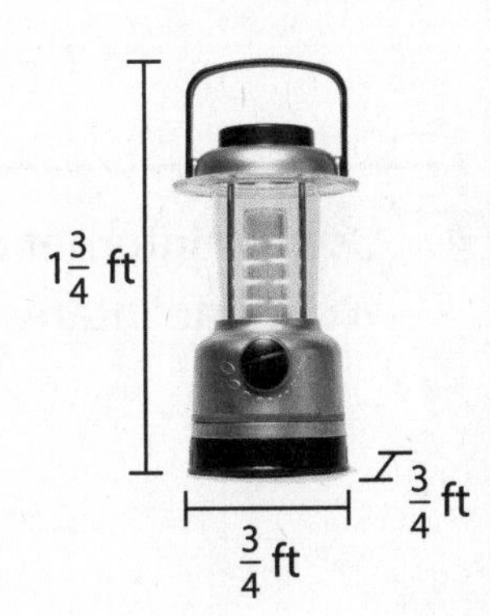

2. On one side of your display, you create an image of the constellation Libra. Each square on the grid is 1 square inch.

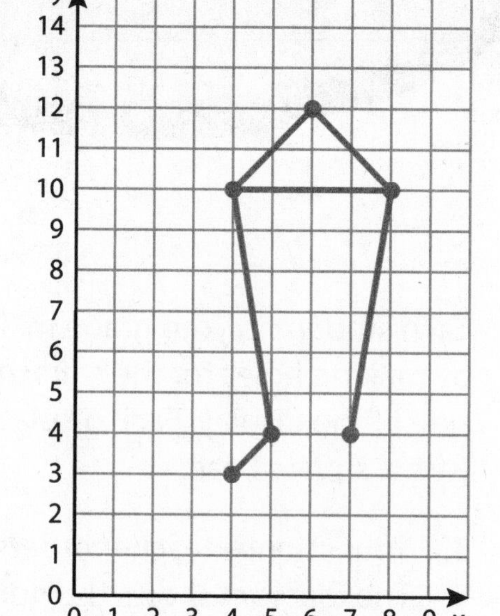

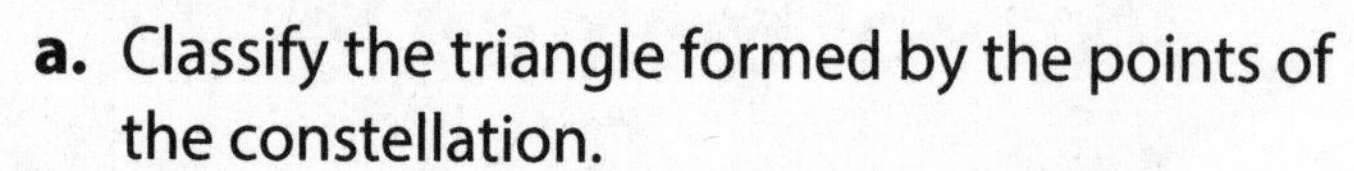

a. Classify the triangle formed by the points of the constellation.

b. What are the coordinates of the points of the constellation?

c. What is the height of the constellation on your display?

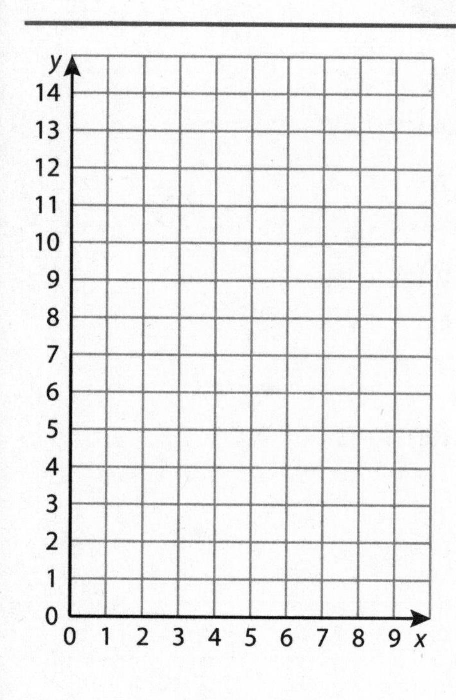

3. You use the coordinate plane to create the image of the Big Dipper.

a. Plot the points $A(6, 2)$, $B(8, 2)$, $C(7, 6)$, $D(5, 5)$, $E(7, 9)$, $F(6, 12)$, and $G(4, 14)$.

b. Draw lines connecting the points of quadrilateral *ABCD*. Draw $\overline{CE}$, $\overline{EF}$, and $\overline{FG}$.

c. Is quadrilateral *ABCD* a trapezoid? How do you know?

4. Use the Internet or some other resource to learn more about constellations. Write one interesting thing you learn.

Glossary

acute triangle **[triángulo actángulo]**

A triangle that has three acute angles

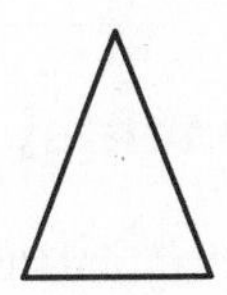

B

base (of a power) **[base]**

The repeated factor in a power

$10 \times 10 \times 10 \times 10 \times 10 = 10^5$ (base: 10)

base (of a prism) **[base]**

The bottom face of a right rectangular prism

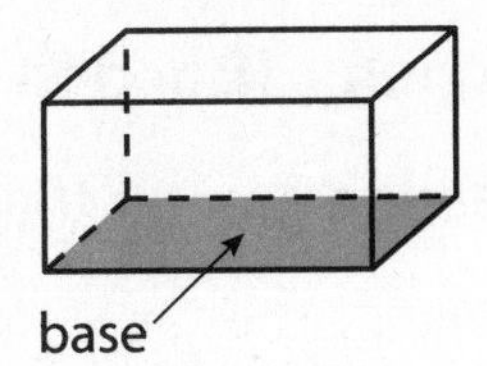

C

common denominator **[denominador común]**

A number that is the denominator of two or more fractions

$\frac{3}{4}$ $\frac{2}{4}$

4 is a common denominator for $\frac{3}{4}$ and $\frac{2}{4}$.

composite figure **[figura compuesta]**

A figure that is made of two or more solid figures

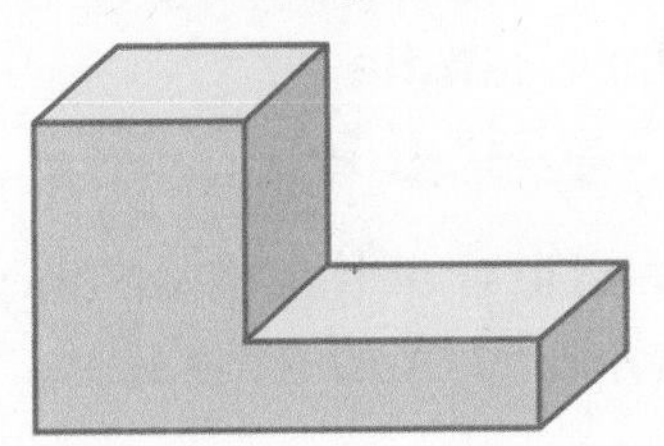

coordinate plane **[plano de coordenadas]**

A plane that is formed by the intersection of a horizontal number line and a vertical number line

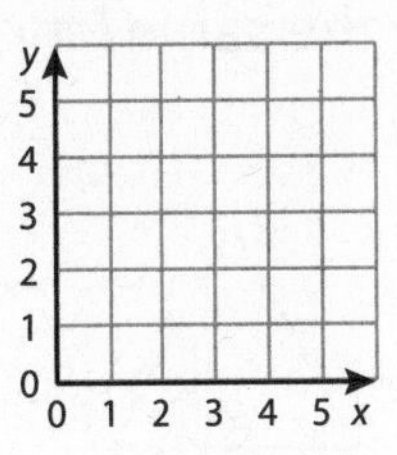

cubic unit **[unidad cúbica]**

A unit used to measure volume

cubic centimeter
cubic inch
cubic foot

data **[datos]**

Values collected from observations or measurements

Day	1	2	3	4	5
Packages Delivered	128	154	137	168	193

equiangular triangle
[triángulo equiángulo]

A triangle that has three angles with the same measure

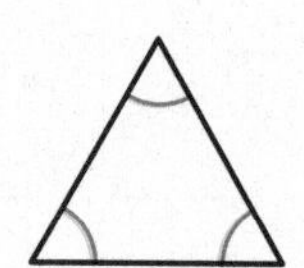

equilateral triangle
[triángulo equilátero]

A triangle that has three sides with the same length

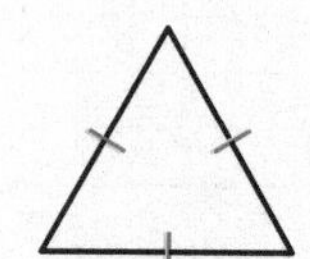

evaluate **[evaluar]**

To find the value of a numerical expression

$$\underbrace{15 + 6 \times 5}_{\text{numerical expression}} = \underbrace{45}_{\text{value}}$$

exponent **[exponente]**

The number of times the base of a power is used as a factor

$$10 \times 10 \times 10 \times 10 \times 10 = 10^5$$

exponent (arrow pointing to 5)

fluid ounces (fl oz) **[onzas fluidas (oz fl)]**

A customary unit used to measure capacity

There are 8 fluid ounces in 1 cup.

improper fraction **[fracción impropia]**

A fraction greater than 1

$$\frac{3}{2}, \frac{6}{3}, \frac{9}{4}$$

inverse operations
[operaciones inversas]

Operations that "undo" each other, such as addition and subtraction, or multiplication and division

Addition $19 + 12 = 31$	⟷	Subtraction $31 - 12 = 19$
Multiplication $18 \times 4 = 72$	⟷	Division $72 \div 4 = 18$

isosceles triangle **[triángulo isósceles]**

A triangle that has two sides with the same length

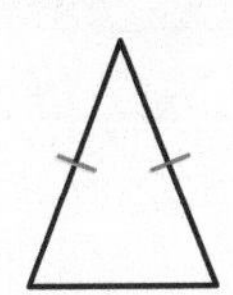

L

line graph **[gráfica de líneas]**

A graph that uses line segments to show how data values change over time

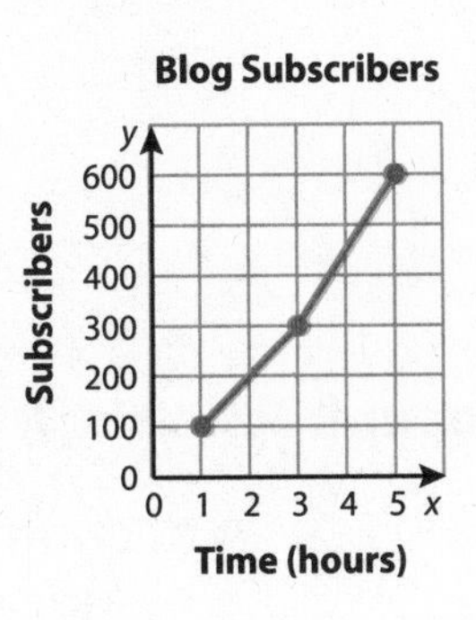

M

milligrams (mg) **[miligramos (mg)]**

A metric unit used to measure mass

One piece of salt weighs about 1 milligram.

N

numerical expression
[expresión numérica]

An expression that contains numbers and operations

$31 + 56$

$19 - 18 \div 6$

O

obtuse triangle
[triángulo obtusángulo]

A triangle that has one obtuse angle

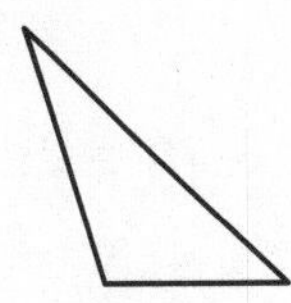

order of operations
[orden de las operaciones]

A set of rules for evaluating expressions

Order of Operations
(with Grouping Symbols)

1. Perform operations in grouping symbols.
2. Multiply and divide from left to right.
3. Add and subtract from left to right.

ordered pair **[par ordenado]**

A pair of numbers that is used to locate a point in a coordinate plane

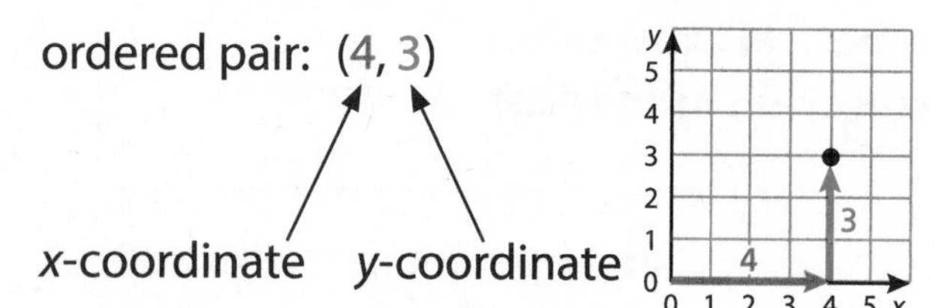

origin **[origen]**

The point, represented by the ordered pair (0, 0), where the x-axis and the y-axis intersect in a coordinate plane

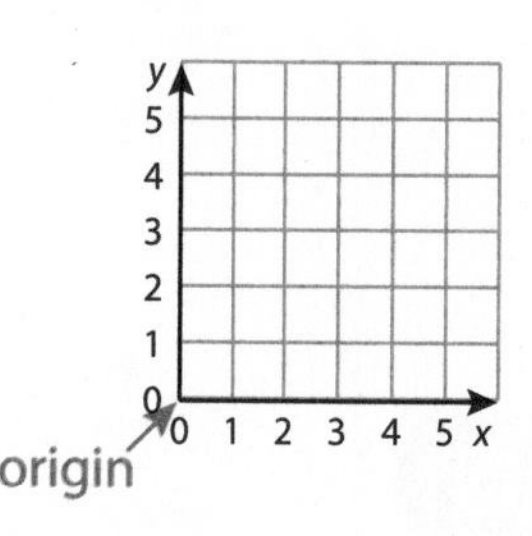

overestimate **[sobrestimar]**

An estimate that is greater than the actual value

38×14

$40 \times 15 = 600$

overestimate

parallelogram **[paralelogramo]**

A quadrilateral that has two pairs of parallel sides

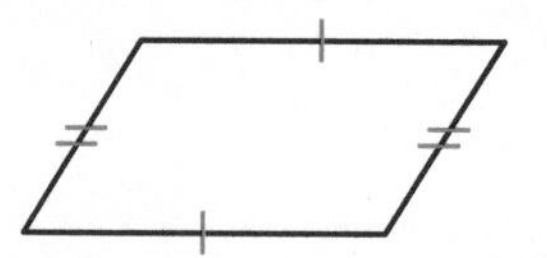

period **[período]**

Each group of three digits separated by commas in a multi-digit number

Thousands Period			Ones Period		
Hundreds	Tens	Ones	Hundreds	Tens	Ones
5	3	8,	7	7	4

power **[potencia]**

A product of repeated factors

$10 \times 10 \times 10 \times 10 \times 10 = 10^5$

power

proper fraction **[fracción propia]**

A fraction less than 1

$\frac{1}{2}, \frac{2}{3}, \frac{3}{4}$

R

rectangle **[rectángulo]**

A parallelogram that has four right angles

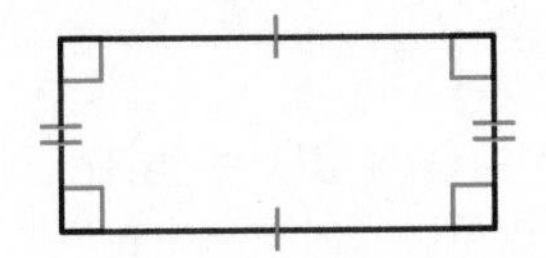

rhombus **[rombo]**

A parallelogram that has four sides with the same length

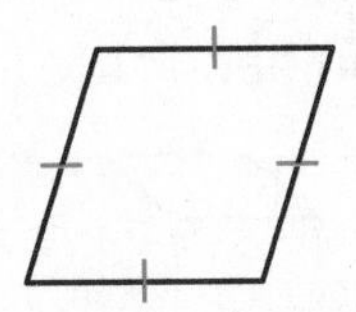

right rectangular prism **[prisma rectangular derecho]**

A solid figure with six rectangular faces

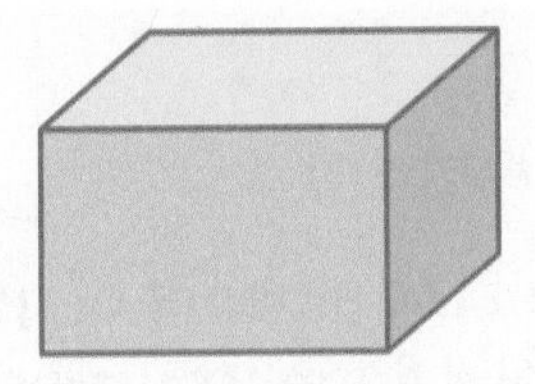

right triangle **[triángulo rectángulo]**

A triangle that has one right angle

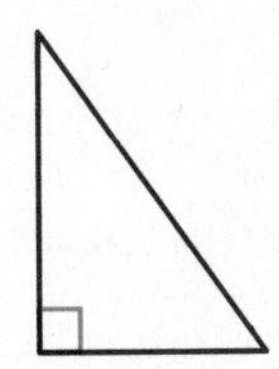

S

scalene triangle **[triángulo escaleno]**

A triangle that has no sides with the same length

simplest form **[mínima expresión]**

When the numerator and denominator of a fraction have no common factors other than 1

$$\frac{2}{6} = \frac{1}{3}$$

↑ simplest form

square **[cuadrado]**

A parallelogram that has four right angles and four sides with the same length

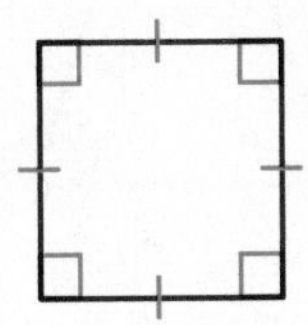

thousandth **[milésimo]**

1 of 1,000 equal parts of a whole

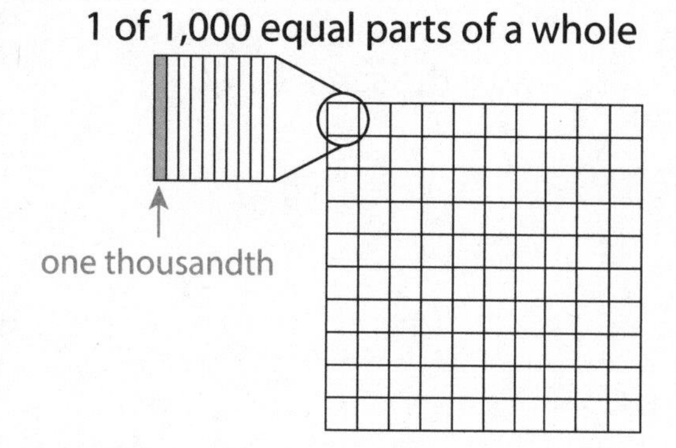

thousandths place **[posición de los milésimos]**

The third place to the right of the decimal point

0.001

↑

thousandths place

trapezoid **[trapecio]**

A quadrilateral that has exactly one pair of parallel sides

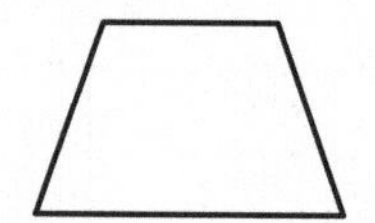

underestimate **[subestimar]**

An estimate that is less than the actual value

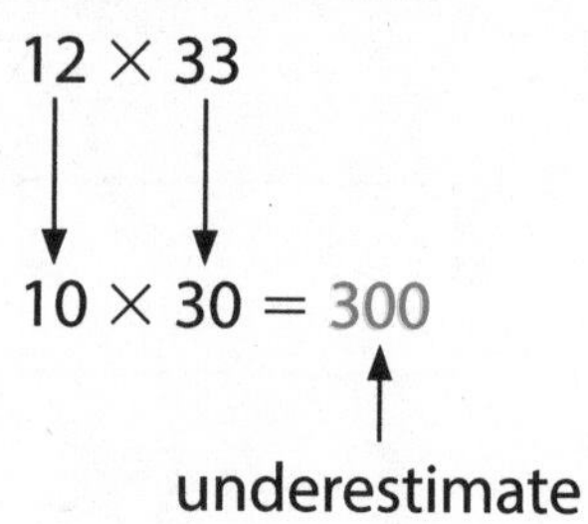

unit cube **[cubo unitaria]**

A cube that measures one unit on each side

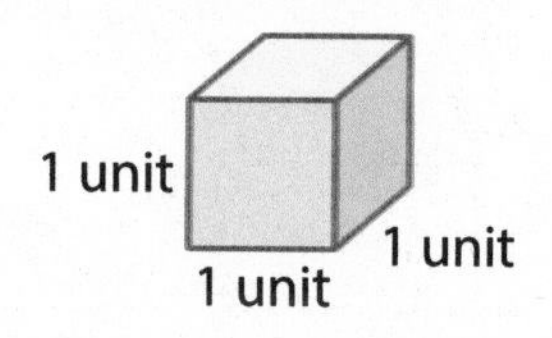

volume **[volumen]**

A measure of the amount of space that a solid figure occupies

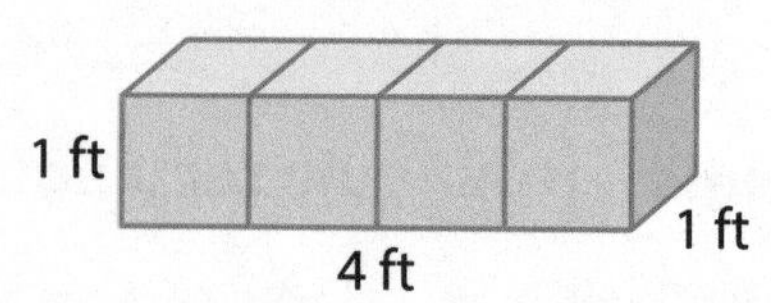

The volume of the figure is 4 cubic feet.

x-axis [eje x]

The horizontal number line in a coordinate plane

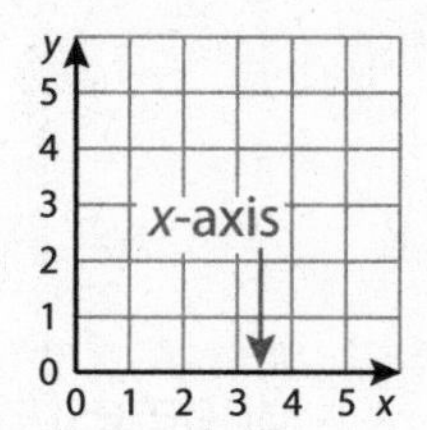

x-coordinate [coordenada x]

The first number in an ordered pair, which gives the horizontal distance from the origin along the x-axis

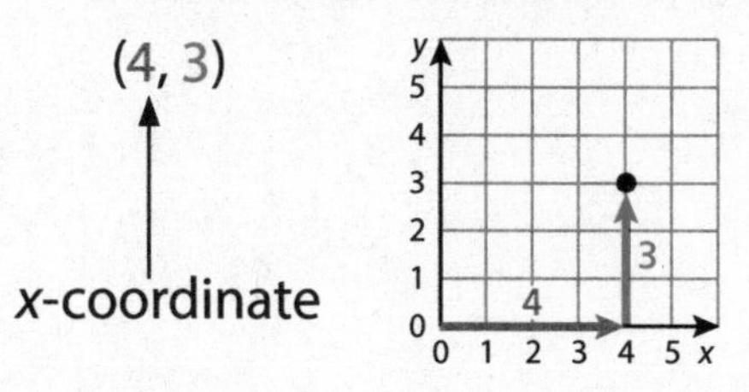

y-axis [eje y]

The vertical number line in a coordinate plane

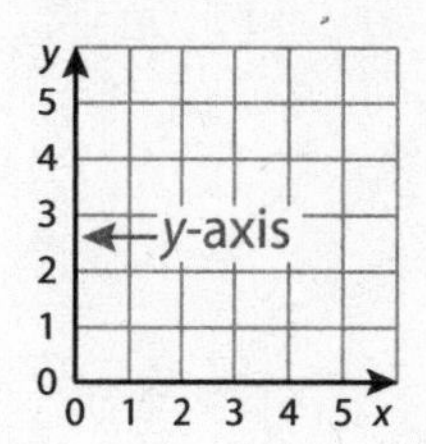

y-coordinate [coordenada y]

The second number in an ordered pair, which gives the vertical distance from the origin along the y-axis

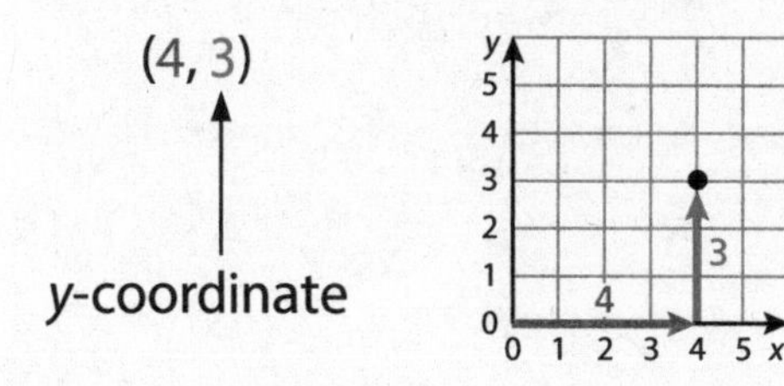

Glossary

Index

C

E

F

Index

M

Index

N

S

T

Index

U

V

W

Reference Sheet

Symbols

=	equals	$\overset{A}{\bullet}$	point A	$\angle ABC$	angle ABC
>	greater than	$\overleftrightarrow{AB}$	line AB	°	degree(s)
<	less than	$\overrightarrow{AB}$	ray AB	°C	degrees Celsius
⊥	is perpendicular to	$\overline{AB}$	line segment AB	°F	degrees Fahrenheit
∥	is parallel to			(x, y)	ordered pair

Length

Metric

1 centimeter (cm) = 10 millimeters (mm)

1 meter (m) = 100 centimeters

1 kilometer (km) = 1,000 meters

Customary

1 foot (ft) = 12 inches (in.)

1 yard (yd) = 3 feet

1 mile (mi) = 1,760 yards

Mass

1 gram (g) = 1,000 milligrams (mg)

1 kilogram (kg) = 1,000 grams

Weight

1 pound (lb) = 16 ounces (oz)

1 ton (T) = 2,000 pounds

Capacity

Metric

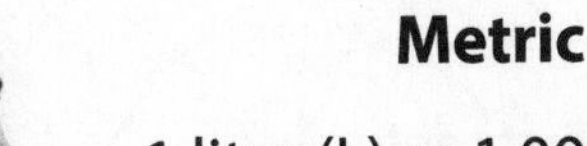

1 liter (L) = 1,000 milliliters (mL)

Customary

1 cup (c) = 8 fluid ounces (fl oz)

1 pint (pt) = 2 cups

1 quart (qt) = 2 pints

1 gallon (gal) = 4 quarts

Reference Sheet

Formulas

Area of a Rectangle

$A = \ell \times w$

Perimeter of a Rectangle

$P = (2 \times \ell) + (2 \times w)$

Volume of a Rectangular Prism

$V = \ell \times w \times h$

or

$V = B \times h$, where $B = \ell \times w$

Triangles

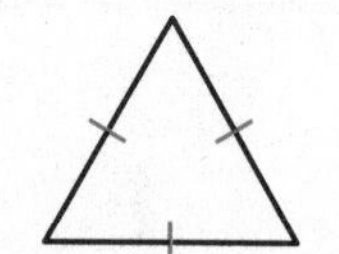

equilateral triangle

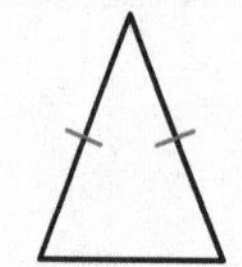

isosceles triangle

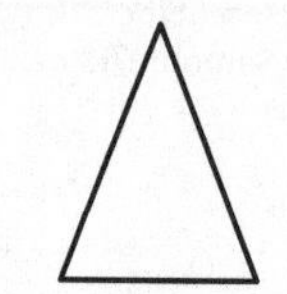

acute triangle

obtuse triangle

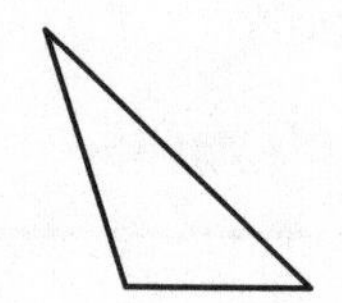

scalene triangle

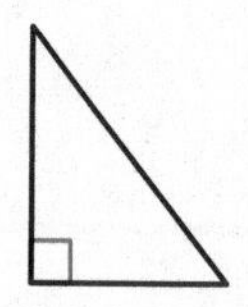

right triangle

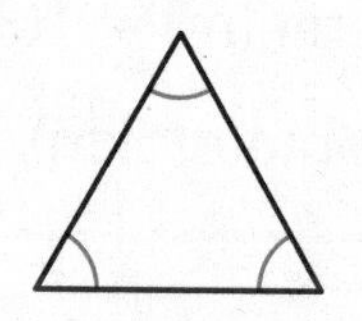

equiangular triangle

Quadrilaterals

The Venn diagram shows the relationship among quadrilaterals.

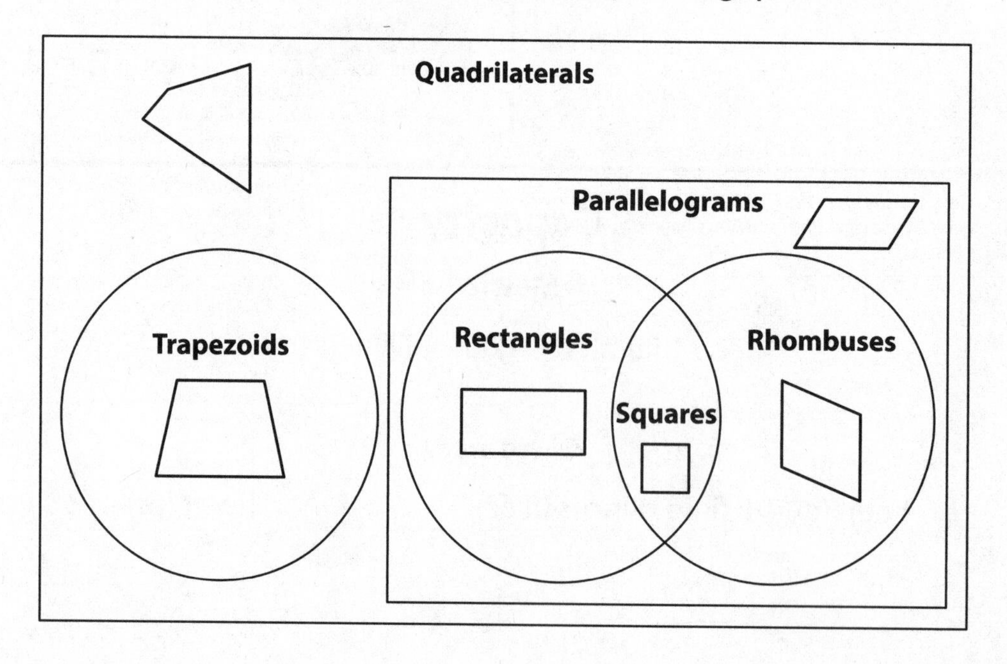

Credits

Front matter

i enmyo /Shutterstock.com; **vii** Steve Debenport/E+/Getty Images

Chapter 1

1 Grafissimo/E+/Getty Images; **6** wrangel/iStock/Getty Images Plus; **11** PicturePartners/iStock/Getty Images Plus; **12** ChrisGorgio/iStock/Getty Images Plus; **18** *left* mlharing/iStock/Getty Images Plus; *right* irin717/iStock/Getty Images Plus; **24** *top* Nearbirds/iStock/Getty Images Plus, dosrayitas//DigitalVision Vectors/Getty Images; *center* WestLight/iStock/Getty Images Plus; *bottom* Chris Gorgio/iStock/Getty Images Plus; **29** reptiles4all/iStock/Getty Images Plus; **32** *top* slovegrove/iStock/Getty Images Plus; *bottom* studiocasper/E+/Getty Images; **36** ebstock/iStock/Getty Images Plus; **38** glassfish/iStock/Getty Images Plus; **41** zanskar/iStock/Getty Images Plus; **42** *top* MCCAIG/E+/Getty Images; *bottom* VitalisG/iStock/Getty Images Plus; **45** *top from left to right* AlesVeluscek/E+/Getty Images; flammulated/iStock/Getty Images Plus; leksele/iStock/Getty Images Plus; photomaru/iStock/Getty Images Plus; *bottom* Samohin/iStock/Getty Images Plus

Chapter 2

51 SerrNovik/iStock/Getty Images Plus; **62** *top* domin_domin/E+/Getty Images; gunaonedesign/iStock/Getty Images Plus; *bottom* mipan/iStock/Getty Images Plus; **64** Ian Dagnall/Alamy Stock Photo; **68** MicrovOne/iStock/Getty Images Plus, Quarta_/iStock/Getty Images Plus; **74** Aerial view of DeFuniak Springs - DeFuniak Springs, Florida . 19--. Color postcard, 4 x 6 in. State Archives of Florida, Florida Memory. <https://www.floridamemory.com/items/show/163817>, accessed 12 December 2017. **77** *top* LoopAll/Shutterstock.com; *center* ollaweila/iStock/Getty Images Plus

Chapter 3

81 sfe-co2/iStock Editorial/Getty Images Plus; **86** *top left* AndreyKrav/iStock Editorial/Getty Images Plus; *top right* JJMaree/iStock Editorial/Getty Images Plus; *Exercise 12 left* TobiasBischof/iStock/Getty Images Plus; *Exercise 12 right* GlobalP/iStock/Getty Images Plus; **92** *top left* Aneese/iStock/Getty Images Plus; *top right* MrLonelyWalker/iStock/Getty Images Plus; *bottom* shanesabin/iStock/Getty Images Plus; **94** Antagain/iStock/Getty Images Plus, Vectortone/iStock/Getty Images Plus; **98** ImpalaPhoto/iStock/Getty Images Plus; **100** photomaru/iStock/Getty Images Plus; **104** *top* N8tureGrl/iStock/Getty Images Plus; *bottom* Nerthuz/iStock/Getty Images Plus; **110** *right* Simeon_les/iStock/Getty Images Plus; *left* GlobalP/iStock/Getty Images Plus; **112** MarivszBlach/iStock/Getty Images Plus; **116** *top* belterz/E+/Getty Images; *bottom* AlbertoChagas/iStock/Getty Images Plus; **118** Siempreverde22/iStock/Getty Images Plus; **120** Schnapps2012/iStock/Getty Images Plus; **121** *top* swedeandsour/E+/Getty Images; *bottom* Shell_114/iStock/Getty Images Plus; **122** *top* eduardrobert/iStock/Getty Images Plus; *center left* macrovector/iStock/Getty Images Plus; *center right* mehmettorlak/iStock/Getty Images Plus; *bottom* erierika/iStock/Getty Images Plus; *bottom left* REKINC1980/E+/Getty Images; **123** *top* Tatabrada/iStock/Getty Images Plus; *bottom* chengyuzheng/iStock/Getty Images Plus; **124** AlexLMX/iStock/Getty Images Plus; **125** Oleh_Slobodeniuk/E+/Getty Images; **135** JackF/iStock/Getty Images Plus; **136** Ron Chapple studios/Hemera/Getty Images Plus

Chapter 4

137 tonda/iStock/Getty Images Plus; **142** leonello/iStock/Getty Images Plus; **144** kojihirano/iStock/Getty Images Plus; **148** *top* Yuri_Arcurs/DigitalVision/Getty Images; *center* DavidByronKeener/iStock/Getty Images Plus; *bottom* goir/iStock/Getty Images Plus; **150** Warpaintcobra/iStock/Getty Images Plus; **154** *top* cmeder/iStock/Getty Images Plus; *bottom* DenisTangneyJr/E+/Getty Images; **156** Hey Darlin/iStock/Getty Images Plus; **160** *top* djvstock/iStock/Getty Images Plus; *bottom* ewg3D/iStock/Getty Images Plus; **162** alexey_boldin/iStock/Getty Images Plus; **165** Harvepino/iStock/Getty Images Plus; **166** *top* ginosphotos/iStock/Getty Images Plus; *bottom* Arnaud_Martinez/iStock/Getty Images Plus; **169** jbeck32661/iStock/Getty Images Plus; **170** colematt/iStock/Getty Images Plus; flyinggiraffestudio/DigitalVision Vectors/Getty Images; artisticco/iStock/Getty Images Plus; DNY59/E+/Getty Images

Chapter 5

173 Dragon Images/Shutterstock.com; **177** Wepix/E+/Getty Images; **178** *top* cmannphoto/E+/Getty Images; *bottom* FRANKHILDEBRAND/iStock/Getty Images Plus; **180** *left* kimberrywood/iStock/Getty Images Plus; *right* Text credit: Ben Garrod, Dr. Diego Pol, Dr. Jose Luis Carballido, Prof. John Hitchinson, and Dr. Luis Chiappe, "Attenborough And The Giant Dinosaur: The story of the discovery of the largest known dinosaur, Facts and Figures," 1 December 2016, BBC Media Centre, accessed 15 November 2017 from http://www.bbc.co.uk/mediacentre/latestnews/2016/attenborough-and-the-giant-dinosaur#heading-facts-and-figures. Photo credit: ©AMNH/D. Finnin; **183** flisak/iStock/Getty Images Plus; **184** Dotted Yeti/Shutterstock.com; **189** BirdImages/E+/Getty Images; **190** *top* Zerbor/iStock/Getty Images Plus; *Exercise 10* IR_Stone/iStock/Getty Images Plus; Tolga TEZCAN/E+/Getty Images; *Exercise 11* seregam/iStock/Getty Images; *bottom* TokenPhoto/E+/Getty Images; **195** bloodua/iStock/Getty Images Plus; **196** scanrail/iStock/Getty Images Plus; **201** Photodisc/Photodisc/Getty Images; **202** *top* GlobalP/iStock/Getty Images Plus; *bottom* ma-k/iStock/Getty Images Plus; **204** *left* Askold Romanov/iStock/Getty Images Plus; *right* ideabug/iStock/Getty Images Plus, duckycards/E+/Getty Images; **207** aapsky/Shutterstock.com; **208** *top* yurimeg/iStock/Getty Images Plus; *bottom* rodrigobarretophoto/iStock/Getty Images Plus; **210** pelooyen/iStock/Getty Images Plus; **213** SchulteProductions/iStock/Getty Images Plus; **214** *center* cirodelia/iStock/Getty Images Plus; *bottom left* MistikaS/E+/Getty Images; DonNichols/iStock/Getty Images Plus; *bottom right* Nikada/Cuba/Getty Images; **216** Tatiana Popova/Shutterstock.com; **219** *left* Hoang Mai Thach/Shutterstock.com; *right* vinaithong/iStock/Getty Images Plus; **220** *top* kaanates/iStock/Getty Images Plus; *bottom left* malerapaso/E+/Getty Images; *bottom right* TokenPhoto/E+/Getty Images; **222** *Exercise 10* GlobalP/iStock/Getty Images Plus; *Exercise 12* clubfoto/iStock/Getty Images Plus; *bottom* bluestocking/E+/Getty Images; **223** Vectorig/DigitalVision Vectors/Getty Images; **224** rimglow/iStock/Getty Images Plus; **225** *top* Ryan McVay/Photodisc/Getty Images; *bottom* DNY59/E+/Getty Images; **229** *top* jxfzsy/iStock/Getty Images Plus; *center* imv/iStock/Getty Images Plus; *bottom* rimglow/iStock/Getty Images Plus; **230** enjoynz/DigitalVision Vectors/Getty Images, artisticco/iStock/Getty Images Plus, pukrufus/DigitalVision Vectors/Getty Images

Chapter 6

235 acilo/iStock Unreleased/Getty Images Plus; **240** Mega Pixel/Shutterstock.com; **242** adempercem/iStock/Getty Images Plus; **246** *top* gremlin/iStock/Getty Images Plus; *center* artisteer/iStock/Getty Images Plus; *bottom* SunChan/iStock/Getty Images Plus; **248** tobiasjo/iStock/Getty Images Plus; **252** Believe_In_Me/iStock/Getty Images Plus; **254** Askold Romanov/iStock/Getty Images Plus; **258** *top* panimoni/iStock/Getty Images Plus; *bottom* Newbird/E+/Getty Images; **260** deyangeorgiev/iStock/Getty Images Plus; **263** GlobalP/iStock/Getty Images Plus; **264** GlobalP/iStock/Getty Images Plus; **269** inthevisual/iStock/Getty Images Plus; **270** KQS/Alamy Stock Photo; **272** hatman12/iStock/Getty Images Plus; **275** MiroNovak/iStock/Getty Images Plus; **276** *top right* Digital Vision/Photodisc/Getty Images; *bottom* Jgalione/E+/Getty Images; **278** *top* kkymek/iStock/Getty Images Plus; *bottom* bobey100/E+/Getty Images; **281** japrz/iStock/Getty Images Plus; **282** blublaf/Vetta/Getty Images; **285** DaddyBit/iStock/Getty Images Plus; **287** aniroot/iStock/Getty Images Plus; **288** CobraCZ/iStock/Getty Images Plus; **290** dblight/E+/Getty Images; **291** *top* lzf/iStock/Getty Images Plus; *bottom* Henrik5000/E+/Getty Images; **296** GlobalP/iStock/Getty Images Plus

Chapter 7
297 Omgimages/iStock/Getty Images Plus; **302** *top* Neustockimages/E+/Getty Images; *center* FernandoAH/E+/Getty Images; *bottom* asafta/iStock Editorial/Getty Images Plus; **307** vnlit/iStock/Getty Images Plus; **308** *top* wabeno/iStock/Getty Images Plus; *center* Serg_Velusceac/iStock/Getty Images Plus; **309** Ratth/iStock/Getty Images Plus; **314** *top* Mustang_79/iStock/Getty Images Plus; *bottom* MRaust/iStock/Getty Images Plus; **320** kickers/iStock/Getty Images Plus; **326** Goodluz/iStock/Getty Images Plus; **328** TokenPhoto/E+/Getty Images; **332** unknown1861/iStock/Getty Images Plus; **334** DougBennett/iStock/Getty Images Plus; **338** *top* Viktar/iStock/Getty Images Plus; *bottom* julichka/iStock/Getty Images Plus; **344** *top* timstarkey/iStock/Getty Images Plus; *bottom* macrovector/iStock/Getty Images Plus; **346** *top* janssenkruseproductions/iStock/Getty Images Plus; *Exercise 14 left* alkir/iStock/Getty Images Plus; *Exercise 14 center* macrovector/iStock/Getty Images Plus; *Exercise 14 right* GlobalP/iStock/Getty Images Plus; **347** DNY59/E+/Getty Images; **348** *left* ewg3D/iStock/Getty Images Plus; *right* Maluson/iStock/Getty Images Plus; **349** *top* abadonian/iStock/Getty Images Plus; 4x6/iStock/Getty Images Plus; *bottom* GaryAlvis/iStock/Getty Images Plus; macrovector/Getty Images Plus; Shimanovskaya/iStock/Getty Images Plus; **350** hudiemm/E+/Getty Images; **352** *left* fergregory/iStock/Getty Images Plus; Rodrusoleg/iStock/Getty Images Plus; *right* iJacky/iStock/Getty Images Plus; **353** kali9/iStock/Getty Images Plus; **354** enjoynz/DigitalVision Vectors/Getty Images; artisticco/iStock/Getty Images Plus; Pukrufus/Digital Vision Vectors/Getty Images; **358** *left* Yurdakul/E+/Getty Images; *right* BreakingTheWalls/iStock/Getty Images Plus; **361** ONiONAstudio/iStock/Getty Images Plus; **363** Vaniatos/iStock/Getty Images Plus; **364** Brand X Pictures/iStock/Getty Images Plus

Chapter 8
365 JamesBrey/Vetta/Getty Images; **369** pialhovik/iStock/Getty Images Plus; **370** groveb/E+/Getty Images; **372** jskiba/E+/Getty Images; **375** FatCamera/iStock/Getty Images Plus; **376** Malex92/iStock/Getty Images Plus; **378** *Exercise 14 left* 3dalia/iStock/Getty Images Plus; *Exercise 14 right* VikaSuh/iStock/Getty Images Plus; *Exercise 15* IvonneW/iStock/Getty Images Plus; **381** nimis69/iStock/Getty Images Plus; **382** *top* Gala98/iStock/Getty Images Plus; *bottom* onetouchspark/iStock/Getty Images Plus; **384** AndreaAstes/iStock/Getty Images Plus; **387** Boonchuay1970/iStock/Getty Images Plus; **388** *top* skegbydave/iStock/Getty Images Plus; *bottom* dszc/E+/Getty Images; **393** haryigit/iStock/Getty Images Plus; **394** *top left* Dr. Norbert Lange/Shutterstock.com; *top right* Beatriz Otero Rivera/iStock/Getty Images Plus; *bottom* foaloce/iStock/Getty Images Plus; **396** GMVozd/E+/Getty Images; **399** Krle/iStock/Getty Images Plus; **400** *top* andrea crisante/Shutterstock.com; *bottom* Antagain/iStock/Getty Images Plus; **402** *top* Comstock/Stockbyte/Getty Images; *bottom* KRyder17/iStock/Getty Images Plus; **405** BWFolsom/iStock/Getty Images Plus; **406** Subsociety/E+/Getty Images; **408** Koraysa/iStock/Getty Images Plus; **409** milehightraveler/E+/Getty Images; **410** 3DMAVR/iStock/Getty Images Plus; **411** Lorado/E+/Getty Images Plus; **412** *top* 3quarks/iStock/Getty Images Plus; *bottom* Beboy_ltd/iStock/Getty Images Plus; **413** Rawpixel/iStock/Getty Images Plus; **414** rimglow/iStock/Getty Images Plus; **415** Sarah8000/E+/Getty Images; **420** andriikoval/iStock/Getty Images Plus

Chapter 9
421 KimberlyDeprey/E+/Getty Images; **425** pada smith/iStock/Getty Images Plus; **426** *top* wundervisuals/iStock/Getty Images Plus; *bottom* hedgehog94/iStock/Getty Images Plus; **428** *left* magpiejst/iStock/Getty Images Plus; *right* M-I-S-H-A/iStock/Getty Images Plus; **432** *top* ALLEKO/iStock/Getty Images Plus; *bottom* GlobalP/iStock/Getty Images Plus; **434** *top* OvertheHill/iStock/Getty Images Plus; *bottom* GomezDavid/E+/Getty Images; **437** MikeLane45/iStock/Getty Images Plus; **438** dorioconnell/E+/Getty Images; **440** *top* yuhirao/iStock/Getty Images Plus; *bottom* adogslifephoto/iStock/Getty Images Plus; **444** *top* republica/iStock/Getty Images Plus; *bottom* arissanjaya/iStock/Getty Images Plus, atoss/iStock/Getty Images Plus; **446** *left* Parrotstarr/iStock/Getty Images Plus; *right* GlobalP/iStock/Getty Images Plus; **449** Smitt/iStock/Getty Images Plus; **450** *top* ShutterWorx/E+/Getty Images; *bottom* YuriyGreen/iStock/Getty Images Plus; **456** borisz/DigitalVision Vectors/Getty Images; Annykos/iStock/Getty Images Plus; johnwoodcock/DigitalVision Vectors/Getty Images; kostenkodesign/iStock/Getty Images Plus; Qvasimodo/iStock/Getty Images Plus; ADDeR_0n3/DigitalVision Vectors/Getty Images; MicrovOne/iStock/Getty Images Plus; colematt/iStock/Getty Images Plus; **458** artisticco/iStock/Getty Images Plus; **462** i-Stockr/iStock/Getty Images Plus; **468** *top* Wavebreakmedia/iStock/Getty Images Plus; *bottom* mladn61/iStock/Getty Images Plus; **470** cynoclub/iStock/Getty Images Plus; **471** jeremkin/iStock/Getty Images Plus; **473** rimglow/iStock/Getty Images Plus

Chapter 10
477 Steve Debenport/E+/Getty Images; **482** *top* pamela_d_mcadams/iStock/Getty Images Plus; *bottom* Anna Kucherova/iStock/Getty Images Plus; **488** *top* omersukrugoksu/iStock/Getty Images Plus; *bottom* Diana Taliun/iStock/Getty Images Plus; **490** WilliamSherman/iStock/Getty Images Plus; **494** *top* Creativeye99/iStock/Getty Images Plus; *center* kipuxa/iStock/Getty Images Plus; **496** tofumax/iStock/Getty Images Plus; **500** *left* Vasko/E+/Getty Images; *right* drewhadley/E+/Getty Images; **502** bernashafo/Shutterstock.com; **503** JacobVanHouten/E+/Getty Images; **505** chictype/E+/Getty Images; **509** *top* Wittayayut/iStock/Getty Images Plus; *center* ©iStockphoto.com/Vladyslav Otsiatsia; **512** ISerg/iStock/Getty Images Plus

Chapter 11
513 mnbb/iStock/Getty Images Plus; **514** *top* EasyBuy4u/E+/Getty Images; *center* Stockbyte/Stockbyte/Getty Images; *right* Enskanto/iStock/Getty Images Plus; **517** esdeem /Shutterstock.com; **518** JTSorrell/iStock/Getty Images Plus; **520** 3DSculptor/iStock/Getty Images Plus; **523** ryasick/E+/Getty Images; **524** *top* scorpp/iStock/Getty Images Plus; *Exercise 18 left* trigga/E+/Getty Images; *right* lightkitegirl/iStock/Getty Images Plus; **526** *top* SarahPage/iStock/Getty Images Plus; *bottom* choness/iStock/Getty Images Plus; **529** *bottom* cinoby/iStock/Getty Images Plus; **530** *top* LOVE_LIVE/E+/Getty Images; *bottom* Alexan2008/iStock/Getty Images Plus; **532** *top* OSTILL/iStock/Getty Images Plus; *Exercise 13 top* 3dalia/iStock/Getty Images Plus; *bottom* CoreyFord/iStock/Getty Images Plus; *bottom* EllenMoran/E+/Getty Images; **533** *right* fotoVoyager/E+/Getty Images; *left* Triduza/iStock/Getty Images Plus; **535** GlobalP/iStock/Getty Images Plus; **536** *top* mrPliskin/iStock/Getty Images Plus; *Exercise 17* ronstik/Shutterstock.com; YuriyZhuravov/Shutterstock.com; Petrenko Andriy/Shutterstock.com; Rafa Irusta/Shutterstock.com; *bottom* rusm/iStock/Getty Images Plus; **538** suprunvitaly/iStock/Getty Images Plus; **539** 4kodiak/iStock/Getty Images Plus; **541** Val_Iva/iStock/Getty Images Plus, andegro4ka/iStock/Getty Images Plus; **542** *top* alex-mit/iStock/Getty Images Plus; *bottom* Laboko/iStock/Getty Images Plus; **545** Sadeugra/E+/Getty Images; **547** Alter_photo/iStock/Getty Images Plus; **549** anythings/Shutterstock.com; **551** ra3rn/iStock/Getty Images Plus; **552** MickyWiswedel/iStock/Getty Images Plus; **554** *left* Mordolff/iStock/Getty Images Plus; *right* Floortje/E+/Getty Images; **555** *right* TokenPhoto/E+/Getty Images; *left* ufuk arslanhan/iStock/Getty Images Plus; **557** *top* Sitade/E+/Getty Images Plus; *bottom* Grafissimo/E+/Getty images; **558** ALINA/DigitalVision Vectors/Getty Images; **561** Volosina/iStock/Getty Images Plus; **562** *top* tropper2000/iStock/Getty Images Plus; *bottom* PaulCowan/iStock/Getty Images Plus; **564** cturtletrax/iStock/Getty Images Plus; **565** BreakingTheWalls/iStock/Getty Images Plus; **566** jmb_studio/iStock/Getty Images Plus; **567** mrtom-uk/iStock/Getty Images Plus; **568** antpkr/Stock/Getty Images Plus

Chapter 12
569 Avalon_Studio/E+/Getty Images, hknoblauch/iStock/Getty Images Plus; **571** hidesy/iStock/Getty Images Plus; **580** Yicai/iStock/Getty Images Plus; **589** AlexeyVS/iStock/Getty Images Plus; **592** monkeybusinessimages/iStock/Getty Images Plus; **595** pixhook/E+/Getty Images; **598** scanrail/iStock/Getty Images Plus; **602** Chiyacat/iStock/Getty Images Plus; **603** baibaz/iStock/Getty Images Plus; **604** PhotoMelon/iStock/Getty Images Plus; **606** ozgurdonmaz/iStock/Getty Images Plus; **608** jenifoto/iStock/Getty Images Plus; **610** *top* frantic00/Shutterstock.com; *bottom* gradyreese/iStock/Getty Images Plus; **613** ChrisGorgio/iStock/Getty Images Plus; **614** IconicBestiary/iStock/Getty Images Plus; **618** *left* Main_sail/iStock/Getty Images Plus; *right* m_pavlov/iStock/Getty Images Plus

Chapter 13
619 vm/E+/Getty Images; **624** Pazhyna/iStock/Getty Images Plus; **630** Jamesmcq24/iStock/Getty Images Plus; **636** Nobilior/iStock/Getty Images Plus; **642** *top* Vasko/E+/Getty Images; *bottom* nata_zhekova/iStock/Getty Images Plus; **651** imaginima/iStock/Getty Images Plus; **656** *left* Horned_Rat/DigitalVision Vectors/Getty Images; *right* Gudella/iStock/Getty Images Plus;

Chapter 14
657 querbeet/iStock/Getty Images Plus; **674** lucielang/iStock/Getty Images Plus; **677** DiyanaDimitrova/iStock/Getty Images Plus; **680** James Knopf/Hemera/Getty Images Plus; **682** Richard G. Bingham II/Alamy Stock Photo; **683** sihuo0860371/iStock/Getty Images Plus; **685** kvkirillov/iStock/Getty Images Plus

Cartoon Illustrations: MoreFrames Animation
Design Elements: oksanika/Shutterstock.com; icolourful/Shutterstock.com; Valdis Torms

Kumon

SUCKS